Microbial Safety and Beneficial Microorganisms in Foods

Microbial Safety and Beneficial Microorganisms in Foods

Guest Editor

Theodoros Varzakas

MDPI

Basel • Beijing • Wuhan • Barcelona • Belgrade • Novi Sad • Cluj • Manchester

Guest Editor
Theodoros Varzakas
Department of Food Science
and Technology
University of Peloponnese
Antikalamos
Greece

Editorial Office
MDPI AG
Grosspeteranlage 5
4052 Basel, Switzerland

This is a reprint of the Special Issue, published open access by the journal *Microorganisms* (ISSN 2076-2607), freely accessible at: https://www.mdpi.com/journal/microorganisms/special_issues/2W4J5KP1LA.

For citation purposes, cite each article independently as indicated on the article page online and as indicated below:

Lastname, A.A.; Lastname, B.B. Article Title. *Journal Name* **Year**, *Volume Number*, Page Range.

ISBN 978-3-7258-8329-5 (Hbk)
ISBN 978-3-7258-8330-1 (PDF)
https://doi.org/10.3390/books978-3-7258-8330-1

Contents

About the Editor

Theodoros Varzakas

Theodoros Varzakas has served as a Full Professor at the Department of Food Science and Technology, University of the Peloponnese, Greece, since 2019, specializing in issues of food technology, food processing/engineering, food quality and safety. He is also a Section Editor-in-Chief in the journal *Foods* in Food Security and Sustainability (2020-), as well as an ex-Editor-in-Chief of *Current Research in Nutrition and Food Science* (2015–2019), and a series Editor for *Contemporary Food Engineering, CRC*. He is also a reviewer and member of the editorial boards of many international journals. Professor Theodoros Varzakas has written more than 200 research papers and chapters in books and has presented more than 160 papers and posters at national and international conferences. He has written and edited 14 books in Greek and 16 in English on sweeteners, biosensors, food engineering, and food processing, published by *CRC*. His participation in many European and national research programs as coordinator or scientific member is also of note. His h-index is 50, and the number of citations is approx. 10,000. According to updates for September 2021, 2022, October 2023, and August 2024, and the September 2025 update for "Updated science-wide author databases of standardized citation indicators" Varzakas Theodoros is in the top 2% of scientists worldwide.

microorganisms

Editorial

Editorial for Special Issue "Microbial Safety and Beneficial Microorganisms in Foods"

Theodoros Varzakas

Department of Food Science and Technology, University of the Peloponnese, 24100 Antikalamos, Greece; t.varzakas@uop.gr

The role of microorganisms in food was first acknowledged a long time ago. Detrimental effects on food systems have been reported from pathogenic microorganisms, such as *Salmonella* and *Escherichia coli* causing severe foodborne illnesses. On the other hand, beneficial microorganisms have been shown to enhance food stability, safety, and nutritional value, and they are particularly found in fermented foods and biopreservation.

This Special Issue provides a platform for researchers to enrich their knowledge on microbial safety and the beneficial use of microorganisms in food. In this Special Issue, six research contributions have been included, along with four review papers.

In the first contribution, we addressed antimicrobial resistance (AMR) and the role of artisanal dairy fermentations (Tyrovolia–Kopanisti) in enriching resistance phenotypes and serving as sensitive sentinels of AMR dynamics.

In the second contribution, it was shown that the fermentative performance of amaranth seeds improved through sprouting and probiotic fermentation, incorporating potential health-promoting properties.

The safety profile of *Lentilactobacillus buchneri*, which has neuroprotective effects, needs to be evaluated through both phenotypic and genotypic analyses for suitability in probiotic applications, as discussed by Kim et al. (contribution 3).

A rapid, robust, and scalable tool for precise probiotic strain tracking and accurate quantification of *Bifidobacterium animalis* subsp. *lactis* has been addressed, enabling critical quality control and meeting regulatory compliance needs (contribution 4).

Waste bread (WB) as a novel, cost-effective, and nutrient-rich substrate for microbial growth has been highlighted in contribution 5. It acts as a fermentation substrate to enhance amylase production, offering sustainable development.

The last research contribution (contribution 6) deals with fucoidan as a natural antibacterial agent for controlling *Salmonella enterica* subsp. *Enterica* Serovar Typhimurium biofilms in seafood processing. Nowadays, enhancement of food safety and minimization of contamination are essential strategies to improve quality of life.

The first review (contribution 7) deals with probiotics selection and precise strain selection, novel encapsulation technologies for viability, and strict safety assessments as described by EFSA and the United States FDA. Unlocking next-generation probiotics depends on omics technologies, synthetic biology, and more detailed microbiome–host interaction studies.

Contribution 8 examines the One Health framework, in terms of the drivers of AMR across sectors, focusing on surveillance systems and regulatory policies aiming at ensuring the microbiological safety of food systems.

Elbehiry et al. (contribution 9) discussed cutting-edge control strategies, such as natural antifungal agents, essential oils, biocontrol methods, and non-thermal technologies like cold plasma and UV-C treatment, along with safe microbial control approaches in fresh produce.

Finally, contribution 10 provided an overview of post-harvest food decontamination methods against some of the most important bacterial foodborne pathogens, focusing on the advantages and challenges of phages and phage-added packaging materials.

Acknowledgments: We would like to thank all authors and reviewers for their excellent contributions to this Special Issue. We are also grateful to all members of the *Microorganisms* Editorial Office for providing us with this opportunity and for their continuous support.

Conflicts of Interest: The author declares no conflict of interest.

List of Contributions:

1. Rozos, G.; Fotou, K.; Gerokomou, V.; Nikolaou, K.; Dadamogia, A.; Hatzizisis, L.; Skoufos, I.; Tzora, A.; Bezirtzoglou, E.; Voidarou, C. One Health Investigation of Stage-Dependent Antimicrobial Resistance Patterns Across Intermediate and Ripened Dairy Matrices: The Tyrovolia–Kopanisti Paradigm. *Microorganisms* **2026**, *14*, 712. https://doi.org/10.3390/microorganisms14030712.
2. Vasile, M.A.; Balan, N.; Grigore-Gurgu, L.; Bahrim, G.E.; Cotârleț, M. Probiotic Lactic Acid Bacteria Fermentation Modulates the Bioactive Properties of Sprouted and Unsprouted Amaranth Seed. *Microorganisms* **2026**, *14*, 340. https://doi.org/10.3390/microorganisms14020340.
3. Kim, S.; Jeong, H.; Lee, N.-K.; Kang, D.-K.; Paik, H.-D.; Park, Y.-S.; Lee, J.H. Comprehensive Safety Assessment of *Lentilactobacillus buchneri* KU200793 as a Potential Probiotic. *Microorganisms* **2025**, *13*, 2067. https://doi.org/10.3390/microorganisms13092067.
4. Mao, D.; Zhao, L.; Zhao, B.; Xu, H.; Zhang, Q. Development of a Strain-Specific Detection and Quantification Method for *Bifidobacterium animalis* subsp. *lactis* HN019 Using WGS-SNP Analysis and qPCR. *Microorganisms* **2025**, *13*, 1596. https://doi.org/10.3390/microorganisms13071596.
5. Ben Mabrouk, S.; Ben Hadj Hmida, B.; Sallami, W.; Dhaouadi, S.; Varzakas, T.; Smaoui, S. Harnessing Waste Bread: From Potential Use in Microbial Growth and Enzyme Production to Techno-Economic Assessment. *Microorganisms* **2025**, *13*, 1571. https://doi.org/10.3390/microorganisms13071571.
6. Roy, A.; Roy, P.K.; Cho, S.R.; Park, S.Y. Effects of Fucoidan on the Inhibition of Biofilm Formation of *Salmonella enterica* Subsp. *enterica* Serovar Typhimurium on Seafoods and Its Molecular Antibiofilm Mechanisms. *Microorganisms* **2025**, *13*, 914. https://doi.org/10.3390/microorganisms13040914.
7. Uhegwu, C.C.; Anumudu, C.K. Probiotic Potential of Traditional and Emerging Microbial Strains in Functional Foods: From Characterization to Applications and Health Benefits. *Microorganisms* **2025**, *13*, 2521. https://doi.org/10.3390/microorganisms13112521.
8. Elbehiry, A.; Marzouk, E.; Abalkhail, A.; Edrees, H.M.; Ellethy, A.T.; Almuzaini, A.M.; Ibrahem, M.; Almujaidel, A.; Alzaben, F.; Alqrni, A.; et al. Microbial Food Safety and Antimicrobial Resistance in Foods: A Dual Threat to Public Health. *Microorganisms* **2025**, *13*, 1592. https://doi.org/10.3390/microorganisms13071592.
9. Elbehiry, A.; Marzouk, E.; Alzaben, F.; Almuaither, A.; Abead, B.; Alamri, M.; Almuzaini, A.M.; Abu-Okail, A. Emerging Technologies and Integrated Strategies for Microbial Detection and Control in Fresh Produce. *Microorganisms* **2025**, *13*, 1447. https://doi.org/10.3390/microorganisms13071447.
10. Braz, M.; Pereira, C.; Freire, C.S.R.; Almeida, A. A Review on Recent Trends in Bacteriophages for Post-Harvest Food Decontamination. *Microorganisms* **2025**, *13*, 515. https://doi.org/10.3390/microorganisms13030515.

 https://doi.org/10.3390/microorganisms14051032

microorganisms

MDPI

Article

One Health Investigation of Stage-Dependent Antimicrobial Resistance Patterns Across Intermediate and Ripened Dairy Matrices: The Tyrovolia–Kopanisti Paradigm

Georgios Rozos [1,*], Konstantina Fotou [1], Vaia Gerokomou [1], Konstantina Nikolaou [1], Aikaterini Dadamogia [1], Lampros Hatzizisis [2], Ioannis Skoufos [2], Athina Tzora [1], Eugenia Bezirtzoglou [3] and Chrysoula (Chrysa) Voidarou [1,*]

1 Laboratory of Animal Health, Food Hygiene and Quality, Department of Agriculture, School of Agriculture, University of Ioannina, 47150 Arta, Greece; kfotou@uoi.gr (K.F.); v.gerokomou@uoi.gr (V.G.); knikolaou@uoi.gr (K.N.); kdadamogia@gmail.com (A.D.); tzora@uoi.gr (A.T.)
2 Laboratory of Animal Science, Nutrition and Biotechnology, Department of Agriculture, School of Agriculture, University of Ioannina, 47150 Arta, Greece; lamprosxatz@uoi.gr (L.H.); jskoufos@uoi.gr (I.S.)
3 Laboratory of Hygiene and Environmental Protection, Department of Medicine, Democritus University of Thrace, Dragana, 68100 Alexandroupolis, Greece; empezirt@yahoo.gr
* Correspondence: grozos@uoi.gr (G.R.); xvoidarou@uoi.gr (C.V.)

Abstract

Antimicrobial resistance (AMR) emerges and circulates across interconnected human, animal, food, and environmental reservoirs; however, food fermentation systems remain underexplored as indicators of local AMR pressure, even though artisanal dairy fermentations may function as natural sentinels of AMR. In this study, we used an artisanal dairy fermentation chain as a One Health model to investigate whether environmentally exposed lactobacilli can reflect stage-associated shifts in resistance. A total of 1.085 isolates representing 16 *Lactobacillus* species were recovered from the same artisanal dairy matrix at two fermentation stages: day 5 ("Tyrovolia"; n = 518) and day 30 ("Kopanisti"; n = 567). Susceptibility to 14 antibiotics was evaluated by broth micro-dilution, and *L. acidophilus* was further screened for selected resistance genes. Overall resistance increased significantly from 69.88% (362/518) at day 5 to 77.25% (438/567) at day 30 ($p = 0.0059$), while multidrug resistance rose from 37.57% to 60.73% of resistant isolates ($p < 0.001$). Across the 224 species–antibiotic combinations examined, 129 (57.58%) showed an increased upper MIC limit at day 30, and resistance increased significantly for 9 of the 14 antibiotics tested, with the largest rises observed for metronidazole (RR = 7.67), chloramphenicol (RR = 5.74), quinupristin/dalfopristin (RR = 4.11), vancomycin (RR = 2.78), and trimethoprim (RR = 2.43). In contrast, erythromycin and oxytetracycline resistance declined significantly at the ripened stage. In *L. acidophilus*, 21 resistance genes were detected in 14/70 day-5 isolates and 19 genes in 13/71 day-30 isolates, but marked genotype–phenotype discordance was observed, including matrix-dependent expression patterns for *tet*M, *erm*B, and *bla*TEM. Collectively, these findings show that environmentally exposed artisanal dairy fermentations can enrich resistance phenotypes and may serve as sensitive sentinels of AMR dynamics at the human–animal–environment interface.

Keywords: *Lactobacillus*; AMR; "Tyrovolia"; "Kopanisti"; *tet*M; *tet*K; *erm*B; *bla*TEM; *cat*

1. Introduction

Antimicrobial resistance (AMR) is widely recognized as one of the major global health challenges of the Anthropocene, undermining the prevention and treatment of bacterial infections in humans and animals and threatening the safety and sustainability of food systems. Resistance has increased worldwide, particularly among clinically important bacterial pathogens, leading to rising morbidity, mortality, and economic burden. Current projections suggest that, without effective mitigation, the global annual economic losses attributable to AMR may increase substantially by 2050, from hundreds of billions of USD today to over one trillion USD [1–3]. At the same time, AMR-associated mortality is expected to rise markedly, with estimates suggesting that annual deaths may reach several million worldwide by mid-century [4–6].

AMR is not limited to pathogenic bacteria. Resistance determinants are also found in commensal and environmental microorganisms, including members of the indigenous microbiota of humans, animals, and diverse environmental niches. Among these, lactobacilli are of particular interest because they are abundant in food-associated ecosystems and may contribute to the persistence and dissemination of resistance genes under both selective and non-selective pressures [7,8]. Although the taxonomy of this group has been substantially revised in recent years, lactobacilli remain ecologically versatile and widespread in plant-associated environments, animal-associated habitats, and a broad range of fermented foods [9–12].

AMR in lactobacilli has been investigated in isolates from fermented dairy products as well as from human, animal, and environmental microbiomes. However, important gaps remain. Much of the available evidence comes either from industrially produced fermented foods, where controlled starter cultures and standardized hygienic conditions may limit environmental influence, or from artisanal products examined only at the final edible stage. As a result, earlier fermentation stages, which may better reflect dynamic environmental inputs during processing and handling, have been insufficiently studied. Consequently, the extent to which fermentation stage and open-environment exposure shape the AMR phenotype of lactobacilli in artisanal dairy systems remains unclear.

Within the One Health framework, AMR emergence and dissemination are driven by interconnected reservoirs and transmission pathways linking humans, animals, food, and the environment [13–15]. Artisanal fermented dairy products provide a particularly relevant system for examining these interactions, because production often takes place in rural household settings with close environmental connectivity. Milk processing may occur in cellars, storerooms, or nearby livestock-associated spaces, where contact with household members, farm workers, animals, and airborne dust may facilitate microbial exchange. In such settings, stage-based sampling may help reveal whether resistance profiles shift during fermentation and handling, consistent with the acquisition, enrichment, or selection of resistance traits.

In the present study, we use a two-stage artisanal fermentation system, "Tyrovolia" (an intermediate fermented dairy matrix) transitioning to "Kopanisti" (the ripened product), as a model to test the hypothesis that artisanal fermentation stages can function as sensitive "sensors" of environmental AMR pressure. Specifically, we investigate (i) how lactobacilli populations are modified across the intermediate-to-ripened transition and (ii) whether their antimicrobial susceptibility profiles shift across stages when tested against a broad panel of antibiotics. By integrating process-stage sampling with a One Health perspective, this work aims to improve understanding of how artisanal dairy environments may influence AMR dynamics in food-associated lactobacilli and, by extension, contribute to AMR circulation across connected human–animal–environment interfaces.

 https://doi.org/10.3390/microorganisms14030712

2. Materials and Methods

In order to assess environmentally induced antimicrobial resistance (AMR), a total of 120 samples of the traditional artisanal cheese "Tyrovolia" (5–7-day curd) and 120 samples of the traditional artisanal Greek cheese "Kopanisti" (≥30-day curd) were collected from local producers directly at their facilities on the island of Mykonos, Greece. Although these cheeses are considered culinarily distinct, they originate from the same curd sampled at different time points during fermentation. "Tyrovolia" is a fresh product represented by a 5–7-day curd, whereas "Kopanisti" is produced from the remaining portion of the same curd, which is further fermented for at least 30 days with the addition of salt. In both products, according to artisanal practice, milk is not heated to pasteurization temperatures but rather to substantially lower temperatures. The two sampling points were selected to represent two distinct stages of the same artisanal fermentation process. "Tyrovolia" (5–7-day curd) corresponds to an early fermentation stage, during which the microbiota is expected to more closely reflect the original milk microbiota and animal-associated microbial inputs. In contrast, "Kopanisti" (≥30-day curd) represents a later stage, after prolonged fermentation, salt addition, and greater exposure to artisanal environmental conditions. Comparison of these two stages allowed an initial evaluation of whether antimicrobial resistance profiles change over the course of fermentation. The selected time points were defined empirically, as similar stage-based studies on these traditional products are not currently available. Under normal sampling conditions, samples were immediately placed in sterile polyethylene bags, sealed, and transported to the laboratory in ice-packed cooler boxes. Upon arrival, samples were stored at 2–4 °C and maintained under refrigeration until analysis. *Note*: Samples corresponding to the "Kopanisti" stage have been partially described in a previous publication [16]. In the present study, they are re-examined as part of an expanded dataset designed to allow comparative evaluation across production stages within a unified analytical framework. The current work therefore does not merely reproduce earlier findings but broadens the study by including additional samples and newly integrated analyses conducted under the same overall experimental design.

2.1. Phenotypic Identification Presumptive Lactobacilli Isolates

To isolate lactic acid bacteria (LAB), the rind of each curd was aseptically removed. From the remaining portion, 10 g was weighed and homogenized for 2 min in pre-warmed (40 °C) peptone water (10 g L^{-1} casein peptone, 5 g L^{-1} NaCl, 20 g L^{-1} trisodium citrate dihydrate; pH 7.0) using a stomacher blender. The homogenate was serially diluted and cultivated on (i) modified decarboxylating agar supplemented with histidine (MDAH) and (ii) MRS medium (de Man, Rogosa and Sharpe). After anaerobic incubation at 37 °C for 24 h, 20 μL (0.02 mL) of the appropriate dilutions were spread-plated on MRS agar and incubated anaerobically for an additional 48 h. Well-isolated colonies were selected and preliminarily screened by Gram staining and standard phenotypic assays (catalase, oxidase, and motility). Isolates meeting the criteria for presumptive LAB were subsequently subjected to species-level identification using matrix-assisted laser desorption/ionization time-of-flight mass spectrometry (MALDI-TOF MS).

Morphologically distinct, well-separated colonies were selected from each sample and subcultured by successive streaking on fresh MRS agar plates until purity was confirmed. Pure isolates were cryopreserved by suspending cells in MRS broth supplemented with 10% (*v*/*v*) sterile glycerol and stored at −80 °C for subsequent analyses.

2.2. Species Identification

Presumptive lactobacilli were re-streaked onto MRS agar and incubated for 24 h prior to identification. Isolates were identified by MALDI-TOF MS using the Bruker Biotyper

https://doi.org/10.3390/microorganisms14030712

Microflex platform (Bruker Daltonik, Bremen, Germany). Spectra were acquired according to the manufacturer's recommendations, processed with MBT Compass software (v4.1), and matched against the Bruker reference library (BDAL, Rev. 11; 10,833 entries). For each isolate, the ten highest-ranking matches and their log(score) values (0.00–3.00) were recorded. Identification reliability was interpreted using Bruker criteria: log(score) < 1.70, unreliable; 1.70–1.99, genus-level; 2.00–2.29, probable species-level (highly probable genus-level); and ≥2.30, highly probable species-level identification. Results were confirmed in two independent experiments [17–19].

2.3. *Phenotypic Antimicrobial Susceptibility Testing (AST)—Determination of Minimal Inhibitory Concentration*

All 1085 isolates were included in the antimicrobial susceptibility/resistance testing, and no specific selection process was applied. Lactobacilli were not selected from a broader pool of isolates; instead, they constituted the target group of microorganisms in this study, and all recovered isolates were subjected to phenotypic susceptibility testing. This approach was based on the ecological ubiquity of lactobacilli (see Supplementary Table S7), which makes them highly relevant as potential indicators of environmental influence on the artisanal dairy matrix, including the possible carriage of resistance genes.

Minimum inhibitory concentrations (MICs) were determined by broth microdilution using 96-well microtiter plates [20–22]. All antimicrobial powders were purchased from Sigma-Aldrich (St. Louis, MO, USA). Test inocula were prepared from fresh cultures suspended in sterile 0.9% NaCl. The standardized suspension was subsequently diluted 1:500 in LSM broth (containing 90% Iso-sensitest broth and 10% MRS.

MICs were assessed for the following antimicrobial agents (final concentration ranges): ampicillin (0.03–64 μg mL^{-1}), ampicillin/sulbactam (0.03–32 μg mL^{-1}), teicoplanin (0.03–32 μg mL^{-1}), clindamycin (0.03–8 μg mL^{-1}), erythromycin (0.03–64 μg mL^{-1}), gentamicin (0.03–500 μg mL^{-1}), streptomycin (0.03–500 μg mL^{-1}), chloramphenicol (0.03–256 μg mL^{-1}), oxytetracycline (0.03–256 μg mL^{-1}), fusidic acid (0.03–256 μg mL^{-1}), metronidazole (0.25–500 μg mL^{-1}), trimethoprim (0.03–500 μg mL^{-1}), quinupristin/dalfopristin (0.03–16 μg mL^{-1}), and vancomycin (0.03–256 μg mL^{-1}). The concentration ranges of the antimicrobial agents were selected individually for each compound, taking into account their expected inhibitory ranges and, where applicable, generally accepted CLSI breakpoint frameworks. However, standardized CLSI/EUCAST interpretive criteria are not consistently available for environmental and non-clinical lactobacilli.

For each well, 50 μL of the diluted inoculum was combined with 50 μL of the corresponding antimicrobial dilution (prepared in LSM), yielding the final concentration ranges stated above. Plates were incubated anaerobically at 37 °C for 48 h, after which MIC endpoints were read as the lowest antimicrobial concentration preventing visible growth. *Escherichia coli ATCC 25922* and *Staphylococcus aureus ATCC 25923* were included as quality-control strains.

Interpretation of MIC data for lactobacilli remains challenging because widely used EUCAST clinical breakpoints and CLSI interpretive criteria are not consistently available and are not always fully compatible with the growth requirements and testing conditions of all *Lactobacillus* spp. Therefore, susceptibility categorization was based on a population-based interpretation of MIC distributions within each species, in order to distinguish the predominant wild-type (WT) population from isolates showing evidence of reduced susceptibility. Accordingly, isolates were classified as non-wild-type (NWT), indicative of acquired resistance and hereafter considered putatively resistant, when their MIC values formed a distinct subpopulation separated from the main WT cluster. Evidence for an NWT subpopulation was inferred from two distributional patterns: (i) discontinuous MIC distributions, in which one or more isolates grew only at higher MIC dilutions that were not contiguous with the modal WT

 https://doi.org/10.3390/microorganisms14030712

range, suggesting a discrete less-susceptible group; and (ii) bimodal or shouldered distributions, in which growth was observed across consecutive MIC dilutions, but the number of isolates increased again at the upper end of the MIC range, suggesting the emergence of a second, less-susceptible cluster rather than simple tailing of the WT distribution. Because the strains were isolated from artisanal matrices exposed to environmental influences, this subpopulation-based approach was considered more appropriate than reliance on reference values developed mainly for clinical or industrial strains.

2.4. PCR Analysis—Detection of Antibiotic Resistance Genes

From the isolate collection, strains affiliated with the *Lactobacillus acidophilus* species group were prioritized for downstream screening of acquired antimicrobial-resistance (AMR) determinants. This group was selected because it was consistently recovered among the LAB isolates and represents a relevant lineage for assessing the potential carriage of transferable resistance genes within dairy-associated microbiota. Genotypic screening was performed using simplex PCR assays to detect resistance genes associated with four major antimicrobial classes: tetracyclines, β-lactams, chloramphenicol, and macrolides (erythromycin). These targets were chosen for three complementary reasons. First, the corresponding agents (or related compounds) are widely used in both human and veterinary practice, creating opportunities for the selection and maintenance of resistance traits across connected ecosystems. Second, these antimicrobial classes include drugs that are clinically important for human health, making the detection of resistance determinants relevant from a public-health perspective. Third, resistance genes linked to these classes are frequently located on mobile genetic elements and therefore have an increased likelihood of horizontal transfer between bacterial hosts. In the context of fermented dairy products, identifying such determinants is also informative for evaluating whether resistance signatures reflect environmental or production-chain inputs (e.g., farm, processing environment, or raw-material microbiota) rather than intrinsic properties of the LAB themselves.

PCR assays were performed to detect selected antimicrobial resistance genes associated with the phenotypes under investigation. For tetracycline resistance, two complementary genetic mechanisms were examined: ribosomal protection and active efflux. Ribosomal protection was initially screened using a universal primer set targeting ribosomal protection determinants and, in positive isolates, was further verified by amplification with *tet*M-specific primers in order to identify the determinant at the gene level. Active efflux was assessed by detection of the *tet*K gene, which encodes a major facilitator superfamily (MFS)-type tetracycline efflux pump. In addition, the presence of *bla*TEM, *cat*, and *ermB* was investigated as markers of resistance to β-lactams, chloramphenicol, and macrolides, respectively. More specifically, *bla*TEM encodes a β-lactamase capable of hydrolyzing β-lactam antibiotics, *cat* encodes chloramphenicol acetyltransferase responsible for enzymatic drug inactivation, and *erm*B encodes a 23S rRNA methyltransferase that mediates target-site modification and is commonly associated with the MLS_B-type resistance phenotype. Primer sequences, expected amplicon sizes, and PCR cycling and annealing conditions, together with the corresponding methodological references [23–30], are presented in Supplementary Table S1.

Genomic DNA was prepared from freshly grown colonies to minimize DNA degradation and to ensure consistent template quality. Briefly, a single well-isolated colony from each isolate was harvested and processed using the DNeasy® UltraClean® Microbial Kit (QIAGEN) according to the manufacturer's instructions. Purified DNA was eluted as provided by the kit protocol and stored at −20 °C until further analysis. A no-template control (NTC; RNase-free water in place of DNA) was included in every run to control for reagent contamination and carry-over during reaction setup. Amplification was per-

 https://doi.org/10.3390/microorganisms14030712

formed on a CFX96™ thermocycler (Bio-Rad, Hercules, CA, USA). PCR products were separated on 2% (*w*/*v*) agarose gels at 100 V, stained, and visualized, and fragment sizes were estimated by comparison with an appropriate DNA molecular weight marker. The PCR cycling conditions consisted of an initial denaturation at 94 °C for 5 min, followed by 35 cycles of denaturation at 94 °C for 30 s, annealing at the primer-specific temperature (Supplementary Table S1) for 30 s, and extension at 72 °C for 45 s. A final extension step at 72 °C for 7 min completed the protocol.

2.5. Questionnaire Survey on Farm Antimicrobial Use

Prior to sampling, the attending veterinarian administered a brief structured questionnaire to the participating producers in order to document routine antimicrobial use in their sheep flocks. The survey recorded the most common clinical indications for treatment, the estimated frequency of antimicrobial administration, and the antibiotic agents and routes of administration most commonly used for therapeutic purposes.

2.6. Statistical Analysis Section

Chi square test of independence was used to evaluate changes in the number of resistant strains during fermentation. The Pearsons correlation coefficient test was used to assess various correlations (e.g., the number of AMR strains to MDR strains and others). Relative Risk of AMR was calculated for each antibiotic as well as for every species (wherever possible). Fischer's exact test was used for the assessment of the combined analysis of phenotypic resistance and corresponding resistance genes in *L. acidophilus* isolates. Multilinear regression was used to determine predictors for AMR and MDR. In all tests significance was considered for $p < 0.05$.

3. Results

Table 1 summarizes the antimicrobial panel used in this study, organized by antibiotic class, primary cellular target/mode of action, and the major resistance mechanisms typically associated with each compound. This framework is used throughout the Results to support interpretation of the susceptibility profiles observed in lactobacilli isolates across the "Tyrovolia"-to-"Kopanisti" transition. In particular, grouping antibiotics by shared targets (cell-wall synthesis, protein synthesis, nucleic-acid integrity, and folate metabolism) allows the resistance phenotypes presented in the subsequent tables to be read not merely as isolate-by-isolate outcomes, but as mechanistically coherent patterns—for example, phenotypes consistent with β-lactamase activity or altered penicillin-binding proteins (β-lactams), target-site modification and efflux (macrolides/lincosamides), efflux- and ribosomal protection–associated tetracycline resistance, aminoglycoside-modifying enzymes, or glycopeptide-associated peptidoglycan precursor remodeling. Accordingly, Table 1 serves as a reference map for the stage-dependent shifts in antimicrobial susceptibility reported below and helps contextualize potential routes of resistance acquisition, enrichment, or selection within an artisanal, environmentally exposed fermentation system.

Table 1. Antimicrobial agents included in the susceptibility testing panel of the present study, grouped according to antimicrobial class, principal mode of action, and major reported resistance mechanisms.

Substance	Class	Mode of Action	Mechanisms of Resistance
Ampicillin	β-lactams	Binds to enzymes involved in peptidoglycan synthesis, inhibiting cell wall formation	-β-lactamases inactivate the antibiotic -changes in cell wall protein enzymes

 https://doi.org/10.3390/microorganisms14030712

Table 1. *Cont.*

Substance	Class	Mode of Action	Mechanisms of Resistance
Sulbactam	β-lactamase inhibitor	Inhibits the β-lactamase enzymes	Extended spectrum β-lactamases (classes A–D)
Erythromycin	Macrolides	Binds to the 23S rRNA molecule of the 50S ribosomal subunit and blocks protein synthesis by inhibiting the transpeptidation/translocation step and also by inhibiting the assembly of the 50S subunit	-target site modification by methylation of a specific nucleotide of the 23S rRNA (*erm* genes) -active drug efflux (*mef*A, *mef*E genes) -cross resistance via the MSLB phenotype
Clindamycin	Lincosamides	-prevents peptide bond formation by binding to the 23S rRNA molecule of the 50S ribosomal subunit. It impedes the assembly of the subunit as well as the translation process.	-target modification: (i) mutation in the 23S rRNA (ii) ribosomal methylations (*erm* genes) -active efflux -target protection by specific proteins -antibiotic inactivation by specific enzymes
Oxytetracycline	Tetracyclines	Binds to the 30S ribosomal subunit and interferes with amino acid transfer	-Inducible efflux -binding site change
Chloramphenicol	Phenicols	Binds to the L16 protein of the 50S ribosomal subunit and inhibits the elongation of peptide chains by suppressing the peptidyl transferase.	-Enzymatic inactivation: (i) CATs: chloramphenicol acetyltransferases (ii) CPTs: chloramphenicol phosphotransferases -efflux pumps -permeability barriers by alterations in outer membrane proteins -target site mutations
Gentamycin	Aminoglycosides	Binds to the 30S ribosomal subunit causing misread of code and thus disrupting membrane permeability and the production of mistranslation of proteins	-Phosphorylation -Adenylation -Acetylation
Streptomycin		Inhibition of protein synthesis at the aminoacyl transfer site of the 16S part of the 30S ribosomal subunit. It binds the formyl -methionyl-tRNA to the 30S subunit.	-Enzymatic inactivation (genes encoding such enzymes are *str*A and *str*B) -mutations in genes encoding ribosomal proteins lead to alterations in the ribosomal subunit (*rps*L and *rrs* genes) -changes in membrane permeability -efflux pumps

 https://doi.org/10.3390/microorganisms14030712

Table 1. *Cont.*

Substance	Class	Mode of Action	Mechanisms of Resistance
Vancomycin	Glycopeptides	Bind to the D-Ala-D-Ala terminus of peptidoglycan precursors preventing cell wall synthesis.	Modification of the bacterial cell wall's peptidoglycan structure: the terminal Alanine is replaced by other dipeptides Serine or Lactate: D-Ala-D-Ser or D-Ala-D-Lac (*van*A and *van*B gene clusters and VanC protein)
Teicoplanin			-As vancomycin above -mutations in the genes encoding D-Ala-D-Ala ligase -VanJ novel protein of the bacterial cell wall reduces the affinity of the drug to its target
Fusidic acid	Tetracyclic steroid (fusidanes)	Binds to the elongation factor G (EF-G) on the ribosome preventing the translocation of the peptide chain inhibiting thus the protein synthesis.	-mutation in the *fus*A gene which encodes EF-G and/or mutation in the *fus*E gene which encodes the ribosomal protein L6 -efflux pumps -enzymatic inactivation -altered drug permeability
Metronidazole	Nitroimidazoles	After intracellular reduction of its molecule, metronidazole interacts with DNA causing strand rupture, helix destabilization and cell death	-reduced uptake or increased efflux -increased DNA repair -altered pyruvate metabolism (necessary to the anaerobic reduction of the drug) -mutation in genes like *rdx*A encoding proteins involved in the activation pathways of metronidazole
Quinupristin/dalfopristin	Streptogramins	-bind on different sites of the 50S ribosomal subunit: Dalfopristin inhibits the early phase of protein synthesis while Quinupristin inhibits the late phase.	-enzymatic inactivation, e.g., acetyltransferases can inactivate Dalfopristin (genes *vat*A, *vat*B and *vat*C). -active efflux (genes *vga*A and *vga*B) -target modification
Trimethoprim	Diaminopyridine	Inhibits dihydrofolate reductase (DHFR), an enzyme necessary for folic acid synthesis in the bacterial cell. Folic acid is crucial to bacterial DNA and protein synthesis.	-mutations in DHFR genes (*dfr*)leads to synthesis of altered DHFR molecule resistant to trimethoprim.

Table 2 summarizes the producers' responses regarding routine therapeutic antibiotic use in their sheep flocks, including the most frequently reported infection categories, their estimated prevalence, and the antimicrobial agents commonly administered. Respiratory infections were reported as the most frequent condition (4–5%), typically treated with tetracycline via subcutaneous (SC) or intramuscular (IM) administration. Udder, post-partum, and traumatic infections were reported at lower frequencies (1–3%) and were consistently managed with a penicillin/dihydrostreptomycin combination administered intramuscu-

 https://doi.org/10.3390/microorganisms14030712

larly. Overall, the interview data indicate that the on-farm therapeutic antimicrobial exposure relevant to this study was dominated by tetracyclines and β-lactam/aminoglycoside combination therapy, providing contextual background for the susceptibility profiles of lactobacilli isolates presented in the subsequent sections.

Table 2. Results of interviews with farm owners regarding the use of antibiotics for common infections in livestock.

Most Common Infections	Prevalence Rate (%)	Antibiotics Used Systematically for Therapeutic Purposes
Respiratory	4–5	Tetracycline SC or IM
Udder	1–2	Penicillin/Dihydrostreptomycin IM
Post partum	2–3	Penicillin/Dihydrostreptomycin IM
Traumatic	1–2	Penicillin/Dihydrostreptomycin IM

Abbreviations: SC, subcutaneous; IM, intramuscular.

Table 3 provides an overview of the lactobacilli isolated from the two fermentation stages examined in this study and summarizes, by species, the frequency of resistant phenotypes and multidrug resistance. In total, 518 isolates were recovered from the 5-day curd ("Tyrovolia") and 567 isolates from the 30-day curd ("Kopanisti"), comprising 16 species. Resistance phenotypes were detected across the antimicrobial panel; however, resistance was not uniformly distributed among species. Notably, all *L. johnsonii* and *L. casei* subsp. *pseudoplantarum* isolates from the 5-day curd were susceptible to all tested antibiotics, and the same complete susceptibility was observed for *L. casei* subsp. *pseudoplantarum* isolates recovered from the 30-day curd. Overall, the proportion of resistant isolates was significantly higher in the 30-day curd (438/567; 77.25%) than in the 5-day curd (362/518; 69.88%) ($\chi^2 = 7.58$, $p = 0.0059$). In addition, multidrug-resistant (MDR) isolates (resistance to >2 antibiotic classes) were detected at a significantly higher frequency in the 30-day curd (266/438; 60.73% of resistant isolates) compared with the 5-day curd (138/362; 37.57%) ($\chi^2 = 40.53$, $p < 0.001$). Together, these findings indicate a stage-associated increase in both overall resistance and MDR phenotypes during the transition from the intermediate curd to the ripened product.

Table 4 breaks down multidrug resistance (MDR) at the species level and shows which exact antibiotics each MDR isolate resisted. In contrast to Table 3, which summarizes MDR counts, Table 4 lists the specific resistance combinations observed within each *Lactobacillus* species and indicates how many distinct MDR phenotypes occurred per species. This format allows a direct comparison between the 5-day and 30-day curds to determine whether the higher MDR frequency at day 30 is driven by (i) more MDR isolates within the same species, (ii) a broader range of MDR profiles within species, or (iii) different species contributing disproportionately to MDR.

Table 5 summarizes, for each of the 14 antibiotics tested, the number (and percentage) of resistant lactobacilli isolates recovered from the 5-day and 30-day curds and quantifies stage-associated change using relative risk (RR) with 95% confidence intervals (CI). No significant stage effect was observed for clindamycin, gentamicin, or fusidic acid, as the RR values were close to unity and the corresponding CIs included 1. In contrast, resistance to erythromycin, and oxytetracycline decreased significantly in the 30-day curd (RR < 1), indicating a lower probability of detecting resistant isolates at the ripened stage. For the remaining nine antibiotics, resistance increased significantly during the transition from the 5-day to the 30-day curd (RR > 1), consistent with stage-associated enrichment and/or acquisition of resistance phenotypes under artisanal, environmentally exposed conditions.

https://doi.org/10.3390/microorganisms14030712

The strongest increases were observed for metronidazole and chloramphenicol, followed by quinupristin/dalfopristin, which exhibited the highest RR values and narrow CIs that did not overlap 1.

Table 3. Species-level distribution of lactobacilli isolates and antimicrobial resistance outcomes across fermentation stages (5–7-day "Tyrovolia" curd vs. ≥30-day "Kopanisti" curd).

Species	5th Day Curd					30th Day Curd				
	n [a]	R [b]	A/B [c]	Ph/T [d]	MDR [e]	(n)	R	A/B	Ph/T	MDR
L. helveticus	19	13	0	0	0	31	27	4	3	7
L. acidophilus	70	48	6	6	36	71	66	6	7	43
L. sakei	43	33	4	3	15	24	19	5	4	7
L. paraplantarum	48	48	8	10	35	48	43	8	8	37
L. brevis	12	7	3	2	5	30	11	2	2	0
L. delbrueckii subsp bulgaricus	58	47	7	8	17	84	80	7	11	54
L. johnsonii	19	0	0	0	0	49	26	4	3	19
L. curvatus	36	30	3	3	0	42	33	7	6	30
L. salivarius	19	11	1	1	0	12	12	5	3	0
L. plantarum	36	18	2	1	0	54	47	6	6	36
L. rhamnosus	38	19	1	1	0	30	24	6	4	13
L. delbrueckii subsp lactis	48	39	5	6	12	36	18	4	3	6
L. fermentum	24	17	6	3	6	36	19	3	3	7
L. pentosus	29	24	8	4	12	25	7	3	1	7
L. casei subsp casei	13	8	2	2	0	12	6	1	1	0
L. casei subsp pseudoplantarum	6	0	0	0	0	6	0	0	0	0
Total	518	362	-	-	138	567	438	-	-	266

[a]: n, number of isolates; [b]: R, number of resistant isolates; [c]: A/B, number of antibiotic classes in which resistance was recorded for the isolates of each species, indicating the range of antimicrobial categories involved; [d]: Ph/T, number of distinct resistance phenotypes observed within each species, defined as different combinations of susceptibility/resistance profiles among the isolates; [e]: MDR, multidrug-resistant isolates (resistance to >2 antibiotic classes). Totals refer to all isolates recovered from the 5-day ("Tyrovolia") and 30-day ("Kopanisti") curds.

Table 4. Species-specific multidrug-resistant (MDR) phenotypes of lactobacilli recovered from the 5-day "Tyrovolia" curd and the ≥30-day "Kopanisti" curd.

Species	5-Day Curd	30-Day Curd
L. helveticus	**0 strains**	6 × Chlor-Met-Q/D-Oxy (4) [a] 1 × Chlor-Met-Q/D (3) **7 strains(22.58%)/2 phenotypes/4 antibiotics**
L. acidophilus	12 × Amp-Ery-Oxy-Gen-Met (5) 17 × Amp-Ery-Oxy (3) 6 × Amp-Ery-Oxy-Gen-Met (5) 1 × Amp-Ery-Oxy-Gen (4) **36 strains (51.43%) [b]/4 phenotypes/6 antibiotics**	12 × Amp-Sulb/Amp-Chlor-Van-Met-Tri (5) 24 × Amp-Sulb/Amp-Van-Met-Tri (4) 7 × Amp-Sulb/Amp-Met-Tri (3) **43 strains(60.56%)/3 phenotypes/6 antibiotics**
L. sakei	15 × Amp-Clin-Chlor (3) **15 strains (34.88%)/1 phenotype/3 antibiotics**	5 × Chlor-Tri-Pen-Fus-Amp (5) 2 × Chlor-Tri-Amp (3) **7 strains(29.17%)/2 phenotypes/5 antibiotics**

https://doi.org/10.3390/microorganisms14030712

Table 4. *Cont.*

Species	5-Day Curd	30-Day Curd
L. paraplantarum	6 × Amp-Ery-Tri-Fus (4) 6 × Amp-Sulb/Amp-Ery-Tri-Fus (4) 1 × Clin-Ery-Tri (3) 3 × Clin-Ery-Amp (3) 2 × Clin-Ery-Amp-Sulb/Amp (3) 17 × Clin-Ery-Chlor (3) **35 strains (72.91%)/6 phenotypes/8 antibiotics**	17 × Ery-Clin-Chlor-Str-Van-Tei-Met-Fus (7) 3 × Ery-Clin-Chlor-Str-Van-Tei-Met (6) 9 × Ery-Clin-Chlor-Str-Van-Tei (5) 6 × Ery-Clin-Van-Tei (4) 1 × Ery-Fus-Van-Chlor (4) 1 × Clin-Str-Tei-Met (4) **37 strains (77.08%)/6 phenotypes/7 antibiotics**
L. brevis	5 × Str-Fus-Tri (3) **5 strains (41.67%)/1 phenotype/3 antibiotics**	**0 strains**
L. delbrueckii subsp. *bulgaricus*	10 × Amp-Ery-Oxy-Q/D (4) 6 × Amp-Ery-Oxy-Tri-Q/D (5) 1 × Amp-Ery-Oxy-Q/D (4) **17 strains (29.31%)/3 phenotypes/5 antibiotics**	23 × Amp-Chlor-Gen-Van-Met-Q/D (6) 1 × Amp-Chlor-Van-Q/D (4) 1 × Amp-Gen-Met (3) 11 × Chlor-Van-Q/D (3) 7 × Amp-Chlor-Gen-Van-Q/D (5) 5 × Amp-Chlor-Gen-Van-Q/D (5) 6 × Chlor-Van-Gen-Q/D (4) **54 strains (64.29%)/7 phenotypes/7 antibiotics**
L. johnsonii	**0 strains**	19 × Amp-Chlor-Str-Met (4) **19 strains (38.78%)/ 1 phenotype/4 antibiotics**
L. curvatus	**0 strains**	10 × Oxy-Met-Amp-Van (4) 2 × Oxy-Met-Amp-Q/D (4) 7 × Oxy-Met-Amp (3) 11 × Amp-Oxy-Met (3) **30 strains (71.43%)/4 phenotypes/5 antibiotics**
L. salivarius	**0 strains**	**0 strains**
L. plantarum	**0 strains**	30 × Amp-Clin-Tei-Met-Fus-Q/D (6) 3 × Amp-Tei-Met-Fus-Q/D (5) 1 × Tei-Met-Fus-Q/D (4) 2 × Tei-Met-Fus (30 **36 strains (66.67%)/4 phenotypes/6 antibiotics**
L. rhamnosus	**0 strains**	6 × Amp-Sulb/Amp-Clin-Chlor-Gen-Van-Met (6) 6 × Amp-Sulb/Amp-Clin-Chlor-Gen (4) 1 × Amp-Sulb/Amp-Chlor-Gen **13 strains(43.33%)/3 phenotypes/6 antibiotics**
L. delbrueckii subsp. *lactis*	6 × Amp-Sulb/Amp-Oxy-Gen-Fus (4) 6 × Amp-Sulb/Amp-Gen-Fus (3) **12 strains (25.00%)/2 phenotypes/5 antibiotics**	6 × Amp-Oxy-Gen-Met (4) **6 strains (16.67%)/ 1 phenotype/4 antibiotics**
L. fermentum	6 × Amp-Sulb/Amp-Str-Van-Fus (4) **6 strains (25.00%)/1 phenotype/6 antibiotics**	7 × Oxy-Str-Q/D (3) **7 strains (28.00%)/1 phenotype/3 antibiotics**
L. pentosus	9 × Amp-Sulb/Amp-Ery-Clin-Oxy-Str-Van-Tri (7) 1 × Amp-Sulb/Amp-Ery-Clin-Oxy-Van-Tri (6) 2 × Amp-Sulb/Amp-Ery-Oxy-Van-Tri (5) **12 strains (33.33%)/3 phenotypes/8 antibiotics**	7 × Chlor-Tri-Met (3) **7 strains (28.00%)/1 phenotype/3 antibiotics**

https://doi.org/10.3390/microorganisms14030712

Table 4. *Cont.*

Species	5-Day Curd	30-Day Curd
L. casei subsp. *casei*	**0 strains**	**0 strains**
L. casei subsp. *pseudoplantarum*	**0 strains**	**0 strains**

Notes: [a]: indicates the number of isolates expressing each MDR phenotype × the corresponding phenotype, with the number in parentheses indicating the number of antibiotic classes involved; [b]: indicates the percentage of MDR isolates relative to the total number of isolates of the corresponding species at that fermentation stage; Abbreviations (antibiotics): Amp, ampicillin; Sulb/Amp, sulbactam/ampicillin; Ery, erythromycin; Clin, clindamycin; Oxy, oxytet-racycline; Chlor, chloramphenicol; Gen, gentamicin; Str, streptomycin; Van, vancomycin; Tei, teicoplanin; Fus, fusidic acid; Met, metronidazole; Q/D, quinupristin/dalfopristin; Tri, trimethoprim; MDR, multidrug-resistant.

Table 5. Antibiotic-specific resistance frequencies in lactobacilli at the early and ripened fermentation stages (5–7-day "Tyrovolia" curd vs. ≥30-day "Kopanisti" curd), and relative risk (RR) of resistance at the ripened stage.

Antibiotic	(n) of Resistant Strains 5-Day Curd [a]	(n) of Resistant Strains 30-Day Curd	(RR) [b]	CI [c] 95%
Ampicillin	190 (36.68%)	318 (55.99%)	1.82	1.62–2.10
Sulbactam/ Ampicillin	53 (10.23%)	91 (16.05%)	1.57	1.14–2.15
Erythromycin	125 (24.13%)	42 (7.41%)	0.31	0.22–0.42
Clindamycin	73 (14.09%)	84 (14.81%)	1.05	0.79–1.41
Oxytetracycline	112 (21.62%)	60 (10.58%)	0.49	0.37–0.65
Chloramphenicol	35 (6.76%)	220 (38.80%)	5.74	4.10–8.04
Gentamycin	78 (15.06%)	72 (12.69%)	0.84	0.63–1.14
Streptomycin	31 (5.98%)	74 (13.05%)	2.18	1.46–3.26
Vancomycin	48 (9.27%)	146 (25.74%)	2.78	2.05–3.76
Teicoplanin	35 (6.76%)	84 (14.81%)	2.19	1.51–3.19
Fusidic acid	69 (13.32%)	66 (11.64%)	0.87	0.64–1.19
Metronidazole	30 (5.79%)	252 (44.45%)	7.67	5.36–10.99
Quinupristin/ Dalfopristin	30 (5.79%)	135 (23.81%)	4.11	2.82–5.99
Trimethoprim	32 (6.18%)	85 (14.99%)	2.43	1.65–3.58
Total (*)	362	438		

Notes: [a]: Values are given as n (%) resistant isolates for each fermentation stage. Percentages are calculated relative to the total isolates recovered from each stage (5-day curd: n = 518; 30-day curd: n = 567); (*): The total does not refer to the product of the addition of the resistant strains of this table because many of them are MDR and would be counted more than once. [b]: RR was calculated as the proportion resistant at day 30 divided by the proportion resistant at day 5 (RR > 1 indicates increased resistance at day 30; RR < 1 indicates decreased resistance at day 30); [c]: CI, 95% confidence interval for RR. Differences were considered statistically significant when the 95% CI did not include 1.

Table 6 presents the range of MIC values, the MIC50 and MIC 90 values as well as the experimental cut-off values of *L. acidophilus* isolates from the 5th and the 30th day matrix. In general, most values of MIC 50 and MIC90 increased under the environmental impact.

In Table 7 the detected genes are depicted with the relevant phenotypes. There is a distinct difference between the expression of these genes between the 5th and the 30th day matrixes.

 https://doi.org/10.3390/microorganisms14030712

Table 6. MIC distributions and experimental cut-off values (mg/L) for *L. acidophilus* isolates recovered from the 5–7-day "Tyrovolia" curd and the ≥30-day "Kopanisti" curd.

	5th Day Curd				30th Day Curd			
Antibiotic	**MIC Range**	**MIC_{50}**	**MI_{90}**	**Cut Off**	**MIC Range**	**MIC_{50}**	**MIC_{90}**	**Cut Off**
Ampicillin	0.03–1	0.5	0.5	0.12	0.12–1	0.5	1	-
Sulbactam/Ampicillin	0.12–0.5	0.12	0.5	-	0.06–2	0.5	1	0.12
Erythromycin	0.03–0.25	0.06	0.25	0.12	0.03–0.5	0.12	0.5	-
Clindamycin	0.12–0.5	0.25	0.5	-	0.12–0.5	0.12	0.25	-
Oxytetracycline	0.12–1	0.5	1	0.25	0.25–16	2	8	-
Chloramphenicol	0.12–0.5	0.25	0.5	-	0.12–8	0.5	2	2
Gentamycin	0.12–0.5	0.25	0.5	0.25	0.25–2	1	2	-
Streptomycin	1–8	4	8	-	0.5–32	4	16	-
Vancomycin	0.12–0.5	0.5	0.5	-	0.25–1	1	1	0.5
Teicoplanin	128–256	256	256	-	16–128	64	128	-
Fusidic acid	16–64	32	64	-	16–64	2	16	-
Metronidazole	32–500	64	256	128	64–≥500	≥500	≥500	-
Trimethoprim	0.25–1	0.5	1	-	2–256	16	256	4
Quinupristin/dalfopristin	0.12–0.5	0.12	0.5	-	0.06–0.5	0.12	0.5	-

Notes: MIC, minimum inhibitory concentration; MIC_{50}, concentration inhibiting 50% of isolates; MIC_{90}, concentration inhibiting 90% of isolates.

Table 7. Detected resistance genes and corresponding phenotypes in *L. acidophilus* isolates recovered from the 5–7-day and ≥30-day curd matrices.

Gene	**5th Day Matrix**		**30th Day Matrix**	
	n	**Phenotype**	**n**	**Phenotype**
*tet*M	9 (12.86%) *	9 resistant to oxytetracycline	5 (7.04%) *	5 susceptible to oxytetracycline
*tet*K	6 (8.57%)	5 resistant + 1 susceptible to oxytetracycline	0	-
cat	0	-	4 (5.63%)	2 resistant + 2 susceptible to chloramphenicol
blaTEM	3 (4.29%)	All 3: Ampicillin resistant Sulbactam/Ampicillin susceptible	7 (9.86%)	6 resistant to Ampicillin and Sulbactam/Ampicillin & 1 Ampicillin susceptible and Sulbactam/Ampicillin resistant
*erm*B	3 (4.29%)	3 resistant to erythromycin	3 (4.22%)	3 susceptible to erythromycin
Total	21	-	19	-

(*): Percentage of the total *L. acidophilus* isolates recovered from the corresponding matrix. Abbreviations: *bla*TEM, gene encoding a beta-lactamase; erm*B*, gene encoding a 23S rRNA methyltransferase; *tet*M, gene encoding a ribosomal protection protein; *tet*K, gene encoding a tetracycline efflux pump; *cat*, gene encoding chloramphenicol acetyltransferase.

 https://doi.org/10.3390/microorganisms14030712

The combined analysis of antimicrobial susceptibility and PCR-based gene detection in *L. acidophilus* revealed only partial agreement between phenotypic resistance and the investigated resistance determinants (Table 8). While phenotypic resistance was observed for multiple antibiotics, the corresponding genes were not uniformly detected among resistant isolates. More specifically, the genes *bla*TEM, *erm*B, *tet*M, *tet*K, and *cat* were present in only a proportion of isolates exhibiting resistance to the related antibiotic classes, indicating incomplete genotype–phenotype correspondence. This pattern may reflect the involvement of additional, non-investigated resistance genes, mutations affecting target sites, regulatory effects on gene expression, or other intrinsic and acquired resistance mechanisms. Therefore, the molecular screening performed herein supports the phenotypic findings only in part and highlights the complexity of antimicrobial resistance in *L. acidophilus* isolates.

Table 8. Combined analysis of phenotypic resistance and corresponding resistance genes in *Lactobacillus acidophilus* isolates (n = 141).

Determining Resistance to Antibiotic Phenotype	Resistance Gene	Phenotypically Resistant, n/N (%)	Gene-Positive Isolates, n/N (%)	Resistant and Gene-Positive, n/N (%)	Gene-Positive Among Resistant, n (%)	Fisher's Exact *p*-Value
Ampicillin	*bla*TEM	104/141 (73.8)	10/141 (7.1)	9/141 (6.4)	9/104 (8.7)	0.455
Sulbactam/ Ampicillin	*bla*TEM	43/141 (30.5)	10/141 (7.1)	6/141 (4.3)	6/43 (14.0)	0.067
Erythromycin	*erm*B	36/141 (25.5)	6/141 (4.3)	3/141 (2.1)	3/36 (8.3)	0.174
Oxytetracycline	*tet*M	36/141 (25.5)	14/141 (9.9)	9/141 (6.4)	9/36 (25.0)	0.001
Oxytetracycline	*tet*K	36/141 (25.5)	6/141 (4.3)	5/141 (3.5)	5/36 (13.9)	0.004
Chloramphenicol	*cat*	12/141 (8.5)	4/141 (2.8)	2/141 (1.4)	2/12 (16.7)	0.036

Abbreviations: *bla*TEM, beta-lactamase gene; *erm*B, 23S rRNA methyltransferase gene; *tet*M, ribosomal protection gene; *tet*K, tetracycline efflux pump gene; *cat*, chloramphenicol acetyltransferase gene. Footnotes: Gene-positive indicates PCR detection of the corresponding resistance determinant. Phenotypically resistant indicates isolates scored as resistant in the binary dataset. Associations between phenotype and genotype were evaluated using Fisher's exact test. p-values < 0.05 were considered statistically significant. For oxytetracycline resistance, both *tet*M and *tet*K were evaluated separately because they represent distinct resistance mechanisms.

To further substantiate the findings of the present study, the detailed antimicrobial susceptibility data are provided in the Supplementary Materials, where they provide a more granular view of the resistance patterns observed among the isolated *Lactobacillus* strains. Specifically, Tables S2 and S3 report the MIC ranges and the corresponding experimental cut-off values for isolates recovered on the 5th and 30th day, respectively, thus allowing direct evaluation of how susceptibility profiles shifted across the two fermentation stages. Building on this comparison, Table S4 summarizes changes in the upper limits of the MIC ranges between day 5 and day 30, thereby highlighting the direction and extent of these shifts and facilitating the identification of antibiotics for which resistance trends became more pronounced during ripening. In parallel, Table S5 presents the EFSA (2018) cut-off values for the relevant *Lactobacillus* groups and species, providing the regulatory benchmark against which the present data can be interpreted [31]. Finally, Table S6 compares the number of resistant strains classified according to EFSA breakpoints with those classified using the experimental cut-off values established in this study, thereby illustrating the degree to which resistance estimates depend on the interpretive framework applied. Taken together, these supplementary tables do not merely provide additional raw data but substantively reinforce the main results by documenting the stage-dependent dynamics of

 https://doi.org/10.3390/microorganisms14030712

antimicrobial susceptibility and by showing that the apparent prevalence of resistance may vary according to the criteria used for classification.

4. Discussion

A central premise of the present work is that artisanal, open-environment dairy fermentations operate at a One Health interface, where microorganisms can be exchanged among livestock, humans, companion animals, and surrounding environmental niches during routine handling and ripening. Accordingly, interpretation of stage-dependent shifts in the lactobacilli community—and in associated antimicrobial resistance phenotypes—should consider not only "starter-like" dairy-adapted taxa, but also taxa plausibly introduced from, or selectively enriched by, the broader production environment. To support this ecological interpretation, Table 6 compiles the lactobacilli taxa isolated in the present study and summarizes their reported occurrence across potential reservoirs described in the literature, including livestock-associated sites (gastrointestinal tract, oral cavity, udder, skin, vagina, nasal cavity), human-associated niches (skin, oral cavity, gastrointestinal tract, nasal cavity), companion animals, and abiotic matrices such as vegetation/rhizosphere and soil. This synthesis is not intended to attribute a definitive origin to any isolate, because species-level occurrence across multiple niches limits source inference based on taxonomy alone, but rather to provide a biologically plausible framework for understanding how environmental exposure during the "Tyrovolia"-to-"Kopanisti" transition could reshape microbial load and community composition, and thereby influence the resistance landscape through introduction, persistence, and/or selective enrichment of specific taxa. In this context, Supplementary Table S7 serves as an interpretive "road map", helping to contextualize the observed stage-dependent patterns in microbial composition and antimicrobial resistance within a realistic One Health framework.

Resistance of lactobacilli to antibiotics has been widely investigated owing to the genus's ubiquitous presence in nature and, above all, its extensive application as a probiotic in animal and human nutrition [32–35]. Intrinsic, or innate, resistance is generally attributed to chromosomal determinants, as demonstrated in glycopeptide resistance, including vancomycin resistance, in *L. sakei*, *L. plantarum*, and *L. rhamnosus* [36,37]. By contrast, acquired resistance is typically linked to mobile genetic elements, such as plasmids and transposons, whose horizontal transfer can give rise to resistant strains and promote the further spread of resistance [38]. Intrinsic resistance in lactobacilli has also been documented for trimethoprim [8], aminoglycosides (gentamicin and streptomycin), metronidazole, and fluoroquinolones [21,39–41].

However, a notable gap in the scientific literature is the investigation of environmentally exposed lactobacilli and their capacity to adapt to environmental conditions that may favor increased antimicrobial resistance. Intensive farming represents a typical environment of this kind because of its extensive use of antibiotics, although it is by no means the only one. Artisan production may also provide prolonged environmental exposure, as these products interact more directly with their surroundings, unlike industrial production, where fermentation processes are generally carried out under controlled conditions. In artisanal settings, products may be exposed to humans (visitors or residents), roaming pets or farm poultry, and fluctuations in weather conditions. Temporary overcrowding within a limited area may likewise contribute to the dissemination of resistance, particularly when combined with heavy foot traffic in animal-rearing or artisanal production premises. Collectively, these factors may increase the likelihood that lactobacilli acquire resistance determinants or adapt through elevated minimum inhibitory concentrations (MICs).

Our findings suggest that environmental exposure significantly affects the resistance profiles of lactobacilli. As shown in Tables S1 and S2, the upper limits of the MIC ranges

 https://doi.org/10.3390/microorganisms14030712

increased significantly from day 5 to day 30, regardless of whether discrete subpopulations were present or emerged within a species. Of the 224 possible combinations of *Lactobacillus* species and antibiotics (16 species × 14 antibiotics), 129 (57.58%) showed an increase in the upper limit of the MIC range at day 30, 33 (14.73%) showed a decrease, and 58 (25.89%) remained unchanged. This overall trend is clearly presented in Table S3. The effect was not uniform across species ($p < 0.001$). For example, *L. helveticus* and *L. curvatus* showed increases in the upper limit of the day-30 MIC range for 12 of the 14 antibiotics tested, whereas *L. pentosus* showed decreases for 6 antibiotics, no change for 2, and increases for the remaining 6. Likewise, the effect varied significantly among antibiotics ($p < 0.001$). Vancomycin, for instance, showed an increase in the upper limit of the MIC range in 15 of the 16 *Lactobacillus* species, whereas penicillin G showed a decrease in 7 species and an increase in 8 species. These observations underline the existence of a species–antibiotic-specific relationship. To some extent, this relationship appeared to extend beyond the phenomenon of innate resistance. Even in the case of vancomycin, which is classically associated with intrinsic resistance in certain *Lactobacillus* species, the upper limit of the MIC range still increased in the day-30 matrix for those species. In Tables 3 and 4 the relation between the resistant strains, the multiresistant ones, the phenotypes and the classes of antibiotics against which resistance was recorded, can be seen. This relationship proved to be strongly positive ($R \geq 0.774$, $p < 0.001$) for the various combinations of these variables. Multilinear analysis however showed that the only significant predictor for MDR strains in the 5th day matrix was the number of phenotypes ($R = 0.857$, $p < 0.001$) while for MDR in the 30th day matrix was the number of resistant strains ($R = 0.954$, $p < 0.001$). As a matter of fact, 38.12% of the resistant strains of the 5th day matrix were found MDR, while 60.73% of the 30th day matrix were found MDR (OR = 2.51, CI 95% 1.886–3.341, $p < 0.001$) which practically means that every resistant strain isolated from the 30th day matrix has 2.51 more chances to be MDR than the ones of the 5th day matrix and this finding implies strong impact of the environmental exposure.

Interestingly, resistance was not universal among the lactobacilli examined. A substantial proportion of isolates remained susceptible to all antibiotics tested, accounting for 30.12% of isolates recovered from the day-5 matrix and 22.75% of those recovered from the day-30 matrix. As this difference was not statistically significant, these findings may indicate the persistence of a baseline susceptible subpopulation throughout the experimental period. One possible explanation is the existence of competitive interactions among resistant phenotypes, whereby strains resistant to one antimicrobial may be outcompeted by strains carrying resistance to another. This interpretation may also help explain why five antibiotics yielded a relative risk (RR) below 1. For example, lactobacilli resistant to ampicillin may have been outcompeted by strains resistant to the sulbactam/ampicillin combination, while gentamicin-resistant isolates may have been displaced by streptomycin-resistant counterparts. Although this hypothesis requires further investigation, it suggests that shifts in resistance prevalence may not depend solely on direct selection pressure, but also on ecological interactions and fitness trade-offs within the lactobacillus community.

In the present study, strains were classified as resistant when they formed a clearly distinct subpopulation within isolates of the same species, whereas the highest MIC value observed for the susceptible subpopulation was taken as the experimental cutoff. The results of this classification approach are presented in Table 3 and Tables S2 and S3. In 2018, EFSA established breakpoint values for *Lactobacillus* species for nine antibiotics, eight of which were included in the present study (Table S5) [31]. Because the EFSA breakpoints are higher than the experimental cutoffs derived here, applying the EFSA criteria to our dataset results in substantially fewer strains being categorized as resistant (Table S6).

https://doi.org/10.3390/microorganisms14030712

Nevertheless, even under the EFSA breakpoint framework, the number of resistant strains recovered from the day-30 matrix remained higher than that isolated from the day-5 matrix, further supporting the conclusion that environmental exposure promotes the emergence or enrichment of resistant phenotypes. This observation is particularly important, as it indicates that the environmental effect identified in the present study is robust and does not depend solely on the stricter classification criteria adopted here. In other words, although our experimental cutoffs were more conservative than the currently accepted EFSA thresholds, the same overall pattern remained evident when the official breakpoints were applied. EFSA breakpoints were developed primarily for practical and regulatory purposes, including the safety assessment of probiotic strains intended for animal feed and human consumption. Although some authors have suggested that current EFSA microbiological cutoffs should be re-examined upward for certain *Lactobacillus* species–antibiotic combinations, our findings point in the opposite direction. Specifically, the present data suggests that higher breakpoint values may underestimate the proportion of strains with reduced susceptibility and may therefore obscure early shifts in MIC distributions associated with environmental exposure. From this perspective, lower or more sensitive microbiological cutoffs may better capture the adaptive dynamics of lactobacilli under environmentally relevant conditions [8,31,42,43].

L. acidophilus was selected for molecular analysis of resistance determinants because it was the most prevalent species among the isolates, representing 70 isolates (13.51%) in the day-5 matrix and 71 isolates (12.52%) in the day-30 matrix. In addition, the number of resistant strains was strongly correlated with the total number of isolates ($R = 0.95$, $p < 0.0001$ for the day-5 matrix; $R = 0.949$, $p < 0.001$ for the day-30 matrix), indicating that this species provided the greatest likelihood of detecting resistance genes. The frequencies of the detected genes are presented in Tables 7 and 8. A total of 21 genes were detected in 14 strains from the day-5 matrix, whereas 19 genes were detected in 13 strains from the day-30 matrix. More importantly, clear genotype–phenotype discordance was observed between the two matrices. The *tet*M gene was associated with oxytetracycline resistance in strains from the day-5 matrix, but with phenotypic susceptibility in strains from the day-30 matrix. A similar pattern was found for *erm*B, which corresponded to erythromycin resistance in the day-5 matrix but to susceptibility in the day-30 matrix. Likewise, *bla*TEM was associated in the day-5 matrix with resistance only to ampicillin, whereas in the day-30 matrix it was associated with resistance to all three β-lactam antibiotics tested. These findings indicate that environmental exposure may affect not only the occurrence of resistance determinants but also their expression. The presence of a resistance gene did not consistently translate into phenotypic resistance, and this relationship appeared to depend on the environmental matrix. Similar genotype–phenotype discordance has been reported previously. Hütt et al. (2018) identified *tet*M and *tet*K in lactobacilli that remained susceptible to oxytetracycline [44], while Anisimova et al. (2019) described strains carrying *tet*K and *erm*B despite showing susceptible phenotypes [37]. Such evidence suggests that molecular detection alone is insufficient for reliable prediction of resistance expression. Several mechanisms may explain this discrepancy. Resistance genes may be transcriptionally silent, altered by mutations in regulatory regions, or inactivated by premature stop codons, insertions, or deletions [45,46]. Even so, non-expressed genes may still remain epidemiologically relevant because they can retain the potential for horizontal transfer [47]. This is particularly important in environmentally exposed systems, where opportunities for gene exchange may be enhanced even in the absence of immediate phenotypic expression.

The detected genes also differed in function and mobility. *tet*M encodes a ribosomal protection protein and may be located on either chromosomes or plasmids, whereas

 https://doi.org/10.3390/microorganisms14030712

*tet*K encodes an efflux pump and is most often plasmid-borne, increasing its transfer potential [8,38]. *erm*B encodes an rRNA methylase targeting the 23S ribosomal subunit and is generally chromosomal, although it remains a key determinant of macrolide resistance [29,48]. The cat genes encode chloramphenicol acetyltransferases and are frequently associated with mobile genetic elements such as plasmids, supporting their transferability [49]. Finally, *bla*TEM genes are well known for their high horizontal mobility within bacterial communities [50]. Taken together, these results suggest that environmental exposure may shape antimicrobial resistance in lactobacilli through mechanisms extending beyond simple gene acquisition. In addition to selecting resistance determinants, environmental conditions may influence whether such genes are phenotypically expressed. This highlights the complexity of resistance ecology in lactobacilli and underscores the need to interpret molecular findings alongside phenotypic susceptibility data. The observed changes in resistance patterns do not, by themselves, identify the exact mechanism responsible for these shifts. Although environmental exposure is considered an important factor, especially under artisanal production conditions with limited protection from external microbial inputs, other explanations should also be taken into account. These include microbial competition, ecological selection during fermentation and ripening, and possible horizontal transfer of resistance determinants within bacterial communities. Therefore, the term environmental pressure should be interpreted broadly, encompassing not only direct exposure to environmental microbiota, but also the selective and interactive processes operating during fermentation. Given the open and dynamic nature of artisanal production environments, the observed AMR shifts are likely to reflect the combined effects of environmental exposure, microbial interactions, and fermentation-associated selection rather than a single causal mechanism [51,52].

5. Limitations of This Study

A limitation of the present study lies in the selection of only two fermentation stages to assess antimicrobial resistance dynamics during artisanal cheese production. These time points were chosen to represent two distinct phases of the process and to allow an initial evaluation of resistance patterns over time in relation to environmental exposure. The selection was empirical, as closely comparable stage-based studies were not available. In this context, the 5–7-day curd may still reflect, to a considerable extent, the original milk microbiota and animal-associated inputs, whereas the $\geq$30-day curd is more likely to reflect the cumulative effects of fermentation, handling, and environmental exposure. Nevertheless, inclusion of additional intermediate sampling points would provide a more refined view of the temporal evolution of AMR during fermentation and should be considered in future investigations.

This consideration also relates to a broader interpretive constraint of the study. Comparisons with previous reports should be made with caution, since differences in artisanal production practices, fermentation stage, microbial ecology, and analytical methodology may substantially influence the AMR patterns observed in lactobacilli. Accordingly, although the present findings offer useful insight into the possible dynamics of resistance in artisanal dairy systems, they should be interpreted within the specific ecological and technological context of the production system examined here.

6. Conclusions

The present study demonstrates that artisanal dairy matrices sampled at different fermentation stages provide a useful model for investigating antimicrobial resistance (AMR) in environmentally exposed lactobacilli. The novelty of the present research does not rely upon the isolation of resistant bacteria from artisan products but on the recording of this re-

 https://doi.org/10.3390/microorganisms14030712

sistance in two different time frames and thus proving that resistance is a dynamic not static effect and -particularly for artisan products- reflects the environmental impact. In that sense artisan products can be used as a sentinel system for assessing the environmental burden of AMR. In industrial scale fermentations, the environment in which these fermentations take place is highly monitored by systems checking the air, the circulation of personnel, etc., and thus its impact is minimal as it should be otherwise the fermented products could end with different properties in every lot, depending on the circumstances. Resistance increased overall from day 5 to day 30 across most *Lactobacillus* species and for most antibiotics tested, while molecular analysis revealed the presence of resistance determinants, including genes with transfer potential, as well as genotype–phenotype discordance. These findings suggest that prolonged environmental exposure during artisanal fermentation may contribute not only to the selection or enrichment of resistant phenotypes, but also to variation in the expression of resistance genes. From a One Health perspective, artisanal dairy fermentations may therefore serve as practical indicators of local AMR pressure at the interface of food, animals, humans, and the environment. Further studies integrating phenotypic, genomic, and source-tracking approaches are needed to clarify the mechanisms underlying these shifts and their epidemiological significance.

In this study we propose a paradigm: monitoring the artisan matrix for resistant lactobacilli in two different time frames and assessing the observed difference, to conclude on environmental burden of resistance genes. This paradigm can work in any artisan environment, since in artisan production the controls of industrial production are not applicable.

Supplementary Materials: The following supporting information can be downloaded at: https://www.mdpi.com/article/10.3390/microorganisms14030712/s1, Supplementary Material; Table S1: Primer sequences and PCR assay conditions for screening antibiotic resistance genes; Table S2: MIC values range and experimental cut off values of the Lactobacilli strains isolated from the 5th day curd; Table S3: MIC values range and experimental cut off values of the Lactobacilli strains isolated from the 30th day curd; Table S4: Alterations in the upper limit of the MIC range from the 5th to the 30th day; Table S5: Proposed cut-off MIC values by EFSA (2018) for *Lactobacillus* spp. [31]; Table S6: Resistant strains (n) by EFSA breakpoints and by the experimental cut off values of this study; Table S7: Lactobacilli taxa isolated in this study and their reported occurrence across potential One Health reservoirs (livestock, humans, companion animals, and environmental niches) based on the published literature. Refs. [53–112] are cited in Supplementary Materials.

Author Contributions: Conceptualization, G.R., E.B. and C.V.; methodology, G.R., K.F., K.N.,V.G., I.S., E.B., A.T. and C.V.; software, G.R., K.F., K.N., V.G. and A.D.; validation, G.R., K.F., E.B., A.T. and C.V.; formal analysis, G.R., K.F., K.N., L.H. and V.G.; investigation, G.R., K.F., V.G., K.N., A.D., L.H. and C.V.; resources, G.R., K.F., K.N., A.D. and V.G.; data curation, K.F., G.R., K.N., A.D., L.H. and V.G.; writing—original draft preparation, G.R.; writing—review and editing, E.B., I.S., A.T. and C.V.; visualization, G.R., E.B., I.S., A.T. and C.V.; supervision, E.B., A.T. and C.V.; project administration, E.B., A.T. and C.V.; funding acquisition, E.B., A.T. and C.V. All authors have read and agreed to the published version of the manuscript.

Funding: This research received no external funding.

Institutional Review Board Statement: Not applicable.

Informed Consent Statement: Not applicable.

Data Availability Statement: The original contributions presented in this study are included in the article/Supplementary Material. Further inquiries can be directed to the corresponding authors.

Acknowledgments: The authors have reviewed and edited the output and take full responsibility for the content of this publication.

Conflicts of Interest: The authors declare no conflicts of interest.

https://doi.org/10.3390/microorganisms14030712

References

1. Ahmad, M.; Khan, A.U. Global economic impact of antibiotic resistance: A review. *J. Glob. Antimicrob. Resist.* **2019**, *19*, 313–316 [CrossRef]
2. Pulingam, T.; Parumasivam, T.; Gazzali, A.M.; Sulaiman, A.M.; Chee, J.Y.; Lakshmanan, M.; Chin, C.F.; Sudesh, K. Antimicrobial resistance: Prevalence, economic burden, mechanisms of resistance and strategies to overcome. *Eur. J. Pharm. Sci.* **2022**, *170* 106103. [CrossRef]
3. McDonnell, A.; Countryman, A.; Laurence, T.; Gulliver, S.; Drake, T.; Edwards, S.; Kenny, C.; Lamberti, O.; Morton, A.; Shafira, A., et al. *Forecasting the Fallout from AMR: Economic Impacts of Antimicrobial Resistance in Humans—A Report from the EcoAMR Series*, Licence: CC BY-SA 3.0 IGO; World Organisation for Animal Health: Paris, France; World Bank: Washington, DC, USA, 2024; p. 58. [CrossRef]
4. GBD 2021 Antimicrobial Resistance Collaborators. Global burden of bacterial antimicrobial resistance 1990–2021: A systematic analysis with forecasts to 2050. *Lancet* **2024**, *404*, 1199–1226. [CrossRef]
5. Naddaf, M. 40 million deaths by 2050: Toll of drug-resistant infections to rise by 70. *Nature* **2024**, *633*, 747–748. [CrossRef]
6. Available online: https://www.who.int/news-room/fact-sheets/detail/antimicrobial-resistance (accessed on 30 January 2026)
7. Dionisio, F.; Domingues, C.P.F.; Rebelo, J.S.; Monteiro, F.; Nogueira, T. The Impact of Non-Pathogenic Bacteria on the Spread of Virulence and Resistance Genes. *Int. J. Mol. Sci.* **2023**, *24*, 1967. [CrossRef]
8. Campedelli, I.; Mathur, H.; Salvetti, E.; Clarke, S.; Rea, M.C.; Torriani, S.; Ross, R.P.; Hill, C.; O'Toole, P.W. Genus-Wide Assessment of Antibiotic Resistance in *Lactobacillus* spp. *Appl. Environ. Microbiol.* **2018**, *85*, e01738–18. [CrossRef]
9. Salvetti, E.; Torriani, S.; Felis, G.E. The Genus *Lactobacillus*: A Taxonomic Update. *Probiotics Antimicrob. Proteins* **2012**, *4*, 217–226 [CrossRef]
10. Goldstein, E.J.; Tyrrell, K.L.; Citron, D.M. *Lactobacillus* Species: Taxonomic Complexity and Controversial Susceptibilities. *Clin Infect. Dis.* **2015**, *60*, S98–S107. [CrossRef] [PubMed]
11. Ibrahim, F.; Lebeer, S.; Salvetti, E.; Felis, G. *The Genus Lactobacillus—Across the Past and Future*; CRC Press: Boca Raton, FL, USA, 2024. [CrossRef]
12. Zheng, J.; Wittouck, S.; Salvetti, E.; Franz, C.M.A.P.; Harris, H.M.B.; Mattarelli, P.; O'Toole, P.W.; Pot, B.; Vandamme, P.; Walter, J., et al. A taxonomic note on the genus *Lactobacillus*: Description of 23 novel genera, emended description of the genus *Lactobacillus* Beijerinck 1901, and union of *Lactobacillaceae* and *Leuconostocaceae*. *Int. J. Syst. Evol. Microbiol.* **2020**, *70*, 2782–2858. [CrossRef] [PubMed]
13. Velazquez-Meza, M.E.; Galarde-López, M.; Carrillo-Quiróz, B.; Alpuche-Aranda, C.M. Antimicrobial resistance: One Health approach. *Vet. World* **2022**, *15*, 743–749. [CrossRef] [PubMed]
14. Singh, K.S.; Anand, S.; Dholpuria, S.; Sharma, J.K.; Blankenfeldt, W.; Shouche, Y. Antimicrobial resistance dynamics and the one-health strategy: A review. *Environ. Chem. Lett.* **2021**, *19*, 2995–3007. [CrossRef]
15. Despotovic, M.; de Nies, L.; Busi, S.B.; Wilmes, P. Reservoirs of antimicrobial resistance in the context of One Health. *Curr. Opin. Microbiol.* **2023**, *73*, 102291. [CrossRef]
16. Rozos, G.; Voidarou, C.; Stavropoulou, E.; Skoufos, I.; Tzora, A.; Alexopoulos, A.; Bezirtzoglou, E. Biodiversity and Microbial Resistance of Lactobacilli Isolated From the Traditional Greek Cheese Kopanisti. *Front. Microbiol.* **2018**, *9*, 517. [CrossRef]
17. Tzora, A.; Nelli, A.; Voidarou, C.C.; Fotou, K.; Bonos, E.; Rozos, G.; Grigoriadou, K.; Papadopoulos, P.; Basdagianni, Z.; Giannenas, I.; et al. Impact of an Omega-3-Enriched Sheep Diet on the Microbiota and Chemical Composition of Kefalograviera Cheese *Foods* **2022**, *11*, 843. [CrossRef]
18. Voidarou, C.; Alexopoulos, A.; Tsinas, A.; Rozos, G.; Tzora, A.; Skoufos, I.; Varzakas, T.; Bezirtzoglou, E. Effectiveness of Bacteriocin-Producing Lactic Acid Bacteria and Bifidobacterium Isolated from Honeycombs against Spoilage Microorganisms and Pathogens Isolated from Fruits and Vegetables. *Appl. Sci.* **2020**, *10*, 7309. [CrossRef]
19. Tzora, A.; Nelli, A.; Voidarou, C.; Fthenakis, G.; Rozos, G.; Theodorides, G.; Bonos, E.; Skoufos, I. Microbiota "Fingerprint" of Greek Feta Cheese through Ripening. *Appl. Sci.* **2021**, *11*, 5631. [CrossRef]
20. Dec, M.; Urban-Chmiel, R.; Stępień-Pyśniak, D.; Wernicki, A. Assessment of antibiotic susceptibility in *Lactobacillus* isolates from chickens. *Gut Pathog.* **2017**, *9*, 54. [CrossRef]
21. Dec, M.; Herman-Ostrzyżek, K.; Zomer, A.; Urban-Chmiel, R. Susceptibility of *Lactobacillaceae* Strains to Aminoglycoside Antibiotics in the Light of EFSA Guidelines. *Life* **2025**, *15*, 732. [CrossRef]
22. Mayrhofer, S.; van Hoek, A.H.; Mair, C.; Huys, G.; Aarts, H.J.; Kneifel, W.; Domig, K.J. Antibiotic susceptibility of members of the Lactobacillus acidophilus group using broth microdilution and molecular identification of their resistance determinants. *Int. J Food Microbiol.* **2010**, *144*, 81–87. [CrossRef] [PubMed]
23. Werner, G.; Willems, R.J.; Hildebrandt, B.; Klare, I.; Witte, W. Influence of transferable genetic determinants on the outcome of typing methods commonly used for Enterococcus faecium. *J. Clin. Microbiol.* **2003**, *41*, 1499–1506. [CrossRef] [PubMed]
24. Gad, G.F.; Abdel-Hamid, A.M.; Farag, Z.S. Antibiotic resistance in lactic acid bacteria isolated from some pharmaceutical and dairy products. *Braz. J. Microbiol.* **2014**, *45*, 25–33. [CrossRef]

 https://doi.org/10.3390/microorganisms14030712

25. Gevers, D.; Danielsen, M.; Huys, G.; Swings, J. Molecular characterization of tet(M) genes in Lactobacillus isolates from different types of fermented dry sausage. *Appl. Environ. Microbiol.* **2003**, *69*, 1270–1275. [CrossRef]
26. Anisimova, E.; Gorokhova, I.; Karimullina, G.; Yarullina, D. Alarming Antibiotic Resistance of Lactobacilli Isolated from Probiotic Preparations and Dietary Supplements. *Antibiotics* **2022**, *11*, 1557. [CrossRef]
27. Egervärn, M.; Roos, S.; Lindmark, H. Identification and characterization of antibiotic resistance genes in *Lactobacillus reuteri* and *Lactobacillus plantarum*. *J. Appl. Microbiol.* **2009**, *107*, 1658–1668. [CrossRef] [PubMed]
28. Colom, K.; Pérez, J.; Alonso, R.; Fernández-Aranguiz, A.; Lariño, E.; Cisterna, R. Simple and reliable multiplex PCR assay for detection of blaTEM, bla(SHV) and blaOXA-1 genes in Enterobacteriaceae. *FEMS Microbiol. Lett.* **2003**, *223*, 147–151. [CrossRef] [PubMed]
29. Hummel, A.S.; Hertel, C.; Holzapfel, W.H.; Franz, C.M. Antibiotic resistances of starter and probiotic strains of lactic acid bacteria. *Appl. Environ. Microbiol.* **2007**, *73*, 730–739. [CrossRef] [PubMed]
30. Guo, H.; Pan, L.; Li, L.; Lu, J.; Kwok, L.; Menghe, B.; Zhang, H.; Zhang, W. Characterization of Antibiotic Resistance Genes from Lactobacillus Isolated from Traditional Dairy Products. *J. Food Sci.* **2017**, *82*, 724–730. [CrossRef]
31. EFSAFEEDAPPanel; Rychen, G.; Aquilina, G.; Azimonti, G.; Bampidis, V.; de Lourdes Bastos, M.; Bories, G.; Chesson, A.; Cocconcelli, P.S.; Flachowsky, G.; et al. Guidance on the characterisation of microorganisms used as feed additives or as production organisms. *EFSA J.* **2018**, *16*, 5206. [CrossRef]
32. Sánchez, Ó.J.; Barragán, P.J.; Serna, L. Review of Lactobacillus in the food industry and their culture media. *Rev. Colomb. Biotecnol.* **2019**, *21*, 63–76. [CrossRef]
33. Daliu, P.; Souto, E.B.; Santini, A. Novel applications of Lactobacillus in the food industry. *J. Asian Sci. Res.* **2025**, *15*, 111–121. [CrossRef]
34. Dewi, G.; Johny, A.K. Lactobacillus in Food Animal Production—A Forerunner for Clean Label Prospects in Animal-Derived Products. *Front. Sustain. Food Syst.* **2022**, *6*, 831195. [CrossRef]
35. Rafique, N.; Mamoona, T.; Bashir, S.; Hussain, I.; Hayat, I. Lactobacilli: Application in Food Industry. In *Lactobacillus—A Multifunctional Genus*; IntechOpen: Rijeka, Croatia, 2023. [CrossRef]
36. Zhang, S.; Oh, J.; Alexander, L.M.; Özçam, M.; van Pijkeren, J. d-Alanyl-d-Alanine Ligase as a Broad-Host-Range Counterselection Marker in Vancomycin-Resistant Lactic Acid Bacteria. *J. Bacteriol.* **2018**, *200*, e00607-17. [CrossRef]
37. Anisimova, E.A.; Yarullina, D.R. Antibiotic Resistance of *LACTOBACILLUS* Strains. *Curr. Microbiol.* **2019**, *76*, 1407–1416. [CrossRef] [PubMed]
38. Das, D.J.; Shankar, A.; Johnson, J.B.; Thomas, S. Critical insights into antibiotic resistance transferability in probiotic *Lactobacillus*. *Nutrition* **2020**, *69*, 110567. [CrossRef] [PubMed]
39. Álvarez-Cisneros, Y.M.; Ponce-Alquicira, E. Antibiotic Resistance in Lactic Acid Bacteria. In *Antimicrobial Resistance—A Global Threat*; IntechOpen: Rijeka, Croatia, 2019. [CrossRef]
40. Qu, C.; Wu, Z.; Pan, D.; Cai, Z.; Liu, X. Characterization of *Lactobacillus reuteri* WQ-Y1 with the ciprofloxacin degradation ability. *Biotechnol. Lett.* **2021**, *43*, 855–864. [CrossRef] [PubMed]
41. Armstrong, E.; Hemmerling, A.; Miller, S.; Burke, K.E.; Newmann, S.J.; Morris, S.R.; Reno, H.; Huibner, S.; Kulikova, M.; Liu, R.; et al. Metronidazole treatment rapidly reduces genital inflammation through effects on bacterial vaginosis-associated bacteria rather than lactobacilli. *J. Clin. Investig.* **2022**, *132*, e152930. [CrossRef]
42. Nøhr-Meldgaard, K.; Struve, C.; Ingmer, H.; Koza, A.; Al-Nakeeb, K.; Agersø, Y. Antimicrobial susceptibility testing and tentative epidemiological cut-off values for *Lactobacillaceae* family species intended for ingestion. *Front. Antibiot.* **2023**, *2*, 1162636. [CrossRef]
43. Kahlmeter, G.; Turnidge, J. Wild-type distributions of minimum inhibitory concentrations and epidemiological cut-off values-laboratory and clinical utility. *Clin. Microbiol. Rev.* **2023**, *36*, e0010022. [CrossRef]
44. Hütt, P.; Lapp, E.; Štšepetova, J.; Smidt, I.; Taelma, H.; Borovkova, N.; Oopkaup, H.; Ahelik, A.; Rööp, T.; Hoidmets, D.; et al. Characterisation of probiotic properties in human vaginal lactobacilli strains. *Microb. Ecol. Health Dis.* **2016**, *27*, 30484. [CrossRef]
45. Shao, Y.; Zhang, W.; Guo, H.; Pan, L.; Zhang, H.; Sun, T. Comparative studies on antibiotic resistance in *Lactobacillus casei* and *Lactobacillus plantarum*. *Food Control* **2015**, *50*, 250–258. [CrossRef]
46. Colautti, A.; Rossi, N.; Piazza, C.; Comi, G.; Iacumin, L. Draft genome sequences of 14 *Lacticaseibacillus* spp. strains, representatives of a collection of 200 strains. *Microbiol. Resour. Announc.* **2023**, *12*, e0048523. [CrossRef] [PubMed]
47. Arnold, B.J.; Huang, I.T.; Hanage, W.P. Horizontal gene transfer and adaptive evolution in bacteria. *Nat. Rev. Microbiol.* **2022**, *20*, 206–218. [CrossRef] [PubMed]
48. Zare, H.; Izadi Amoli, R.; Zaboli, F.; Rezapour, M.; Kaboosi, H. The phenotypic and genotypic antibiotic susceptibility of vaginal Lactobacillus with potential probiotic properties isolated from healthy women in northern Iran. *Iran. J. Microbiol.* **2024**, *16*, 227–235. [CrossRef] [PubMed]
49. Moradi, J.; Fathollahi, M.; Halimi, S.; Alvandi, A.; Abiri, R.; Vaziri, S.; Rezaei, A. Characterization of the resistome in *Lactobacillus* genomic sequences from the human gut. *J. Glob. Antimicrob. Resist.* **2022**, *30*, 451–458. [CrossRef]

50. Floris, I.; Battistini, R.; Tramuta, C.; Garcia-Vozmediano, A.; Musolino, N.; Scardino, G.; Masotti, C.; Brusa, B.; Orusa, R.; Serracca, L.; et al. Antibiotic Resistance in Lactic Acid Bacteria from Dairy Products in Northern Italy. *Antibiotics* **2025**, *14*, 375. [CrossRef]
51. Larsson, D.G.J.; Flach, C.F. Antibiotic resistance in the environment. *Nat. Rev. Microbiol.* **2022**, *20*, 257–269. [CrossRef]
52. Samtiya, M.; Matthews, K.R.; Dhewa, T.; Puniya, A.K. Antimicrobial Resistance in the Food Chain: Trends, Mechanisms, Pathways, and Possible Regulation Strategies. *Foods* **2022**, *11*, 2966. [CrossRef]
53. Wang, L.; Fang, M.; Hu, Y.; Yang, Y.; Yang, M.; Chen, Y. Characterization of the most abundant *Lactobacillus* species in chicken gastrointestinal tract and potential use as probiotics for genetic engineering. *Acta Biochim. Biophys. Sin.* **2014**, *46*, 612–619. [CrossRef]
54. Verso, L.L.; Lessard, M.; Talbot, G.; Fernandez, B.; Fliss, I. Isolation and Selection of Potential Probiotic Bacteria from the Pig Gastrointestinal Tract. *Probiotics Antimicrob. Proteins* **2017**, *10*, 299–312. [CrossRef]
55. Donnell, M.M.O.; Harris, H.M.B.; Lynch, D.B.; Ross, R.P.; O'toole, P.W. *Lactobacillus ruminis* strains cluster according to their mammalian gut source. *BMC Microbiol.* **2015**, *15*, 80. [CrossRef]
56. Zhang, J.; McWhorter, A.R.; Khan, S.; Willson, N.-L.; Chousalkar, K.K. Characterization of *Lactobacillus* spp. isolated from layer hens as probiotic candidates. *BMC Veter. Res.* **2025**, *21*, 416. [CrossRef]
57. Plaizier, J.C.; Li, S.; Danscher, A.M.; Derakshani, H.; Andersen, P.H.; Khafipour, E. Changes in microbiota in rumen digesta and feces due to a grain-based subacute ruminal acidosis (SARA) challenge. *Microb. Ecol.* **2017**, *74*, 485–495. [CrossRef]
58. Maldonado, N.C.; de Ruiz, C.S.; Otero, M.C.; Sesma, F.; Nader-Macías, M.E. Lactic acid bacteria isolated from young calves—Characterization and potential as probiotics. *Res. Veter. Sci.* **2012**, *92*, 342–349. [CrossRef] [PubMed]
59. Walter, J. Ecological Role of Lactobacilli in the Gastrointestinal Tract: Implications for Fundamental and Biomedical Research. *Appl. Environ. Microbiol.* **2008**, *74*, 4985–4996. [CrossRef] [PubMed]
60. Han, H.; Ogata, Y.; Yamamoto, Y.; Nagao, S.; Nishino, N. Identification of lactic acid bacteria in the rumen and feces of dairy cows fed total mixed ration silage to assess the survival of silage bacteria in the gut. *J. Dairy Sci.* **2014**, *97*, 5754–5762. [CrossRef]
61. Sirisopapong, M.; Shimosato, T.; Okrathok, S.; Khempaka, S. Assessment of lactic acid bacteria isolated from the chicken digestive tract for potential use as poultry probiotics. *Anim. Biosci.* **2023**, *36*, 1209–1220. [CrossRef]
62. Lin, W.-C.; Ptak, C.P.; Chang, C.-Y.; Ian, M.-K.; Chia, M.-Y.; Chen, T.-H.; Kuo, C.-J. Autochthonous Lactic Acid Bacteria Isolated From Dairy Cow Feces Exhibiting Promising Probiotic Properties and in vitro Antibacterial Activity Against Foodborne Pathogens in Cattle. *Front. Veter. Sci.* **2020**, *7*, 239. [CrossRef]
63. Abdou, A.M.; Fouad, E.A.; Alam, S.S.; Hakim, A.S. Isolation and identification of probiotic lactobacilli from non-ruminant animals. *Inter. J. Vet. Sci.* **2019**, *8*, 349–354.
64. Hattab, J.; Marruchella, G.; Pallavicini, A.; Gionechetti, F.; Mosca, F.; Trachtman, A.R.; Lanci, L.; Gabrielli, L.; Tiscar, P.G. Insights into the Oral Bacterial Microbiota of Sows. *Microorganisms* **2021**, *9*, 2314. [CrossRef]
65. Lim, J.-A.; Cha, J.; Choi, S.; Kim, J.-H.; Kim, D. Early Colonization of the Intestinal Microbiome of Neonatal Piglets Is Influenced by the Maternal Microbiome. *Animals* **2023**, *13*, 3378. [CrossRef]
66. Izhar, M.Z.; Nawaz, M.; Yaqub, T.; Avais, M.; Anjum, A.A. In vitro characterization of probiotic potential of *Lactobacillus plantarum* CM49 against selected cattle mastitogens. *BMC Microbiol.* **2024**, *24*, 310. [CrossRef]
67. Monareng, N.J.; Ncube, K.T.; van Rooi, C.; Modiba, M.C.; Mtileni, B. A Systematic Review on Microbial Profiling Techniques in Goat Milk: Implications for Probiotics and Shelf-Life. *Int. J. Mol. Sci.* **2025**, *26*, 5551. [CrossRef]
68. Toquet, M.; Gómez-Martín, Á.; Bataller, E. Review of the bacterial composition of healthy milk, mastitis milk and colostrum in small ruminants. *Res. Veter. Sci.* **2021**, *140*, 1–5. [CrossRef]
69. Derakhshani, H.; Fehr, K.B.; Sepehri, S.; Francoz, D.; De Buck, J.; Barkema, H.W.; Plaizier, J.C.; Khafipour, E. Invited review: Microbiota of the bovine udder: Contributing factors and potential implications for udder health and mastitis susceptibility. *J. Dairy Sci.* **2018**, *101*, 10605–10625. [CrossRef] [PubMed]
70. Swartz, J.D.; Lachman, M.; Westveer, K.; O'Neill, T.; Geary, T.; Kott, R.W.; Berardinelli, J.G.; Hatfield, P.G.; Thomson, J.M.; Roberts, A.; et al. Characterization of the Vaginal Microbiota of Ewes and Cows Reveals a Unique Microbiota with Low Levels of Lactobacilli and Near-Neutral pH. *Front. Veter. Sci.* **2014**, *1*, 19. [CrossRef] [PubMed]
71. Poole, R.K.; Soffa, D.R.; McAnally, B.E.; Smith, M.S.; Hickman-Brown, K.J.; Stockland, E.L. Reproductive Microbiomes in Domestic Livestock: Insights Utilizing 16S rRNA Gene Amplicon Community Sequencing. *Animals* **2023**, *13*, 485. [CrossRef] [PubMed]
72. Frétin, M.; Martin, B.; Rifa, E.; Isabelle, V.-M.; Pomiès, D.; Ferlay, A.; Montel, M.-C.; Delbès, C. Bacterial community assembly from cow teat skin to ripened cheeses is influenced by grazing systems. *Sci. Rep.* **2018**, *8*, 200. [CrossRef]
73. Bouchard, D.S.; Seridan, B.; Saraoui, T.; Rault, L.; Germon, P.; Gonzalez-Moreno, C.; Nader-Macias, F.M.E.; Baud, D.; François, P.; Chuat, V.; et al. Lactic Acid Bacteria Isolated from Bovine Mammary Microbiota: Potential Allies against Bovine Mastitis. *PLoS ONE* **2015**, *10*, e0144831. [CrossRef]
74. Kagkli, D.M.; Vancanneyt, M.; Hill, C.; Vandamme, P.; Cogan, T.M. Enterococcus and Lactobacillus contamination of raw milk in a farm dairy environment. *Int. J. Food Microbiol.* **2007**, *114*, 243–251. [CrossRef]

 https://doi.org/10.3390/microorganisms14030712

5. Delanghe, L.; De Boeck, I.; Van Malderen, J.; Gehrmann, T.; Allonsius, C.N.; Bron, P.A.; Claes, I.; Hagendorens, M.; Leysen, J.; Wittouck, S.; et al. The inner elbow skin microbiome contains *Lactobacillus* among its core taxa and varies with age, season and lifestyle. *Microbiome Res. Rep.* **2024**, *3*, 43. [CrossRef]
6. Skowron, K.; Bauza-Kaszewska, J.; Kraszewska, Z.; Wiktorczyk-Kapischke, N.; Grudlewska-Buda, K.; Kwiecińska-Piróg, J.; Wałecka-Zacharska, E.; Radtke, L.; Gospodarek-Komkowska, E. Human Skin Microbiome: Impact of Intrinsic and Extrinsic Factors on Skin Microbiota. *Microorganisms* **2021**, *9*, 543. [CrossRef]
7. Kõll, P.; Mändar, R.; Marcotte, H.; Leibur, E.; Mikelsaar, M.; Hammarström, L. Characterization of oral lactobacilli as potential probiotics for oral health. *Oral Microbiol. Immunol.* **2008**, *23*, 139–147. [CrossRef]
8. Badet, C.; Thebaud, N. Ecology of Lactobacilli in the Oral Cavity: A Review of Literature. *Open Microbiol. J.* **2008**, *2*, 38–48. [CrossRef] [PubMed]
9. Caufield, P.; Schön, C.; Saraithong, P.; Li, Y.; Argimón, S. Oral Lactobacilli and Dental Caries: A Model for Niche Adaptation in Humans. *J. Dent. Res.* **2015**, *94*, 110S–118S. [CrossRef] [PubMed]
0. Zavisic, G.; Petricevic, S.; Radulovic, Z.; Begovic, J.; Golic, N.; Topisirovic, L.; Strahinic, I. Probiotic features of two oral *Lactobacillus isolates*. *Braz. J. Microbiol.* **2012**, *43*, 418–428. [CrossRef] [PubMed]
1. Wasfi, R.; El-Rahman, O.A.A.; Zafer, M.M.; Ashour, H.M. Probiotic *Lactobacillus* sp. inhibit growth, biofilm formation and gene expression of caries-inducing *Streptococcus mutans*. *J. Cell. Mol. Med.* **2018**, *22*, 1972–1983. [CrossRef]
2. Kõll-Klais, P.; Mändar, R.; Leibur, E.; Marcotte, H.; Hammarström, L.; Mikelsaar, M. Oral lactobacilli in chronic periodontitis and periodontal health: Species composition and antimicrobial activity. *Oral Microbiol. Immunol.* **2005**, *20*, 354–361. [CrossRef]
3. Huang, R.; Wu, F.; Zhou, Q.; Wei, W.; Yue, J.; Xiao, B.; Luo, Z. Lactobacillus and intestinal diseases: Mechanisms of action and clinical applications. *Microbiol. Res.* **2022**, *260*, 127019. [CrossRef]
4. Dempsey, E.; Corr, S.C. *Lactobacillus* spp. for Gastrointestinal Health: Current and Future Perspectives. *Front. Immunol.* **2022**, *13*, 840245. [CrossRef]
5. De Boeck, I.; Broek, M.F.v.D.; Allonsius, C.N.; Spacova, I.; Wittouck, S.; Martens, K.; Wuyts, S.; Cauwenberghs, E.; Jokicevic, K.; Vandenheuvel, D.; et al. Lactobacilli Have a Niche in the Human Nose. *Cell Rep.* **2020**, *31*, 107674. [CrossRef]
6. Silva-Bea, S.; Francisco-Tomé, M.; Cabrera-Alvargonzález, J.J.; Potel, C.; Álvarez, M.; Pérez, S.; Regueiro, B.; Cabral, M.P. In vivo monitoring of Lactiplantibacillus plantarum in the nasal and vaginal mucosa using infrared fluorescence. *Appl. Microbiol. Biotechnol.* **2022**, *106*, 6239–6251. [CrossRef]
7. Chen, M.; He, S.; Miles, P.; Li, C.; Ge, Y.; Yu, X.; Wang, L.; Huang, W.; Kong, X.; Ma, S.; et al. Nasal Bacterial Microbiome Differs Between Healthy Controls and Those With Asthma and Allergic Rhinitis. *Front. Cell. Infect. Microbiol.* **2022**, *12*, 841995. [CrossRef]
8. Dehcheshmeh, M.M.; Grant, L.; Ebrahimie, E.; Khabiri, A.; Hemmatzadeh, F.; Shipstone, M.; Trott, D.J. Oral probiotic and postbiotic supplementation enhances the abundance of *Lactobacillus acidophilus*, *Lactobacillus johnsonii*, and *Limosilactobacillus reuteri* in both canine skin and gastrointestinal microbiota: Insights from long-read 16S rRNA gene sequencing. *Veter. Res. Commun.* **2025**, *50*, 1–13. [CrossRef]
9. Štempelová, L.; Micenková, L.; Andrla, P.; Strompfová, V.; Lucia. The skin microbiome on healthy and inflammatory altered canine skin determined by next generation sequencing. *Front. Microbiol.* **2025**, *16*, 1528747. [CrossRef]
0. Grześkowiak, Ł.; Endo, A.; Beasley, S.; Salminen, S. Microbiota and probiotics in canine and feline welfare. *Anaerobe* **2015**, *34*, 14–23. [CrossRef] [PubMed]
1. Raju, M.K.; Kapali, B.S.C.; JebaSingh, G.V.J.; Subathra, Y.; Nithya, A.D.M. Isolation, Characterization and Sequencing of Lactobacillus from the Oral and Fecal Samples of Healthy Dogs. *Res. J. Pharm. Technol.* **2018**, *11*, 5061–5065. [CrossRef]
2. Kalaiselvi, G.; Pazhanivel, N.; Bobade, S.S. Characterisation of Canine-derived Lactic Acid Bacteria from Faecal and Oral Sources: Potential for Probiotic Use in Dog Health. In *Microbiology and Biotechnology Research: An Overview*; BP International: Hong Kong, China, 2025; Volume 3, pp. 131–142.
3. Zhao, M.; Li, Y.; Zhang, Y.; Li, G. Genomic analysis and functional properties of *Lactobacillus johnsonii* GJ231 isolated from healthy beagles. *Front. Microbiol.* **2024**, *15*, 1437036. [CrossRef] [PubMed]
4. Jang, H.-J.; Son, S.; Kim, J.-A.; Jung, M.Y.; Choi, Y.-J.; Kim, D.-H.; Lee, H.K.; Shin, D.; Kim, Y. Characterization and Functional Test of Canine Probiotics. *Front. Microbiol.* **2021**, *12*, 625562. [CrossRef]
5. Niemiec, B.A.; Gawor, J.; Tang, S.; Prem, A.; Krumbeck, J.A. The bacteriome of the oral cavity in healthy dogs and dogs with periodontal disease. *Am. J. Veter. Res.* **2022**, *83*, 50–58. [CrossRef]
6. Jang, H.-J.; Kim, J.-A.; Kim, Y. Characterization of feline-originated probiotics *Lactobacillus rhamnosus* CACC612 and Bifidobacterium animalis subsp. lactis CACC789 and evaluation of their host response. *BMC Veter. Res.* **2024**, *20*, 128. [CrossRef]
7. Masuoka, H.; Shimada, K.; Kiyosue-Yasuda, T.; Kiyosue, M.; Oishi, Y.; Kimura, S.; Ohashi, Y.; Fujisawa, T.; Hotta, K.; Yamada, A.; et al. Transition of the intestinal microbiota of cats with age. *PLoS ONE* **2017**, *12*, e0181739. [CrossRef]
8. Bugrov, N.; Rudenko, P.; Lutsay, V.; Gurina, R.; Zharov, A.; Khairova, N.; Molchanova, M.; Krotova, E.; Shopinskaya, M.; Bolshakova, M.; et al. Fecal Microbiota Analysis in Cats with Intestinal Dysbiosis of Varying Severity. *Pathogens* **2022**, *11*, 234. [CrossRef]

99. Older, C.E.; Diesel, A.B.; Lawhon, S.D.; Queiroz, C.R.R.; Henker, L.C.; Hoffmann, A.R. The feline cutaneous and oral microbiota are influenced by breed and environment. *PLoS ONE* **2019**, *14*, e0220463. [CrossRef] [PubMed]
100. Singhal, N.; Singh, N.S.; Mohanty, S.; Kumar, M.; Virdi, J.S. Rhizospheric *Lactobacillus plantarum* (*Lactiplantibacillus plantarum*) strains exhibit bile salt hydrolysis, hypocholestrolemic and probiotic capabilities in vitro. *Sci. Rep.* **2021**, *11*, 15288. [CrossRef] [PubMed]
101. Jaffar, N.S.; Jawan, R.; Chong, K.P. The potential of lactic acid bacteria in mediating the control of plant diseases and plant growth stimulation in crop production—A mini review. *Front. Plant Sci.* **2023**, *13*, 1047945. [CrossRef]
102. Zhang, X.; Liao, H.; Cai, T.; Cai, P.; Wu, X.; Wang, Z.; Ma, H.; Qiu, G.; Zhao, M.; Lu, X.; et al. Features and rhizosphere colonization strategies of *Lactobacillus plantarum* 0308 in soil-tomato systems. *Front. Microbiol.* **2025**, *16*, 1652881. [CrossRef]
103. Skotniczny, M.; Satora, P. Molecular Detection and Identification of Plant-Associated *Lactiplantibacillus plantarum*. *Int. J. Mol. Sci.* **2023**, *24*, 4853. [CrossRef]
104. Mota-Gutierrez, J.; Cocolin, L. Current trends and applications of plant origin lactobacilli in the promotion of sustainable food systems. *Trends Food Sci. Technol.* **2021**, *114*, 198–211. [CrossRef]
105. Amin, M.; Jorfi, M.; Khosravi, A.; Samarbafza, A.; Sheikh, A.F. Isolation and Identification of *Lactobacillus casei* and *Lactobacillus plantarum* from Plants by PCR and Detection of their Antibacterial Activity. *J. Biol. Sci.* **2009**, *9*, 810–814. [CrossRef]
106. Leandro, E.d.S.; Ginani, V.C.; de Alencar, E.R.; Pereira, O.G.; Rose, E.C.P.; Vale, H.M.M.D.; Pratesi, R.; Hecht, M.M.; Cavalcanti, M.H.; Tavares, C.S.O. Isolation, Identification, and Screening of Lactic Acid Bacteria with Probiotic Potential in Silage of Different Species of Forage Plants, Cocoa Beans, and Artisanal Salami. *Probiotics Antimicrob. Proteins* **2020**, *13*, 173–186. [CrossRef] [PubMed]
107. Mundt, J.O.; Hammer, J.L. Lactobacilli on Plants. *Appl. Microbiol.* **1968**, *16*, 1326–1330. [CrossRef] [PubMed]
108. Chen, Y.-S.; Yanagida, F.; Shinohara, T. Isolation and identification of lactic acid bacteria from soil using an enrichment procedure. *Lett. Appl. Microbiol.* **2005**, *40*, 195–200. [CrossRef] [PubMed]
109. Ajaj, D.F.; Hassan, A.A. Isolation and Identification of *Ssome Lactobacillus* spp. Bacteria and Evaluation Their Efficacy in The Management of Damping off Disease on Peas. *IOP Conf. Ser. Earth Environ. Sci.* **2021**, *910*, 012106. [CrossRef]
110. Su, L.; Ma, F.; An, Z.; Ji, X.; Zhang, P.; Yue, Q.; Zhao, C.; Sun, X.; Li, K.; Li, B.; et al. The Metabolites of *Lactobacillus fermentum* F-B9-1 Relieved Dextran Sulfate Sodium-Induced Experimental Ulcerative Colitis in Mice. *Front. Microbiol.* **2022**, *13*, 865925. [CrossRef]
111. Jarocki, P.; Komoń-Janczara, E.; Glibowska, A.; Dworniczak, M.; Pytka, M.; Korzeniowska-Kowal, A.; Wzorek, A.; Kordowska-Wiater, M. Molecular Routes to Specific Identification of the *Lactobacillus casei* Group at the Species, Subspecies and Strain Level. *Int. J. Mol. Sci.* **2020**, *21*, 2694. [CrossRef]
112. Siraj, N.; Sood, K.; Yadav, R. Isolation and Identification of Potential Probiotic Bacteria from Cattle Farm Soil in Dibrugarh District. *Adv. Microbiol.* **2017**, *7*, 265–279. [CrossRef]

 https://doi.org/10.3390/microorganisms14030712

microorganisms

MDPI

Article

Probiotic Lactic Acid Bacteria Fermentation Modulates the Bioactive Properties of Sprouted and Unsprouted Amaranth Seed

Mihaela Aida Vasile, Nicoleta Balan, Leontina Grigore-Gurgu, Gabriela Elena Bahrim * and Mihaela Cotârleț *

Faculty of Food Science and Engineering, "Dunărea de Jos" University of Galați, 111 Domnească Street, 800201 Galați, Romania; aida.vasile@ugal.ro (M.A.V.); nicoleta.balan@ugal.ro (N.B.); gurgu.leontina@ugal.ro (L.G.-G.)
* Correspondence: gabriela.bahrim@ugal.ro (G.E.B.); mihaela.cotarlet@ugal.ro (M.C.)

Abstract

This study aims to investigate the functional and biochemical characteristics of sprouted and unsprouted red and black amaranth flours by fermentation with four probiotic strains (*Lactiplantibacillus plantarum* MIUG BL21, *Lactiplantibacillus pentosus* MIUG BL24, *Lacticaseibacillus rhamnosus* MIUG BL38, and *Lactiplantibacillus paraplantarum* MIUG BL74). Aqueous extracts from freeze-dried fermented products derived from sprouted and raw seed of two *Amaranthus* species (*Amaranthus cruentus*—red amaranth and *Amaranthus hypochondriacus*—black amaranth) were characterised for their acidification and phytochemical profiles by titrimetric, spectrophotometric and chromatographic methods, and their antioxidant activities by ABTS and DPPH assays. Water-soluble proteins were evaluated by SDS-PAGE analysis. Nine phenolic acids (gallic acid, protocathechic acid, syringic acid, ellagic acid, ferulic acid, cinnamic acid, caffeic acid, *p*-coumaric acid, and chlorogenic acid) and twelve flavonoids (epicatechin gallate, hesperitin, quercetin, apigenin, luteolin, naringenin, quercetin 3-glucoside, isorhamnetin, peonidin 3-O rutinoside, epicatechin, keracyanin, and rutin trihydrate) were identified in the extracts of amaranth samples. The titratable acidity ranged from 0.59 to 5.50 mL of 0.1 N NaOH. Total flavonoid content (TFC) varied from 1.09 to 4.67 mg CE/g DW; whereas, total phenolic content (TPC) fluctuated from 1.99 to 5.76 mg GAE/g DW. The spectrum of ABTS and DPPH values was from 17.49 to 56.82% and 0.60 to 35.50%, respectively. More biologically active compounds were found in red amaranth-based samples, both sprouted and unsprouted, compared to black amaranth-based samples. There was a moderate correlation between the TPC and the antioxidant activity. The fermentation of red amaranth with *L. rhamnosus* MIUG BL38 led to a global increase in the protein background intensity, consistent with protein hydrolysis. Overall, sprouting and probiotics fermentation improved the fermentative performance of the amaranth seeds, enabling their effective use as a nutritive food with potential health-promoting properties.

Keywords: red amaranth; *Amaranthus cruentus*; black amaranth; *Amaranthus hypochondriacus*; sprouting; fermentation; *Lactiplantibacillus* spp.; *Lacticaseibacillus* spp.; gluten-free flour

1. Introduction

Amaranth (*Amaranthus* spp.) is an age-old pseudocereal cultivated by the Aztec, Inca, and Mayan civilisations and is now recognised for its distinctive nutritional and functional qualities [1]. Unlike traditional cereals, amaranth is gluten-free, making it a valuable alternative for individuals with celiac disease or gluten sensitivity [2]. Its seeds are rich in high-quality protein (15–18%), containing a high level of lysine, the limiting amino acid in most cereals, as well as dietary fibre, unsaturated fatty acids, and essential minerals such as iron, calcium, and magnesium [3–5].

Amaranth is a nutrient-rich, easily cultivated, and underutilised pseudocereal that can enhance climate change mitigation and food and nutritional security in Africa [6]. Instead, in Europe, amaranth seed production has rapidly increased, where it is considered a niche of pseudocereals market development. Germany is the primary consumer market for the seeds. Still, its popularity is growing in other countries such as the United Kingdom, the Netherlands, Belgium, Sweden, France, and Romania due to its valuable nutritional and functional benefits [7,8].

The complex biochemical profile of amaranth also includes bioactive compounds such as polyphenols, flavonoids, phytosterols, squalene, and antioxidant peptides, which help reduce oxidative stress and prevent cardiovascular, metabolic, and inflammatory diseases [1,2]. Because of these properties, amaranth is regarded as a new generation functional food with both nutritional and therapeutic potential [2,5]. Recently, the food industry has increasingly utilised the potential of this pseudocereal in gluten-free products and fortified nutritional formulas, including bread, biscuits, pasta, fermented drinks, sports products, and dietary formulas for patients with metabolic disorders [4]. At the same time, the high adaptability of amaranth plants to various climatic conditions and their agronomic benefits make it a strategic crop for global food security [3,4].

The importance of protein hydrolysates has grown owing to their bioactive properties and health benefits. They are an essential source of bioactive peptides which exhibit antibacterial, antihypertensive [9], anticoagulant, hypocholesterolemic, anti-diabetic [10], immunomodulatory, anti-inflammatory [11], anticancer, and antioxidant properties [5]. In summary, proteins and peptides derived from amaranth have shown potential as ingredients in functional foods and nutraceuticals, offering health benefits in the prevention of metabolic diseases.

The germination process is one of the most effective natural biotransformation methods for enhancing the nutritional and functional value of pseudocereals. During germination, the activation of endogenous enzymes leads to the breakdown of antinutritional compounds (phytic acid, tannins) and the production of bioactive metabolites, such as phenolic compounds, flavonoids, and antioxidant peptides [12–14].

Many studies have shown that germinating amaranth seeds significantly increases antioxidant activity, amino acid bioavailability, and the digestibility of protein and starch [12,13,15]. Additionally, bacterial fermentation, especially when performed with probiotic strains of lactic acid bacteria, is a biotechnological process that can further improve the functional properties of amaranth flour. During fermentation, microorganisms convert the plant substrate into bioactive compounds, decrease the levels of antinutrients, and release peptides with antioxidant, anti-inflammatory, and hypoglycaemic effects [14,16]. Recent research indicates that fermenting germinated amaranth with *Lactiplantibacillus plantarum* results in higher antioxidant activity and a more favourable phenolic profile compared to unfermented products [15,17].

Therefore, combining germination and fermentation offers a promising synergistic approach to creating functional foods with high biological value. In the current climate of growing interest in sustainable, gluten-free products with proven nutritional benefits, amaranth presents a significant opportunity for developing innovative ingredients aimed at maintaining health and preventing diseases related to oxidative stress [3,17].

This study aimed to evaluate the effect of probiotic lactic acid bacteria fermentation, using four LAB strains, on the functional and biochemical properties of red and black

 https://doi.org/10.3390/microorganisms14020340

amaranth flours, both germinated and ungerminated. The research is expected to identify the optimal combination of lactic acid bacteria strain and amaranth type, in order to obtain a fermented food ingredient with functional potential, intended to promote a healthy diet.

2. Materials and Methods

2.1. Plant Seeds and Amaranth Flour

The seeds of black amaranth (BA) (*Amaranthus hypochondriacus*) were collected from the rural commune of Ditinn, Dalaba prefecture (Guinea), in 2024 [18]. The seeds were divided into 500 g batches, vacuum-packed, and kept at room temperature in the dark [19].

The red amaranth (RA) seeds 'Garnet Red' (*Amaranthus cruentus*) were purchased from a commercial distributor S.C. XTREME ANG MARKETING S.R.L., Bucharest, Romania.

Red amaranth seeds (RA), *Amaranthus caudatus* L. var. Red Garnet, purchased from a local commercial distributor (Bucharest, Romania), was selected due to its high content of phenolic compounds, promoters of the antioxidant activity [20]. Meanwhile, black amaranth seeds (BA) (*Amaranthus hypochondriacus*), an invasive weed, collected from the rural commune of Ditinn, Dalaba prefecture (Guinea), were chosen based on their dietary minerals, vitamins, proteins, and bioactive compounds with potential health benefits [18,21].

Seed germination was performed using the EasyGreen EGL 55 automated germinator (Biovie Co., Langlade, France), following the method described by Ghinea et al. (2021), with modifications [22]. In brief, to prevent microbial contamination, seeds were exposed to UV radiation in a safety cabinet (SafeFAST Elite 209 S, Cornaredo, Italy) for 15 min, followed by washing with 2% sodium hypochlorite solution for 10 min and rinsing with distilled water for 20 min. Amaranth seeds (BA and RA) were allowed to germinate for 70 h at 23 ± 1 °C in semi-dark conditions. A high relative humidity (~90–95%) was ensured by spraying the seeds with water for 15 min every 4 h. Then, the seeds were dried at 40 °C for 24 h in an incubator (Stericell Typ 111, Munich, Germany).

Sprouting efficiency (%) was determined as the ratio between the number of germinated seeds and the total number of incubated seeds [23], calculated by the following equation:

$$\text{Sprouting efficiency } (\%) = \text{ Number of sprouted seeds/Total number of seeds } \times 100 \quad (1)$$

For red amaranth seeds, the sprouting efficiency was 82.83 ± 0.45%, while for black amaranth, it was 72.53 ± 0.72%.

To obtain the flour from unsprouted amaranth seeds (BA and RA), the same disinfecting, washing, and drying steps as described above were carried out. Using an electric grinder (Heiner HCG150SS, Bucharest, Romania), the dried seeds and sprouts were ground, and the resulting flours were stored at 4 °C [18,19].

2.2. Reagents and Chemicals

The following reagents were purchased from Sigma-Aldrich (Hamburg, Germany): DPPH (2,2-Diphenyl-1-picrylhydrazyl), ABTS (2,2′-azino-bis(3-ethylbenzothiazoline-6-sulfonic acid)), aluminium chloride ($AlCl_3$), sodium carbonate (Na_2CO_3), sodium nitrite ($NaNO_2$), potassium chloride (KCl), sodium acetate (CH_3COONa), calcium carbonate ($CaCO_3$), Folin–Ciocalteau and Bradford reagents. Merck Millipore (Darmstadt, Germany) provided culture media, including De Man, Rogosa, and Sharpe (MRS) broth and agar. Ethanol and methanol, which were HPLC-grade solvents, were acquired from Scharlau (Barcelona, Spain).

https://doi.org/10.3390/microorganisms14020340

2.3. Lactic Acid Bacteria Reactivation and Fermentation

Four probiotic strains, *Lactiplantibacillus plantarum* MIUG BL21, *Lactiplantibacillus pentosus* MIUG BL24, *Lacticaseibacillus rhamnosus* MIUG BL38, and *Lactiplantibacillus paraplantarum* MIUG BL74, from the Microorganisms Collection (abbreviated MIUG) at the Integrated Centre for Research, Expertise and Technological Transfer in Food Industry (BioAliment-TehnIA), "Dunărea de Jos" University of Galaţi, Galaţi, Romania, were examined (Table 1) [21]. Two of these strains, *Lp. plantarum* MIUG BL21 (LMG P-33745) and *Lp. paraplantarum* MIUG BL74 (LMG P-33746) were deposited at the Belgian Coordinated Collections of Microorganisms (BCCM), Gent, Belgium.

Table 1. Sources of probiotic strain isolation.

Probiotic Strains	Isolation Source	Accession No.
Lp. plantarum MIUG BL21	Soil	MW617265.1
Lp. pentosus MIUG BL24	Whey	OK325938.1
Lc. rhamnosus MIUG BL38	Chickpea	MT463821.1
Lp. paraplantarum MIUG BL74	Pickles	MT604712.1

Several factors, including technological, functional, and safety-related properties, were considered when the probiotic strains were chosen. All strains were selected based on their previously demonstrated resistance to simulated gastric juice (low pH and bile salt tolerance) and adhesion to intestinal epithelial cells. Additionally, the selected strains showed no antibiotic resistance, no hemolytic activity, and antibacterial activity against common foodborne pathogens [21].

A volume of 2 mL of stock culture was transferred into 9 mL of MRS broth (Merck, Darmstadt, Germany) and cultivated for 48 h at 37 °C, using an incubator (Binder BF4000, Tuttlingen, Germany) to reactivate the stock cultures, which had been preserved in 40% (*w*/*w*) glycerol solution at a temperature of −80 °C.

A sterile loop was used to streak 1.0 μL of the reactivated culture onto MRS agar medium (Merck, Darmstadt, Germany) supplemented with 30 g/L $CaCO_3$ to obtain single colonies [9].

The LAB inoculum was obtained by transferring a single colony onto 50 mL of MRS broth and incubating it stationary, for 48 h at 37 °C. A spectrophotometer (Biochrom, Libra 22, Holliston, MA, USA) was then used to measure and adjust the optical density at 600 nm (OD 600) to a value of about 2.0 (using fresh MRS broth), which indicates a cell concentration of 1×10^8 CFU/mL solution [24].

The fermentation media were previously autoclaved at 121 °C for 15 min, in an autoclave (Panasonic, MLS-3751L, Watertown, MA, USA). A 2% (*v*/*v*) LAB inoculum was added to the fermentation medium, which contained 5% (*w*/*w*) black and red amaranth seed flour, both sprouted and unsprouted in distilled water. Then, the mixture was incubated for 48 h at 37 °C and 100 rpm in an orbital shaker (Lab Companion SI-300, GMI, Minneapolis, MN, USA) [25]. Control samples were obtained using both red and black amaranth flour, including sprouted and unsprouted variants. Following fermentation, the fermented products were freeze-dried with a freeze-drier (Christ Alpha 1–4 LD plus, Osterode am Harz, Germany), at −42 °C and 0.10 mbar. Then, the moisture content was analysed with a moisture analyser (KERN, DAB 100-3, Balingen, Germany).

2.4. pH and Total Titratable Acidity (TTA) Measurement

The pH and the total titratable acidity of the fermented samples were measured with a pH metre (FiveEasy Plus FP20, Mettler Toledo, Greifensee, Switzerland) and an automatic titrator (TitroLine Easy, Schott Instruments, Mainz, Germany). To summarise, for the total

 https://doi.org/10.3390/microorganisms14020340

titratable acidity, a mixture of 10 g of fermented products and 90 mL of distilled water was titrated to an endpoint of 8.50 [26]. The TTA was assessed in millilitres of 0.1 N NaOH [21].

2.5. Antioxidant Activity (AA)

The 2,2-diphenyl-1-picrylhydrazyl (DPPH) and 2,2′-azino-bis(3-ethylbenzothiazoline-6-sulfonate) (ABTS) radicals were used to evaluate the antioxidant activity of the freeze-dried samples [27,28].

Water extracts 1:10 (*w*/*v*) of the freeze-dried fermented products were prepared. After vortexing for 5 s, the mixture was incubated overnight at room temperature with shaking at 200 rpm. It was then sonicated for 30 min at 40 °C in an ultrasonic water bath (DU-32; ARGOLAB, Capri, Italy). The supernatant was separated by centrifuging at 7000 rpm for 10 min at 4 °C (Hettich Universal 320R, Tuttlingen, Germany). These aqueous extracts were globally characterised regarding antioxidant activity, total flavonoid content, total polyphenolic content and water-soluble proteins.

In short, 3900 μL of 0.004% (*w*/*v*) DPPH (in methanol HPLC) combined with 100 μL of the extract was incubated for 90 min in the dark, and then the absorbance was measured at 515 nm.

To perform the ABTS assay, 20 μL of the extract was mixed with 1980 μL of ABTS solution (which had a 0.7 absorbance at OD734), homogenised, and after 15 min in the dark, the mix's absorbance at 734 nm was measured.

According to Equation (2) [28], the antioxidant activity was calculated as the radical scavenging activity (RSA, %):

$$RSA\ (\%) = A_{control}/A_{sample} \times 100 \qquad (2)$$

where

$A_{control}$—the absorbance of the control sample.

A_{sample}—the absorbance of the analysed sample.

2.6. Total Flavonoid Content (TFC)

The aluminium chloride colorimetric assay was employed to determine the total flavonoid content. In summary, a mixture of 250 μL of the extracts, 1250 μL of distilled water and 75 μL of a 5% (*w*/*v*) sodium nitrite solution was homogenised and kept at room temperature for 5 min. In addition, 150 μL of a 10% (*w*/*v*) aluminium chloride solution was added to the samples, and after 6 min of incubation, 500 μL of 1.0 M sodium hydroxide and distilled water were added, up to a total volume of 3000 μL [29]. Using a spectrophotometer (Libra S22 UV-VIS, Biochrom, Cambridge, UK), the mixes' absorbance was measured at 510 nm.

Using a calibration curve (y = 1.856x − 0.0201, R^2 = 0.9951), TFC was calculated as mg catechin equivalents per gram of dry weight (CE)/g DW.

2.7. Total Polyphenolic Content (TPC)

The total polyphenolic content in the samples was evaluated with the modified Folin–Ciocalteu method. Briefly, 500 μL of Folin–Ciocalteu reagent was mixed with 200 μL of the extracts that had been diluted in 7900 μL of distilled water. After 5 min, 500 μL of a 20% (*w*/*v*) sodium carbonate solution was added, and the samples were left to stand in the dark for 60 min. Then the absorbance at a wavelength of 765 nm was evaluated [29].

Using a gallic acid standard curve (y = 1.3612x − 0.0205, R^2 = 0.9838), TPC was considered as mg gallic acid equivalents per gram dry weight (GAE)/g DW.

https://doi.org/10.3390/microorganisms14020340

2.8. Soluble Protein Content (SPC)

The soluble protein content of samples was assessed by the Bradford method. Briefly, 900 μL of Bradford reagent was mixed with quantities of 30 μL of the extracts. After 30 min of dark incubation at room temperature, the absorbance at 595 nm was measured. Using a bovine serum albumin calibration curve ($y = 0.4382x + 0.0127$, $R^2 = 0.9933$), the protein content was measured and expressed as mg/g DW [30].

2.9. Probiotic Cell Counting

The viability of lactic acid bacteria in fermented sprouted and unsprouted amaranth products was established using the classical method by counting the colonies on Man, Rogosa and Sharpe (MRS) agar medium, supplemented with calcium carbonate [18]. Briefly, one gram of freeze-dried fermented products was aseptically homogenised with 9 mL of sterile 0.9% saline solution. Decimal dilutions (10^{-1}–10^{-7}) were made, and 1 mL of each dilution was distributed and homogenised with MRS agar containing 3% (*w*/*v*) $CaCO_3$. The mixtures were cultivated at 37 °C for 48–72 h in aerophilic conditions and distinct colonies were enumerated in plates containing between 30 and 300 colonies and then expressed as CFU/gram of sample.

2.10. Reverse-Phase High-Performance Liquid Chromatography (HPLC) Analysis of Phenolic Compounds

The identification of phenolic compounds followed the method outlined by Mërtiri et al. [27], using an Agilent 1200 HPLC system (Agilent Technologies, Santa Clara, CA, USA) that includes a degasser, quaternary pump, autosampler, column, and multi-wavelength detector. Separation was carried out on a BDS Hypersil C18 column (150 mm × 4.6 mm, 5 μm) with the following settings: 30 °C temperature, 10 μL injection volume, 1 mL/min flow rate. The mobile phases consisted of 100% methanol (solvent A) and 10% formic acid (solvent B). The gradient programme was as follows: 0–20 min, 9% A–91% B; 20–30 min, 35% A–65% B; 30–40 min, 50% A–50% B; 40–45 min, 9% A–91% B. Results are calculated as mean values ± STDEV for duplicate runs, in ng/μL, based on peak area and calibration curves with standard references.

2.11. SDS-PAGE Analysis

The SDS-PAGE assay was performed according to Laemmli [31]. For protein electrophoresis, all the samples were diluted with ultrapure water (1:2), mixed with Laemmli buffer containing beta-mercaptoethanol, and treated for 5 min at 95 °C [32]. The electrophoresis was carried out at 100 V for 120 min. Further, the protein gels were fixed for 90 min, stained with Coomassie Brilliant blue G-250, and destained in 10% acetic acid for 40 min.

2.12. Statistical Analysis

Minitab V19.1 (Minitab LLL, State College, PA, USA) was used to evaluate the data. The one-way ANOVA method was used to analyse the results, which were presented as mean ± standard deviation of three replicates. At a 95% confidence level, the data were examined for homogeneity of variances (Bartlett's test) and normal distribution (Ryan-Joiner test), and then either the Tukey test ($p < 0.05$) or the Games–Howell test ($p < 0.05$). The data was assessed using linear multiple regression and Pearson correlation.

3. Results and Discussion

3.1. pH and Total Titratable Acidity (TTA) of the Amaranth Fermented Products

The RA-fermented products exhibited pH values ranging from 3.16 ± 0.04 to 4.35 ± 0.69, with no significant differences ($p < 0.05$) among them. However, the low-

https://doi.org/10.3390/microorganisms14020340

est pH values were observed in fermented products (FPs) with sprouted RA seeds, as shown in Figure 1. While the BA-fermented products displayed pH values ranging from 3.07 ± 0.10 to 4.89 ± 0.04, with significant differences ($p < 0.05$) between them. Visibly, the probiotics fermentation of sprouted RA flour produced the lowest pH value when compared with the fermentation substrates. Castro-Alba et al. [19] shown that using *Lactobacillus plantarum* 299v® for fermenting unsprouted amaranth flour resulted in a pH drop above 4.0 after 48 h of fermentation.

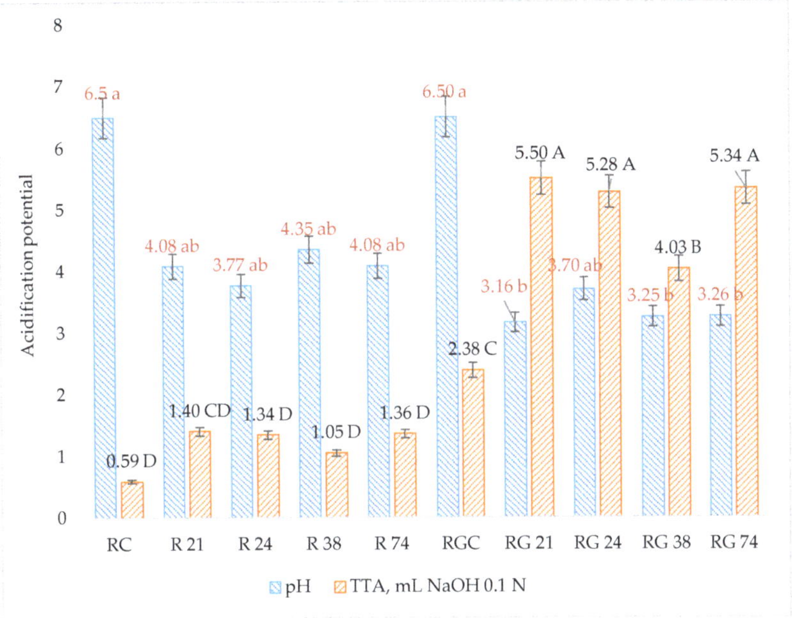

Figure 1. Fermented RA products pH (■) and TTA (■) after 48 h of fermentation. Means that do not share a letter are significantly different based on Tukey test ($p < 0.05$). RC—ungerminated RA control (unfermented), R 21, R 24, R 38, and R 74—ungerminated RA fermented with *Lp. plantarum* MIUG BL21, *Lp. pentosus* MIUG BL24, *Lc. rhamnosus* MIUG BL38, and *Lp. paraplantarum* MIUG BL74. RGC—germinated RA control (unfermented), RG 21, RG 24, RG 38, and RG 74—germinated RA fermented with the strains listed above. The lowercase letters are used to compare the pH values, while the uppercase letters are used to compare the TTA values for red amaranth FPs.

Alternatively, sprouted RA fermented products with *Lp. pentosus* MIUG BL24, *Lp. plantarum* MIUG BL21, and *Lp. paraplantarum* MIUG BL74 showed the highest TTA values (5.28 ± 0.22 to 5.50 ± 0.11 mL of 0.1 N NaOH), with no significant differences ($p < 0.05$) among them (Figure 1).

Similarly, sprouted BA fermented products with *Lp. pentosus* MIUG BL24 and *Lp. plantarum* MIUG BL21 showed the highest TTA values (4.03 ± 0.08 to 4.18 ± 0.01 mL of 0.1 N NaOH), with no significant differences ($p < 0.05$) among them (Figure 2).

Further, the FPs derived from unsprouted seeds listed lower values for the total titratable acidity.

The results are in accordance with Owheruo et al. [33], who reported that the sesame seed flour during sprouting exhibits a decrease in pH with a simultaneous increase in TTA during sprouting, due to naturally occurring microorganisms that convert carbohydrates into an acidic medium through a metabolic mechanism. In a comparable study, employing

 https://doi.org/10.3390/microorganisms14020340

sprouted and fermented *Phalaris canariensis* seeds, when compared to unsprouted seeds, resulted in a greater TTA value when fermented beverages were prepared [34].

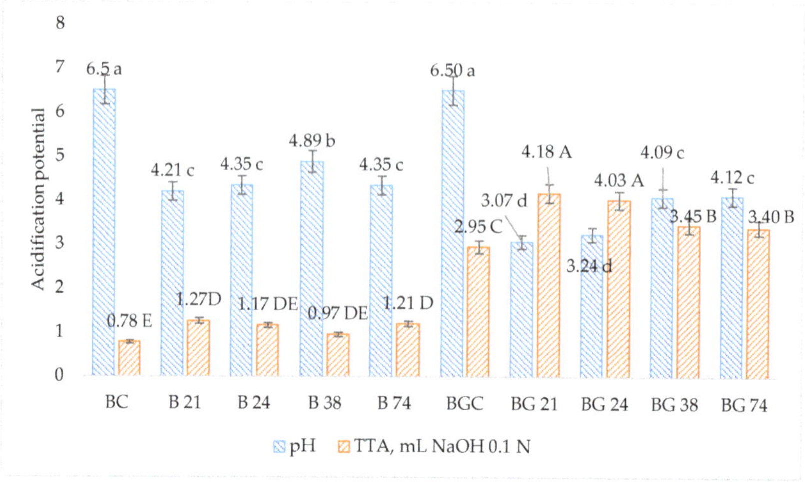

Figure 2. Fermented BA products pH (■) and TTA (■) after 48 h of fermentation. Means that do not share a letter are significantly different based on Tukey test ($p < 0.05$). BC—ungerminated BA control (unfermented), B 21, B 24, B 38, and B 74—ungerminated BA fermented with *Lp. plantarum* MIUG BL21, *Lp. pentosus* MIUG BL24, *Lc. rhamnosus* MIUG BL38, and *Lp. paraplantarum* MIUG BL74. BGC—germinated BA control (unfermented), BG 21, BG 24, BG 38, BG 74—germinated BA fermented with the strains listed above. The lowercase letters are used to compare the pH values, while the uppercase letters are used to compare the TTA values for black amaranth FPs.

Similarly, unsprouted BA dough fermented with *Lp. pentosus* MIUG BL24 demonstrated the highest TTA value of 12.33 ± 0.35 mL of 0.1 N NaOH, according to a recent study by Souare et al. [18].

Venturi et al. [35] reported that in the amaranth-based sourdough fermentation, the pH decreased from 6.20 to 4.35–4.60, while TTA reached higher values (6.45–6.90 mL NaOH 0.1 N), highlighting the differences between fermentations with monoculture of lactic acid bacteria and those with co-culture of lactic acid bacteria and yeast.

Overall, the results confirm that germination increases the availability of fermentable carbohydrates, which in turn increases acidity by stimulating the metabolism of lactic acid bacteria. The main role of LAB is to ferment carbohydrates into organic acids, mainly lactic acid, which increases the acidity and reduces the pH of the end product. Since pH and TTA are related to the antibacterial and antifungal properties that are derived from the organic acid content of fermented products, they are key indicators in food preservation, processing, and protection [36].

3.2. Antioxidant Activity (AA) of the Amaranth Fermented Products

The DPPH and ABTS radical-scavenging methods were used to assess the substantial changes in the fermented amaranth samples' antioxidant activity. The results showed that the DPPH values ranged from 0.60 ± 0.60% to 26.28 ± 0.18% and the ABTS radical scavenging values for red amaranth FPs ranged from 17.49 ± 1.83% to 56.82 ± 5.21% (Figure 3). While the DPPH values ranged from 11.17 ± 2.13% to 35.67 ± 2.05%, the ABTS values for the fermented black amaranth products ranged from 33.03 ± 6.70% to 55.09 ± 9.97% (Figure 4).

https://doi.org/10.3390/microorganisms1402034

The antioxidant potential of black amaranth FPs was greater than that of red amaranth FPs. Furthermore, compared to unsprouted seeds, seed germination leads to an increased antioxidant potential for both amaranth species [20].

The results are in accordance with Souare et al., who reported that black amaranth doughs fermented with the *Lp. pentosus* MIUG BL24 strain and the *Lc. rhamnosus* MIUG BL38 strain exhibited the highest inhibition of the ABTS radical (58.00 ± 1.81% and 58.92 ± 6.05%), nearly 2.2 times higher than the control [18].

Additionally, red amaranth leaves had the highest total antioxidant capacity (TAC) (ABTS+) concentration (34.72 TEAC µg/g DW) compared to green amaranth leaves (25.27 TEAC µg/g DW), according to a recent study [37].

RA fermented products exhibited lower antioxidant activity values when compared to germinated and non-germinated controls. The germinated control had the maximum antioxidant potential such as 56.82 ± 5.21% (ABTS) and 33.08 ± 0.84% (DPPH), respectively ($p < 0.05$). Fermented products coded RG 24, RG 38, R38, and R24 demonstrated comparable RSA values (ABTS assay), with no statistically significant differences among them ($p < 0.05$).

Conversely, the germinated and ungerminated BA controls exhibited the lowest antioxidant potential, as measured by the DPPH assay. The germinated FPs based on black amaranth seed flour fermented with *Lc. rhamnosus* MIUG BL38 (coded BG 38) strain showed the highest antioxidant activity at a value of 55.09 ± 9.97%, by ABTS assay. Additionally, germinated FPs fermented with *Lc. rhamnosus* MIUG BL38, *Lp. pentosus* MIUG BL24, and *Lp. paraplantarum* MIUG BL74 (coded BG 38, BG 24, and BG 74) strains registered similar values of antioxidant potential by DPPH method (35.67 ± 2.05%, 35.50 ± 0.82%, and 35.04 ± 1.30%, DPPH), with no significant statistical differences among them ($p < 0.05$).

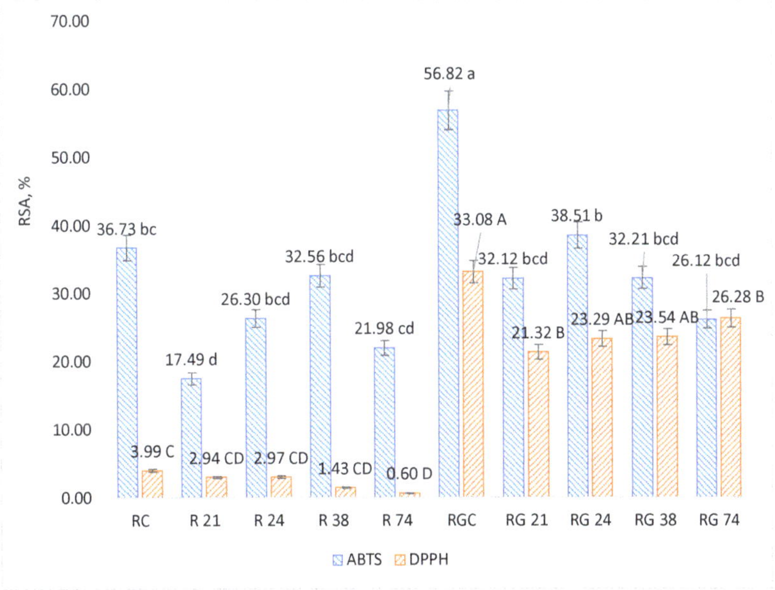

Figure 3. Fermented RA products ABTS (■) and DPPH (■) scavenging activity. Means that do not share a letter are significantly different based on the Tukey test ($p < 0.05$). RC—ungerminated RA control (unfermented), R 21, R 24, R 38, and R 74—ungerminated RA fermented with *Lp. plantarum* MIUG BL21, *Lp. pentosus* MIUG BL24, *Lc. rhamnosus* MIUG BL38, and *Lp. paraplantarum* MIUG BL74. RGC—germinated RA control (unfermented), RG 21, RG 24, RG 38, RG 74—germinated RA fermented with the strains listed above. The lowercase letters are used to compare the ABTS values, while the uppercase letters are used to compare the DPPH values for red amaranth FPs.

https://doi.org/10.3390/microorganisms14020340

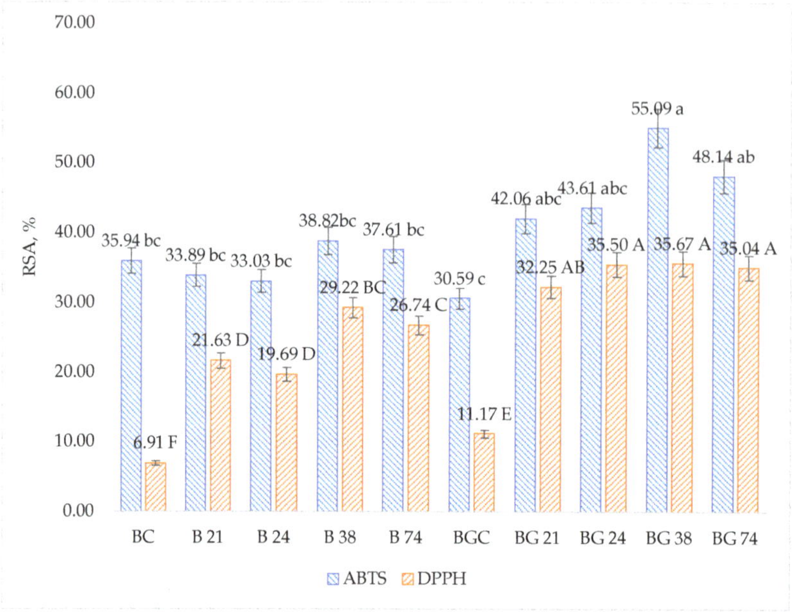

Figure 4. Fermented BA products ABTS (■) and DPPH (■) scavenging activity. Means that do not share a letter are significantly different based on the Tukey test ($p < 0.05$). BC—ungerminated BA control (unfermented), B 21, B 24, B 38, and B 74—ungerminated BA fermented with *Lp. plantarum* MIUG BL21, *Lp. pentosus* MIUG BL24, *Lc. rhamnosus* MIUG BL38, and *Lp. paraplantarum* MIUG BL74. BGC—germinated BA control (unfermented), BG 21, BG 24, BG 38, BG 74—germinated BA fermented with the strains listed above. The lowercase letters are used to compare the ABTS values, while the uppercase letters are used to compare the DPPH values for black amaranth FPs.

The data align with those of Chauhan et al., who showed that optimised sprouted amaranth flour had more antioxidant compounds than raw flour [12]. Additionally, Vento et al. [20] noticed that germination increased antioxidant activity in quinoa and amaranth seeds (13.07% and 9.06%, respectively), as shown by the DPPH approach. Compared to non-germinated seeds, germination and fermentation may improve the antioxidant potential of amaranth seeds, enhancing their health benefits [20].

In a recent study, raw, cooked and fermented seeds and germinated seeds of *Chenopodium quinoa* Willd. var. Tunkahuan and *Amaranthus caudatus* L. var. Alegrìa were compared regarding the total antioxidant capacity. After 48 h without sterilisation, fermentation of germinated seeds inoculated with *Lactobacillus plantarum* demonstrated the maximum antioxidant capacity (55.3 $\pm$ 2.03%) compared to the control (9.06 $\pm$ 1.25%). In contrast, antioxidant activity dropped significantly up to 8.45 $\pm$ 1.1% when germinated seeds were first sterilised and then fermented with *L. plantarum* for 48 h at 30 °C, compared to the control (9.06 $\pm$ 1.25%). Overall, the results showed that both species' antioxidant activity decreased after sterilisation when compared to unsterilized samples [20].

Rich in bioactive substances, amaranth may strengthen the body's defences against free radicals, preserving and enhancing the body's functions [4]. It has been demonstrated that phenolic antioxidants reduce oxidative stress and inflammation, which may help prevent chronic illnesses like diabetes, cancer, and heart disease [3].

Given that DPPH activity supposed electron transfer, while ABTS activity included hydrogen atom transfer, there may be a variation in antioxidant processes [9]. Additionally, various bacterial strains may have different antioxidant processes [38].

 https://doi.org/10.3390/microorganisms14020340

Overall, fermentation and sprouting are promising approaches for enhancing or generating antioxidant compounds and improving their bioaccessibility [39].

3.3. Global Characterisation of Amaranth Fermented Products

3.3.1. Soluble Protein Content (SPC) of Amaranth Samples

Soluble protein contents ranging from 26.15 ± 0.47 to 41.97 ± 8.81 mg/g DW were obtained from RA fermented products, with no significant differences upon them ($p < 0.05$) (Table 2). However, ungerminated BA flour fermented with *Lp. plantarum* MIUG BL21 and *Lp. pentosus* MIUG BL24 (coded B 21 and B 24), showed a higher content of water-soluble proteins, reaching 50.82 ± 1.03 and 48.47 ± 4.44 mg/g DW, respectively.

Table 2. The global characterisation of the RA samples.

Sample	SPC, mg/g DW	TPC, mg GAE/g DW	TFC, mg CE/g DW
RC	26.15 ± 0.47 ^{a}	1.99 ± 0.02 ^{b}	1.09 ± 0.03 ^{d}
R 21	32.39 ± 6.39 ^{a}	3.72 ± 0.94 ^{ab}	2.29 ± 0.30 ^{bcd}
R 24	33.84 ± 8.85 ^{a}	3.85 ± 0.31 ^{a}	2.99 ± 0.29 ^{abc}
R 38	41.97 ± 8.81 ^{a}	3.98 ± 0.13 ^{a}	4.24 ± 0.51 ^{a}
R 74	29.23 ± 6.64 ^{a}	3.24 ± 0.63 ^{ab}	3.56 ± 1.07 ^{ab}
RGC	29.28 ± 0.36 ^{a}	3.83 ± 0.03 ^{ab}	2.51 ± 0.13 ^{bc}
RG 21	35.68 ± 1.63 ^{a}	3.78 ± 0.90 ^{ab}	3.52 ± 0.19 ^{ab}
RG 24	35.34 ± 3.36 ^{a}	4.25 ± 0.11 ^{a}	3.51 ± 0.59 ^{ab}
RG 38	27.24 ± 2.08 ^{a}	3.84 ± 0.55 ^{ab}	2.02 ± 0.19 ^{cd}
RG 74	30.38 ± 0.34 ^{a}	4.02 ± 0.28 ^{a}	2.67 ± 0.54 ^{bc}

Means that do not share a letter, in the same column, are significantly different based on the Tukey test ($p < 0.05$). RC—ungerminated RA control (unfermented), R 21, R 24, R 38, and R 74—ungerminated RA fermented with *Lp. plantarum* MIUG BL21, *Lp. pentosus* MIUG BL24, *Lc. rhamnosus* MIUG BL38, and *Lp. paraplantarum* MIUG BL74. RGC—germinated RA control (unfermented), RG 21, RG 24, RG 38, and RG 74—germinated RA fermented with the strains listed above.

The results displayed in Tables 2 and 3 showed that, when compared to the unfermented control, the soluble protein content improves by fermenting the flour based on ungerminated black and red amaranth seeds. Rather, by fermenting the sprouted red and black amaranth seeds, their soluble protein content is comparable to that of the control samples and does not differ significantly ($p < 0.05$).

Similarly, the doughs fermented with *Lp. pentosus* MIUG BL24 strain and *Lc. rhamnosus* MIUG BL38 strain showed comparable effects on protein hydrolysis during the fermentation of BA flour in a previous research. Although it was lower than the control (1.23 ± 0.03 mg/g DW), the dough fermented with the *Lc. rhamnosus* MIUG BL38 strain had the highest protein content (1.18 ± 0.06 mg/g DW) [18].

The results obtained are consistent with Cruz-Casas et al. [9], who concluded that releasing protein hydrolysates through fermentation is a strain-dependent method. The data, especially in the case of red amaranth, agree with Maldonado-Alvarado et al., who found that germination increased protein content in three types of quinoa (white, red, and black) [40].

Lactic acid fermentation is a relatively underexplored technique for obtaining protein hydrolysates from amaranth. Owing to their unique proteolytic system, which consists of intracellular peptidases, transport mechanisms, and cell envelope proteinases, lactic acid bacteria are crucial for fermentation-based production of protein hydrolysates and bioactive peptides. They contain GRAS-classified genera and easily adapt to plant substrates. Even within the same species, there is a great variation in LAB proteolytic activity due to strain dependence [9].

https://doi.org/10.3390/microorganisms14020340

Table 3. The global characterisation of the BA samples.

Sample	SPC, mg/g DW	TPC, mg GAE/g DW	TFC, mg CE/g DW
BC	23.20 ± 0.67 c	2.10 ± 0.22 f	1.19 ± 0.11 d
B 21	48.47 ± 4.44 ab	5.76 ± 0.12 a	5.28 ± 0.44 a
B 24	50.82 ± 1.03 a	4.37 ± 0.27 b	4.67 ± 0.62 ab
B 38	32.16 ± 0.97 bc	3.08 ± 0.11 e	2.16 ± 0.27 cd
B 74	38.72 ± 3.96 abc	3.86 ± 0.22 bc	3.25 ± 0.98 bc
BGC	22.57 ± 0.69 c	2.54 ± 0.13 f	1.66 ± 0.12 d
BG 21	32.33 ± 6.41 abc	3.48 ± 0.18 cde	2.38 ± 0.80 cd
BG 24	27.44 ± 5.14 abc	3.65 ± 0.28 cd	2.45 ± 0.11 cd
BG 38	24.07 ± 1.22 c	3.42 ± 0.09 cde	2.28 ± 0.45 cd
BG 74	21.46 ± 0.17 c	3.14 ± 0.03 de	1.39 ± 0.21 d

Means that do not share a letter, in the same column, are significantly different based on the Tukey test ($p < 0.05$). BC—ungerminated BA control (unfermented), B 21, B 24, B 38, and B 74—ungerminated BA fermented with *Lp. plantarum* MIUG BL21, *Lp. pentosus* MIUG BL24, *Lc. rhamnosus* MIUG BL38, and *Lp. paraplantarum* MIUG BL74. BGC—germinated BA control (unfermented), BG 21, BG 24, BG 38, and BG 74—germinated BA fermented with the strains listed above.

The literature highlighted that amaranth proteins have lower quantities of prolamins (about 0–13%) than albumins (11–51%), globulins (16–51%), and glutelins (7–36%). In particular, compared to typical grains like wheat, barley, and maize, amaranth has a greater amount of albumins and globulins and lower levels of prolamins and glutelins [10].

3.3.2. Total Polyphenol Content (TPC) of Amaranth Samples

There were notable differences in total phenolic content (TPC) between RA fermented products (Table 2). The FPs coded R 24, R 38, RG 24, and RG 74 had higher TPC values ranging from 3.85 ± 0.31 to 4.25 ± 0.11 mg GAE/g DW, with no significant differences between them ($p < 0.05$). The control sample (RC) had the lowest TPC (1.99 ± 0.02 mg GAE/g DW).

Furthermore, TPC varied considerably amongst the BA variants (Table 3). Ungerminated BA fermented with *Lp. plantarum* MIUG BL21 (coded B 21) showed the highest TPC (5.76 ± 0.12 mg GAE/g DW), followed by B 24 (4.37 ± 0.27 mg GAE/g DW), which were statistically different ($p < 0.05$). The control samples (BC and BGC) had the lowest values, 2.10 ± 0.22 and 2.54 ± 0.13 mg GAE/g DW, respectively. The phenolic compounds concentration of BA samples was generally positively impacted by fermentation and germination.

Recently, Souare et al. [18] reported that fermentation with selected probiotic bacteria reduced the TPC in the doughs when compared to the control. BA ungerminated dough fermented with *Lc. rhamnosus* MIUG BL38 had the highest TPC (2.33 ± 0.20 mg GAE/g DW), suggesting that both the strain and fermentation system have an impact on polyphenol content.

The results obtained are in agreement with those of Vento et al., who found that by fermenting sterilised, ungerminated amaranth seeds with *Lactobacillus plantarum*, a significantly higher total phenolic content (8.55 ± 0.49 mg QE/g DW) than the unfermented control (2.35 ± 0.33 mg QE/g DW) occurred. On the other hand, fermentation of germinated seeds with the same strain resulted in a lower TPC value (5.58 ± 0.54 mg QE/g DW), after 48 h. Overall, these findings indicate that lactic acid fermentation led to a higher content of phenolic compounds in ungerminated amaranth seeds compared with germinated ones [20].

The increased phenolic content observed in fermented ungerminated amaranth seed samples may be attributed to the release of phenolic compounds that are originally bound to the seed cell wall matrix [41]. The enzymatic activity of lactic acid bacteria, combined with the acidic conditions generated during fermentation, likely enhances the release of these bound phenolics into soluble free phenols [42]. Additionally, endogenous enzymes produced during sprouting promote the release of phenolic compounds [43]. Overall, fermentation and germination increased the total phenolic content and improved nutritional value of amaranth through activation of endogenous enzymes. By changing the bioactivity and digestibility of foods, fermentation modulates their nutritional and anti-nutritional properties. Particularly, polyphenols are mostly found in bound forms, coupled with proteins, lipids, or sugars, in unfermented cereals. These compounds can be released into free forms during bioprocessing stages, which increases their bioavailability and antioxidant potential. However, free phenolic content may also drop as a result of degradation or interactions with other elements of the food matrix, depending on the type of bioprocess and biocatalyst properties [20].

During germination, procyanidins and catechins are among the phenolic chemicals whose concentration decreases as a result of condensation and polymerisation, which are aided by enzymes such as polyphenol oxidases. On the other hand, hydrolytic enzymes linked to total phenolic content, such as cellulases, amylases, hemicellulases, glucanases, and polyphenol oxidases, become more active and support the polymerisation and modification of phenolic compounds. Furthermore, germination increases the availability of free phenolic chemicals by breaking cell wall polymers. Additionally, germination triggers secondary metabolic processes, including those linked to phenolic metabolism, such as the oxidative pentose phosphate route, glycolysis, acetate/malonate pathway, shikimate pathway, and phenylpropanoid pathway. This activation increases the content of phenolic compounds and their antioxidant activity in germinated seeds, as does the creation of enzymes such as phenylalanine ammonia-lyase [43].

Through fermentation, due to the breakdown of bonds with grain cell wall components, activities of enzymes such as β-glucosidase, decarboxylases, esterases, hydrolases, and reductases, as well as the metabolic activity of fermenting microorganisms, bound phenolic compounds are converted from their linked or conjugated forms to free ones [44]. The released free aglycones can increase antioxidative activity, and phenolic compounds are more bioaccessible in their free form. On the other hand, because free phenolic compounds can associate with other molecules in the food matrix, either hydrolysed by particular microbial strains, or broken down by microbial enzymes, a decrease in their amount may occur during fermentation [45]. Overall, the combined application of germination and fermentation leads to increased availability and transformation of phenolic acids and flavonoids, contributing to enhanced antioxidant potential.

3.3.3. Total Flavonoid Content (TFC) of the Amaranth Samples

Total flavonoid content (TFC) varied among the RA fermented products. The control sample (RC) showed the lowest TFC (1.09 ± 0.03 mg CE/g DW), while higher values were generally observed in coded samples R 38, R 74, RG 21, and RG 24, which were obtained from germinated and non-germinated RA, fermented with *Lp. plantarum* MIUG BL21, *Lp. pentosus* MIUG BL24, *Lc. rhamnosus* MIUG BL38, and *Lp. paraplantarum* MIUG BL74 strains, indicating that fermentation and sprouting enhanced flavonoid levels, with no significant differences among them ($p < 0.05$).

TFC variations were more noticeable in BA samples. Coded samples B 21 and B 24 exhibited the highest flavonoid content (5.28 ± 0.44 and 4.67 ± 0.62 mg CE/g DW, respectively), whereas the control samples (BC and BGC) had the lowest values. In summary, the amount of flavonoids in BA samples was positively impacted by both fermentation and germination.

https://doi.org/10.3390/microorganisms14020340

A recent study stated that the total flavonoid content of black amaranth dough, based on ungerminated seed flour, fermented with *Lc. paracasei* MIUG BL4 strain, *Lc. paracasei* MIUG BL13 strain, and *Lp. pentosus* MIUG BL24 strain was higher than that of the control, with no statistical difference among them ($p < 0.01$) [18].

Total flavonoid content was found to be similar in raw and cooked amaranth seeds (1.8 mg QE/g DW) but sprouting significantly increased it to 5.0 mg QE/g DW. Most fermentation variants further increased TFC, although decreases were seen after fermentation with *S. cerevisiae* alone and in combination with *L. plantarum*. The highest levels of TFC were obtained through spontaneous fermentation in both ungerminated seeds (6.1 mg QE/g DW) and germinated seeds (9.41 mg QE/g DW), corresponding to increases of nearly 2.6- and 1.9-fold when compared to the controls [20].

Studies show that germination significantly improves total phenolics, total flavonoids, and antioxidant activity in a variety of grains, with nutritional benefits. Soaking, ultrasound, and thermo-alkaline hydrolysis are recommended pre-germination treatments that modify the structure of grains to influence germination outcomes, promoting the accumulation of flavonoids and phenolic acids during germination and increasing antioxidant activity [43].

Increased levels of phenolic compounds and flavonoids as a result of microbial hydrolysis are primarily responsible for the improvement of antioxidant activity during fermentation [38] and germination [43].

Particularly, for RA fermented products, TFC had a strong and moderate, respectively, positive correlation with SPC ($r = 0.632$, $p < 0.05$) and TPC ($r = 0.435$, $p < 0.05$), suggesting that total flavonoid content increased with soluble protein content, as an effect of bioprocessing modification of vegetal structure. TPC showed moderate positive correlations with DPPH ($r = 0.376$, $p < 0.05$), indicating that phenolics may contribute to the increment of the antioxidant activity (Figure 5).

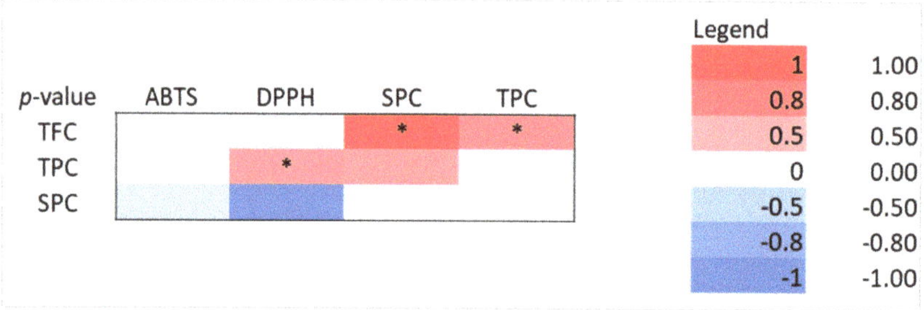

Figure 5. Heatmap indicating the Pearson's correlation coefficient of TFC, TPC, SPC, DPPH, and ABTS for the RA fermented products. The correlation (r) was considered as follows: ±0.8 to ±1.0: strong, ±0.50 to ±0.79: moderate, ±0.30 to ±0.49: weak, ±0.00 to ±0.29: no correlation; *—*p*-values that represent statistical significance ($p < 0.05$).

A strong negative correlation was observed between antioxidant activity (ABTS) and water-soluble proteins ($r = -0.386$, $p < 0.05$), in the case of black amaranth FPs. Other strong correlations, such as those between SPC and TFC or TPC ($r = 0.881$, $p < 0.05$, $r = 0.819$, $p < 0.05$ were statistically significant (Figure 6).

https://doi.org/10.3390/microorganisms14020340

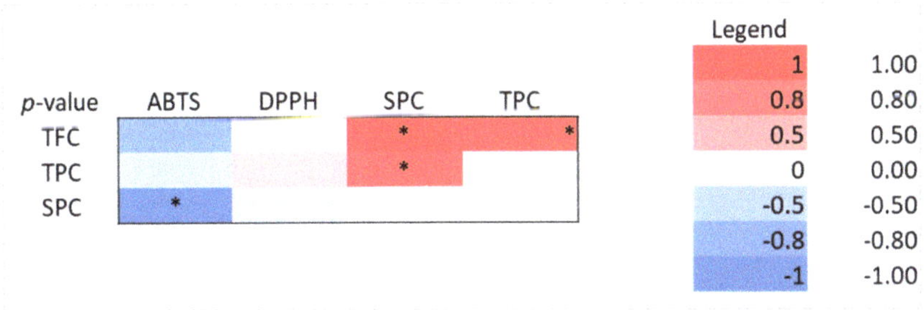

Figure 6. Heatmap indicating the Pearson's correlation coefficient of TFC, TPC, SPC, DPPH, and ABTS for the BA fermented products. The correlation (r) was considered as follows: ±0.8 to ±1.0: strong, ±0.50 to ±0.79: moderate, ±0.30 to ±0.49: weak, ±0.00 to ±0.29: no correlation; *—*p*-values that represent statistical significance ($p < 0.05$).

According to regression analysis, soluble proteins significantly affected ABTS activity ($p = 0.037$), while total phenolic content was the main factor influencing DPPH activity ($p = 0.015$). These findings demonstrate various antioxidant response patterns (Table 4).

Table 4. Regression analysis of ABTS and DPPH versus TFC, TPC, and PC, for BA samples.

Source	DF	Adj SS	Adj MS	F-Value	*p*-Value
Regression/ABTS	3	582.70	194.23	3.34	0.035
SPC	1	280.85	280.85	4.83	0.037
Error	26	1513.18	58.20		
Total	29	2095.98			
Regression/DPPH	3	943.87	314.62	4.20	0.015
TPC	1	838.18	838.18	11.19	0.003
Error	26	1946.99	74.88		
Total	29	2890.87			

DF = degrees of freedom, SS = sums of squares, MS = mean square, F = variance ratio, *p* = probability.

3.4. The Probiotic Cell Enumeration of the Amaranth Samples

The statistical analysis revealed significant differences ($p < 0.05$) between the mean values of probiotic viability (log CFU/g) from FPs based on red and black amaranth, germinated and ungerminated. With values ranging from 7.68 ± 0.11 to 8.15 ± 0.49 log CFU/g, the samples fermented with unsprouted RA coded R21, R24, and R74 had the highest cell concentrations; there were no significant differences between them ($p < 0.05$). When compared to the other samples, ungerminated BA seeds fermented with the *Lp. paraplantarum* MIUG BL74 strain had the highest cell concentration (8.84 ± 0.06 log CFU/g), which was statistically significant. The fermented products based on sprouted red and black amaranth seeds exhibited significantly lower cell concentrations, up to 3.11 ± 0.04 and 5.60 ± 0.24 log CFU/g, respectively (Table 5). Instead, among the sprouted FPs, those based on black and red amaranth (RG 38 and BG 38) exhibited the greatest probiotic viability.

Probiotic viability in FPs derived from ungerminated amaranth fell within ranges previously reported for amaranth- and pseudocereal-based fermentations, with cell counts of 2.60–8.54 log CFU/g in spontaneously fermented amaranth sourdough [46] and approximately 8 log CFU/g in amaranth-based purees [47]. According to these results, unsprouted amaranth flour might be a suitable substrate for the growth of lactic acid bacteria under optimal fermentation conditions [48].

https://doi.org/10.3390/microorganisms14020340

Table 5. Probiotic cell counting (log CFU/g) in fermented products based on red and black amaranth flour, sprouted and unsprouted.

Sample	Log CFU/g	Sample	Log CFU/g
R 21	7.68 ± 0.11^{a}	B 21	8.38 ± 0.18^{AB}
R 24	8.15 ± 0.49^{a}	B 24	8.00 ± 0.12^{B}
R 38	6.87 ± 0.16^{b}	B 38	7.95 ± 0.05^{BC}
R 74	8.15 ± 0.14^{a}	B 74	8.84 ± 0.06^{A}
RG 21	3.11 ± 0.04	BG 21	5.87 ± 0.18^{D}
RG 24	3.18 ± 0.12^{e}	BG 24	5.60 ± 0.24^{D}
RG 38	6.20 ± 0.19^{c}	BG 38	7.52 ± 0.24^{C}
RG 74	3.93 ± 0.09^{d}	BG 74	5.89 ± 0.02^{D}

Mean values that do not share the same letter, in the same column, are statistically different at $p < 0.05$, based on posthoc Tukey test. The lowercase letters are used to compare red amaranth FPs, while the uppercase letters are used to compare black amaranth FPs. Coded samples B 21, B 24, B 38, and B 74/R 21, R 24, R 38, and R 74—ungerminated BA/RA fermented with *Lp. plantarum* MIUG BL21, *Lp. pentosus* MIUG BL24, *Lc. rhamnosus* MIUG BL38, and *Lp. paraplantarum* MIUG BL74 strains. BG 21, BG 24, BG 38, BG 74/RG 21, RG 24, RG 38, and RG 74—germinated BA/RA fermented with the strains listed above.

Additionally, the biochemical changes that occur in amaranth seeds during germination impacted the resilience of probiotic lactic acid bacteria. Similar results regarding the effect of pretreatment (including sprouting) and substrate type on probiotic viability have been reported by Nkhata et al. [49] and Petrova and Petrov [50], who agreed that both the characteristics of the probiotic strain and the degree of processing of the pseudocereal influence the probiotic's survivability following fermentation.

3.5. RP-HPLC Characterisation of Amaranth Samples

For subsequent analyses, fermented products from both sprouted and unsprouted red and black amaranth with the strain *Lc. rhamnosus* MIUG BL38 were chosen, along with the controls. The selection criteria included balanced acidification, enhanced ABTS and DPPH antioxidant activity, high total phenolic and flavonoid contents, and a probiotic threshold of at least 6 log CFU/g.

In the analysed RA fermented products, the nine phenolic acids that were identified were four hydroxybenzoic acids (gallic acid, protocathechic acid, syringic acid, ellagic acid), and five hydroxycinnamic acids (ferulic acid, cinnamic acid, caffeic acid, ρ-coumaric acid, chlorogenic acid), as described in Tables 6 and 7. The fermentation and germination processes revealed twelve flavonoids, as follows, epicatechin gallate, hesperitin, quercetin, apigenin, luteolin, naringenin, quercetin 3-glucoside, isorhamnetin, peonidin 3-O rutinoside, epicatechin, keracyanin, and rutin trihydrate.

The amount of phenolic acids and flavonoids in the examined samples varied significantly. More biological active compounds were found in red amaranth-based samples, both germinated and ungerminated, than in black amaranth-based samples.

The current findings are supported by Sarker and Oba [51], who reported abundant phenolic acids such as salicylic acid, vanillic acid, protocatechuic acid, gallic acid, gentisic acid, β-resorcylic acid, *p*-hydroxybenzoic acid, syringic acid, ellagic acid, sinapic acids, chlorogenic acid, trans-cinnamic acid, m-coumaric acid, caffeic acid, *p*-coumaric acid, and ferulic acid, as well as flavonoids like rutin, hyperoside, isoquercetin, myricetin, quercetin, apigenin, kaempferol, and catechin in the red-coloured amaranth leaves, which were much higher compared to the green-coloured amaranth.

Both the germinated and ungerminated RA-controls displayed a variety of phenolic compounds, with the maximum quantity of epicatechin (4518.68 ± 15.86 ng/µL) and protocatechuic acid (2000.3 ± 2.38 ng/µL) in the germinated control (RGC). Gallic acid was the most prevalent, with the highest concentration of 833.08 ± 1.23 ng/µL in the

https://doi.org/10.3390/microorganisms1402034(

germinated RA sample fermented with *Lc. rhamnosus* MIUG BL38 strain (coded sample RG38), followed by chlorogenic acid (110.67 ± 3.28 ng/µL). However, the FPs included small amounts of phenolic acids such as syringic, protocatechuic, and chlorogenic acids. Instead, only in controls, ferulic, cinnamic, and caffeic acids, quercetin, and apigenin were identified (Table 6).

Table 6. RP-HPLC characterisation of red amaranth samples, both germinated and non-germinated, fermented with probiotic *Lc. rhamnosus* MIUG BL38 strain.

Identified Compounds, ng/µL	RC	R 38	RGC	RG 38
		Phenolic acids		
Gallic acid	42.37 ± 0.59 ^{d}	135.85 ± 1.01 ^{a}	nd	833.08 ± 1.23 ^{a}
Syringic acid	35.55 ± 0.37 ^{e}	6.46 ± 1.57 ^{e}	291.40 ± 4.19 ^{c}	18.19 ± 0.84 ^{e}
p-Cumaric acid	6.25 ± 2.04 ^{i}		nd	
Protocatechuic acid	nd	36.94 ± 3.25 ^{d}	2000.3 ± 2.38 ^{b}	56.01 ± 5.86 ^{d}
Chlorogenic acid	145.12 ± 0.20 ^{a}	50.26 ± 0.97 ^{c}	nd	110.67 ± 3.28 ^{b}
Ferulic acid	6.95 ± 0.78 ^{hi}	nd	6.66 ± 0.69 ^{k}	nd
Cinnamic acid	4.20 ± 0.20 ^{i}	nd	4.87 ± 0.53 ^{m}	nd
Caffeic acid	55.19 ± 0.86 ^{c}	nd	57.07 ± 3.73 ^{f}	nd
Ellagic acid	33.79 ± 0.37 ^{e}	nd	nd	nd
		Flavonoids		
Epicatechin gallate	89.93 ± 0.19 ^{b}	nd	nd	nd
Hesperitin	28.46 ± 0.32 ^{f}	69.80 ± 6.19 ^{b}	195.25 ± 2.99 ^{d}	75.80 ± 8.06 ^{c}
Quercetin	10.88 ± 0.30 ^{g}	nd	11.95 ± 1.18 ^{h}	nd
Apigenin	7.08 ± 0.14 ^{hi}	nd	9.57 ± 3.99 ^{i}	nd
Luteolin	9.51 ± 0.19 ^{gh}	nd	nd	nd
Naringenin	nd	nd	60.02 ± 4.02 ^{e}	nd
Quercetin 3-glucoside	nd	nd	16.08 ± 3.62 ^{g}	nd
Isorhamnetin	5.61 ± 1.56 ^{i}	nd	5.18 ± 0.42 ^{l}	nd
Peonidin 3-O rutinoside	nd	nd	2.85 ± 0.46 ^{n}	nd
Epicatechin	nd	37.04 ± 1.10 ^{d}	4518.68 ± 15.86 ^{a}	103.38 ± 1.08 ^{b}
Keracyanin	nd	nd	nd	17.50 ± 0.59 ^{e}
Rutin trihydrate	10.77 ± 0.38 ^{g}	nd	7.07 ± 0.13 ^{j}	nd

nd—Not detected; Mean values that do not share the same letter, in the same column, are statistically different at $p < 0.05$, based on Tukey and Games–Howell tests. RC—ungerminated RA control (unfermented), R 38—ungerminated RA fermented with *Lc. rhamnosus* MIUG BL38, RGC—germinated RA control (unfermented), RG 38—germinated BA fermented with the strain listed above.

Table 7. RP-HPLC characterisation of black amaranth samples, both germinated and non-germinated, fermented with probiotic *Lc. rhamnosus* MIUG BL38 strain.

Identified Compounds, ng/µL	BC	B 38	BGC	BG 38
		Phenolic acids		
Gallic acid	38.57 ± 0.14 ^{c}	97.16 ± 0.69 ^{c}	47.92 ± 0.06 ^{b}	288.45 ± 2.18 ^{b}
Protocatechuic acid	42.77 ± 0.93 ^{c}	24.31 ± 0.25 ^{d}	49.63 ± 0.35 ^{b}	nd
Syringic acid	12.26 ± 0.58 ^{d}	5.17 ± 0.04 ^{f}	40.73 ± 0.78 ^{c}	11.25 ± 0.27 ^{d}
p-Cumaric acid	5.09 ± 0.17 ^{e}	4.80 ± 0.25 ^{ef}	4.72 ± 0.17 ^{d}	6.00 ± 0.08 ^{e}
Chlorogenic acid	nd	53.55 ± 2.83 ^{cdef}	189.07 ± 4.35 ^{a}	73.15 ± 2.30 ^{c}
		Flavonoids		
Quercetin 3-glucoside	70.99 ± 0.84 ^{b}	nd	nd	nd
Luteolin	9.98 ± 0.11 ^{d}	nd	nd	nd
Epicatechin	nd	177.03 ± 1.16 ^{b}	nd	619.86 ± 11.62 ^{a}
Epigalocatechin	511.64 ± 0.98 ^{a}	1349.76 ± 3.14 ^{a}	nd	nd

nd—Not detected; Mean values that do not share the same letter, in the same column, are statistically different at $p < 0.05$, based on Tukey and Games–Howell tests. BC—ungerminated BA control (unfermented), B 38—ungerminated BA fermented with *Lc. rhamnosus* MIUG BL38, BGC—germinated BA control (unfermented), BG 38—germinated BA fermented with the strain listed above.

https://doi.org/10.3390/microorganisms14020340

Gallic acid levels in the FPs derived from germinated seed were significantly higher in BG38 (288.45 ± 2.18 ng/μL) than in control, coded BGC (47.92 ± 0.06 ng/μL). Fermented products based on ungerminated BA seed showed the same pattern, as seen in coded sample B38, which had a concentration of 97.16 ± 0.69 ng/μL compared to the control's concentration of 38.57 ± 0.14 ng/μL. The main flavonoid in coded sample B38 was epigallocatechin (1349.76 ± 3.14 ng/μL), but epicatechin was mostly found in BG38 (619.86 ± 11.62 ng/μL), suggesting that the distribution of flavonoid compounds varies depending on the sample (Table 7). Overall, the HPLC data showed notable qualitative and quantitative variations in the phenolic acid and flavonoid content of the examined FPs, indicating that the phytochemical profile is significantly influenced by processing (fermentation and germination).

Studies comparing the leaves of red genotypes of *A. tricolour* and the green genotypes of *A. lividus* show that the red genotypes are richer in flavonoids and phenolic acids. In particular, phenolic contents in red accessions ranged from 85 to 312 μg/g DW, which was higher than the range of 71 to 220 μg/g DW in green accessions. Gallic acid, vanillic acid, salicylic acid, protocatechin, caffeic acid, *p*-coumaric acid, ferulic acid, and trans-cinnamic acids constituted the majority of these compounds. Additionally, red accessions were rich in flavonoids such as rutin, quercetin, myricetin, and kaempferol [51,52].

According to a recent study by Suoare et al. [18], black amaranth fermented doughs had higher contents of flavonoids than phenolic acids, with epigallocatechin being the most prevalent, followed by epicatechin and catechin. The levels of epigallocatechin were 7789.88 ± 17.00 ng/μL in the control, 6942.47 ± 5.63 ng/μL in the dough fermented with *Lc. rhamnosus* MIUG BL38, and 4983.16 ± 7.29 ng/μL in the dough fermented with *Lp. pentosus* MIUG BL24. Only the control had minor phenolic acids, such as *p*-coumaric and 4-hydroxybenzoic acids.

Assessing other plant parts (leaves, stalks, seeds, and flowers), sprouts contained the fewest identifiable phenolic chemicals; only gallic acid, protocatechuic acid, and rutin were quantified [53]. Cinnamic acid derivatives (ferulic and chlorogenic acid) were less common and generally at lower concentrations in sprouts of *A. caudatus*.

Phenolic compounds, including flavonoids, hydroxycinnamic acids (ferulic acid, cinnamic acid, caffeic acid, *p*-coumaric acid, chlorogenic acid), and hydroxybenzoic acids (gallic acid, protocathechic acid, syringic acid), are commonly found in *Amaranthus* species [54].

Sprouting (germination) induced nutritionally favourable changes, making amaranth sprouted seeds nutritionally superior compared to non-sprouted seeds. Consequently, they represent promising ingredients for developing foods with enhanced nutrient and antioxidant properties.

3.6. SDS-PAGE of the FPs Extract Analysis

The SDS-PAGE evaluation of water-soluble proteins extracted from black and red amaranth FPs highlighted a few distinct protein bands between the germinated, ungerminated and fermented samples, respectively (Figure 7).

More precisely, in the ungerminated black amaranth control (BC), a pronounced background smear of molecular weight proteins smaller than 15 kDa were observed, which was reduced in the germinated control (BGC). In addition, two distinct protein bands of ~35 kDa and ~27 kDa were maintained across all the samples, including the red amaranth controls. The ~35 kDa band may correspond to the acidic subunits of 11S globulin [55,56], one of the major storage proteins found in amaranth seeds, while the ~27 kDa band represents a smaller globulin fraction, both of which were resistant to germination process and remained visible after germination. Moreover, the germination process led to a reduction in the intensity of proteins bands having a molecular weight between 10 and 15 kDa and a decrease in the background smear, suggesting a partial hydrolysis of storage proteins. Different electrophoretic profiles have been documented in the literature, highlighting variations in amaranth genetic diversity, extraction solvents and processing conditions. For instance, Bojórquez-Velázque et al. [57] showed that the hydrophilic protein fraction of the seeds contains a wide range of proteins, with major bands at ~33 kDa, 37 kDa, and 52 kDa, while the hydrophobic fractions have different bands which can be clustered in three regions (20–24 kDa, 32–35 kDa and 50–70 kDa), with clear species-dependent differences in band presence. Making a comparative proteomic analysis of amaranth seeds, Bojórquez-Velázque et al. [56] found that the hydrophilic fraction contained abundant small- and medium-molecular-weight proteins of 10–15 kDa and 25–40 kDa, respectively. The main proteins from the high-variation region were the 11S globulin, granule-bound starch synthase I and the Vicilin-like seed storage proteins, respectively [56].

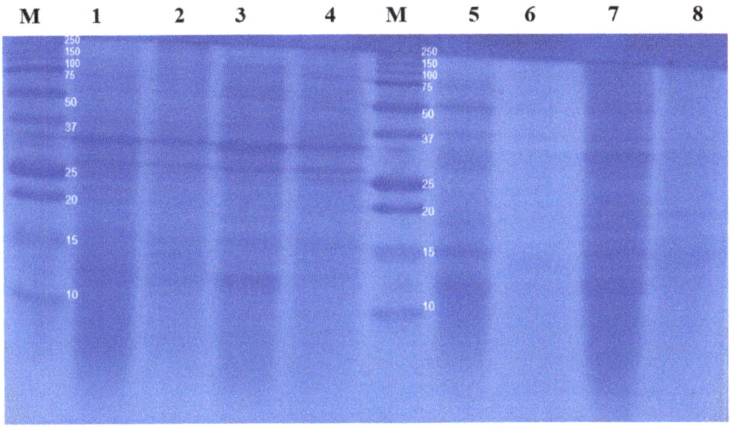

Figure 7. The SDS-PAGE profile of water-soluble proteins from the following samples: lines 1 and 2—ungerminated and germinated black amaranth, BC and BGC (controls); lines 3 and 4—ungerminated and germinated red amaranth, RC and RGC (controls); lines 5 and 6—ungerminated and germinated black amaranth FPs fermented with *L. rhamnosus* MIUG BL38 (B38 and BG38); lines 7 and 8—ungerminated and germinated red amaranth flours fermented with *L. rhamnosus* MIUG BL38 (R38 and RG38); M—Precision Plus Proteins Dual Xtra Prestained Marker (kDa).

Additionally, the fermentation process with *L. rhamnosus* MIUG BL38 probiotic strain further changed the protein's background. In ungerminated black amaranth (B38, line 5), two clear bands of 15 kDa and 50 kDa were emphasised, along with a faint one of 75 kDa. In contrast, when the germinated black amaranth was used as substrate in the fermentation, the protein background became uniform, with no sharp protein bands visible, suggesting that an extensive hydrolysis occurred (line 6). Further, in red ungerminated amaranth, the fermentation with *L. rhamnosus* MIUG BL38 (R38—line 7) led to an overall increase in the protein background intensity, indicating the presence of a greater number of soluble proteins. This pattern suggests that the fermentation process generates a range of polypeptides and peptides through partial proteolysis, rather than complete protein degradation. Furthermore, the fermentation of germinated red amaranth produced a proteins profile similar to that of fermented germinated black amaranth, indicating an extensive hydrolysis of storage proteins in both varieties.

https://doi.org/10.3390/microorganisms14020340

4. Conclusions

The current investigation shows that the functional and biochemical properties of red and black amaranth seeds are effectively improved by the combined use of sprouting and probiotic fermentation. Acidifying potential, antioxidant activity, phenolic and flavonoid profiles, water-soluble protein content and probiotic viability were all greatly influenced by sprouting and fermentation processing. The results are highly strain- and amaranth variety-dependent. Black amaranth-derived fermented products demonstrated stronger antioxidant activity but sprouted and unsprouted red amaranth generally showed higher levels of bioactive compounds. *Lacticaseibacillus rhamnosus* MIUG BL38 had the most balanced technological and functional performance among the strains studied, promoting favourable acidification, increased soluble protein release, improved antioxidant potential, and proper probiotic viability. Overall, the results indicate that red amaranth, *Amaranthus cruentus* sprouts fermented with *Lc. rhamnosus* MIUG BL38 represents a promising gluten-free functional food ingredient with numerous potential health benefits, suitable for incorporation into innovative probiotic and nutraceutical food formulations.

Author Contributions: Conceptualisation, M.C. and M.A.V.; methodology, M.C., M.A.V. and L.G.-G.; software, N.B.; validation, M.C., M.A.V. and G.E.B.; formal analysis, M.A.V. and N.B.; investigation, M.C. and L.G.-G.; resources, M.A.V.; data curation, M.C. and M.A.V.; writing—original draft preparation, M.A.V. and M.C.; writing—review and editing, M.C., M.A.V., L.G.-G. and G.E.B.; visualisation, M.C., M.A.V., L.G.-G. and G.E.B.; supervision, G.E.B.; project administration, M.A.V. and G.E.B.; funding acquisition, M.A.V. All authors have read and agreed to the published version of the manuscript.

Funding: This research was funded by the "Dunărea de Jos" University of Galați, contract no. IG 7971/2025, which benefited the first author, Mihaela Aida Vasile.

Institutional Review Board Statement: Not applicable.

Informed Consent Statement: Not applicable.

Data Availability Statement: The original contributions presented in this study are included in the article. Further inquiries can be directed to the corresponding authors.

Acknowledgments: The Integrated Centre for Research, Expertise and Technological Transfer in Food Industry (BioAliment-TehnIA), Dunărea de Jos University of Galati, Romania is acknowledged for providing technical support. The support regarding the probiotic strains offered by MIUG Collection affiliated with the Microbial Resource Research Infrastructure (MIRRI) is gratefully acknowledged.

Conflicts of Interest: The authors declare no conflicts of interest.

References

1. Baraniak, J.; Kania-Dobrowolska, M. The Dual Nature of Amaranth—Functional Food and Potential Medicine. *Foods* **2022**, *11*, 618. [CrossRef]
2. Park, S.-J.; Sharma, A.; Lee, H.-J. A Review of Recent Studies on the Antioxidant Activities of a Third-Millennium Food: *Amaranthus* spp. *Antioxidants* **2020**, *9*, 1236. [CrossRef] [PubMed]
3. Osei, E.D.; Afedzi, A.E.K.; Amotoe-Bondzie, A.; Ivanišová, E.; Encina-Zelada, C.R.; Czaplicki, S.; Čičová, I.; Jančo, I.; Gálik, B.; Afoakwah, N.A. Nutritional and Bioactive Characterization of *Amaranthaceae* Seeds from Peru, Slovakia, and Poland: A Comparative Study. *Food Sci. Nutr.* **2025**, *13*, e70901. [CrossRef]
4. Toimbayeva, D.; Saduakhasova, S.; Kamanova, S.; Kiykbay, A.; Tazhina, S.; Temirova, I.; Muratkhan, M.; Shaimenova, B.; Murat, L.; Khamitova, D.; et al. Prospects for the Use of Amaranth Grain in the Production of Functional and Specialized Food Products. *Foods* **2025**, *14*, 1603. [CrossRef]
5. Lopez-Martinez, J.M.; Ahmad, I. Amaranth Seeds: A Promising Functional Ingredient for Gastronomy—A Review. *Sarhad J. Agric.* **2024**, *40*, 39–53. [CrossRef]

 https://doi.org/10.3390/microorganisms14020340

Alemayehu, F.R.; Bendevis, M.A.; Jacobsen, S.-E. The Potential for Utilizing the Seed Crop Amaranth (*Amaranthus* spp.) in East Africa as an Alternative Crop to Support Food Security and Climate Change Mitigation. *J. Agron. Crop Sci.* **2015**, *201*, 321–329. [CrossRef]

Aderibigbe, O.R.; Ezekiel, O.O.; Owolade, S.O.; Korese, J.K.; Sturm, B.; Hensel, O. Exploring the Potentials of Underutilized Grain Amaranth (*Amaranthus* spp.) along the Value Chain for Food and Nutrition Security: A Review. *Crit. Rev. Food Sci. Nutr.* **2022**, *62*, 656–669. [CrossRef] [PubMed]

Mătieș, A.; Negrușier, C.; Roșca Mare, O.; Mintaș, O.S.; Zanc Săvan, G.; Odagiu, A.C.M.; Andronie, L.; Păcurar, I. Characterization of Nutritional Potential of *Amaranthus* sp. Grain Production. *Agronomy* **2024**, *14*, 630. [CrossRef]

Cruz-Casas, D.E.; Aguilar, C.N.; Ascacio-Valdés, J.A.; Rodríguez-Herrera, R.; Chávez-González, M.L.; Flores-Gallegos, A.C. Bioactive Protein Hydrolysates Obtained from Amaranth by Fermentation with Lactic Acid Bacteria and *Bacillus* Species. *Heliyon* **2023**, *9*, e13491. [CrossRef]

Zhu, F. Amaranth Proteins and Peptides: Biological Properties and Food Uses. *Food Res. Int.* **2023**, *164*, 112405. [CrossRef]

Rizzello, C.G.; Lorusso, A.; Russo, V.; Pinto, D.; Marzani, B.; Gobbetti, M. Improving the Antioxidant Properties of Quinoa Flour through Fermentation with Selected Autochthonous Lactic Acid Bacteria. *Int. J. Food Microbiol.* **2017**, *241*, 252–261. [CrossRef]

Chauhan, A.; Kumari, N.; Saxena, D.C.; Singh, S. Effect of Germination on Fatty Acid Profile, Amino Acid Profile and Minerals of Amaranth (*Amaranthus* spp.) Grain. *J. Food Meas. Charact.* **2022**, *16*, 1777–1786. [CrossRef]

Aguilar-Felices, E.J.; Romero-Viacava, M.; Enciso-Roca, E.; Herrera-Calderon, O.; Común-Ventura, P.; Yuli-Posadas, R.; Chacaltana-Ramos, L.; Pari-Olarte, B. Antioxidant Activity of the Germinated Seed of Four Varieties of *Amaranthus caudatus* L. from Peru. *Pharmacogn. J.* **2019**, *11*, 588–593. [CrossRef]

Valadez-Vega, C.; Lugo-Magaña, O.; Figueroa-Hernández, C.; Bautista, M.; Betanzos-Cabrera, G.; Bernardino-Nicanor, A.; González-Amaro, R.M.; Alonso-Villegas, R.; Morales-González, J.A.; González-Cruz, L. Effects of Germination and Popping on the Anti-Nutritional Compounds and the Digestibility of *Amaranthus hypochondriacus* Seeds. *Foods* **2022**, *11*, 2075. [CrossRef]

Janmohammadi, H.; Hosseintabar-Ghasemabad, B.; Oliyai, M.; Alijani, S.; Gorlov, I.F.; Slozhenkina, M.I.; Mosolov, A.A.; Suarez Ramirez, L.; Seidavi, A.; Laudadio, V.; et al. Effect of Dietary Amaranth (*Amaranthus hybridus chlorostachys*) Supplemented with Enzyme Blend on Egg Quality, Serum Biochemistry and Antioxidant Status in Laying Hens. *Antioxidants* **2023**, *12*, 456. [CrossRef]

García-Vicente, E.J.; Martín, M.; Rey-Casero, I.; Pérez, A.; Martínez, R.; Bravo, M.; Alonso, J.M.; Risco, D. Effect of Feed Supplementation with Probiotics and Postbiotics on Strength and Health Status of Honey Bee (*Apis mellifera*) Hives during Late Spring. *Res. Vet. Sci.* **2023**, *159*, 237–243. [CrossRef]

Hernández-García, Y.; Melgar-Lalanne, G.; Téllez-Medina, D.I.; Ruiz-May, E.; Salgado-Cruz, M.d.l.P.; Andrade-Velásquez, A.; Dorantes-Álvarez, L.; López-Hernández, D.; Santiago Gómez, M.P. Scavenging Peptides, Antioxidant Activity, and Hypoglycemic Activity of a Germinated Amaranth (*Amaranthus hypochondriacus* L.) Beverage Fermented by Lactiplantibacillus Plantarum. *J. Food Biochem.* **2022**, *46*, e14139. [CrossRef]

Souare, M.L.; Diallo, A.O.S.; Balan, N.; Vasile, M.A.; Traore, L.; Bahrim, G.E.; Cotârleț, M.; Dumitru, C.N. An Assessment of the Functional Properties of Black Amaranth Flour During Fermentation with Probiotic Lactic Acid Bacteria. *Fermentation* **2025**, *11*, 414. [CrossRef]

Castro-Alba, V.; Lazarte, C.E.; Perez-Rea, D.; Carlsson, N.; Almgren, A.; Bergenståhl, B.; Granfeldt, Y. Fermentation of Pseudocereals Quinoa, Canihua, and Amaranth to Improve Mineral Accessibility through Degradation of Phytate. *J. Sci. Food Agric.* **2019**, *99*, 5239–5248. [CrossRef]

Vento, M.; Della Croce, C.M.; Bellani, L.; Tassi, E.L.; Echeverria, M.C.; Giorgetti, L. Effect of Sprouting, Fermentation and Cooking on Antioxidant Content and Total Antioxidant Activity in Quinoa and Amaranth. *Int. J. Mol. Sci.* **2024**, *25*, 10972. [CrossRef]

Cotârleț, M.; Pihurov, M.; Păcularu-Burada, B.; Vasile, A.M.; Bahrim, G.E.; Grigore-Gurgu, L. Selection of New Lactobacilli Strains with Potentially Probiotic Properties. *Ann. Univ. Dunarea Jos Galati Fascicle VI—Food Technol.* **2023**, *47*, 27–50. [CrossRef]

Ghinea, I.O.; Ionica Mihaila, M.D.; Blaga (Costea), G.-V.; Avramescu, S.M.; Cudalbeanu, M.; Isticioaia, S.-F.; Dinica, R.M.; Furdui, B. HPLC-DAD Polyphenolic Profiling and Antioxidant Activities of Sorghum Bicolor during Germination. *Agronomy* **2021**, *11*, 417. [CrossRef]

Busquère, A.; Lefebvre, D.; Galaup, P.; Tricoulet, L.; Musset, C.; Lacroux, E.; Merah, O. Osmopriming Increases Seed Germination of *Amaranthus cruentus* (L.). *Seeds* **2025**, *4*, 37. [CrossRef]

Kamal, H.; Mudgil, P.; Bhaskar, B.; Fisayo, A.F.; Gan, C.-Y.; Maqsood, S. Amaranth Proteins as Potential Source of Bioactive Peptides with Enhanced Inhibition of Enzymatic Markers Linked with Hypertension and Diabetes. *J. Cereal Sci.* **2021**, *101*, 103308. [CrossRef]

Li, J.; Zhao, W.; Pan, X.; Lao, F.; Liao, X.; Shi, Y.; Wu, J. Improvement of Antioxidant Properties of Jujube Puree by Biotransformation of Polyphenols via *Streptococcus thermophilus* Fermentation. *Food Chem. X* **2022**, *13*, 100214. [CrossRef]

Nionelli, L.; Wang, Y.; Pontonio, E.; Immonen, M.; Rizzello, C.G.; Maina, H.N.; Katina, K.; Coda, R. Antifungal Effect of Bioprocessed Surplus Bread as Ingredient for Bread-Making: Identification of Active Compounds and Impact on Shelf-Life. *Food Control* **2020**, *118*, 107437. [CrossRef]

27. Mërtiri, I.; Coman, G.; Cotârlet, M.; Turturică, M.; Balan, N.; Râpeanu, G.; Stănciuc, N.; Mihalcea, L. Phytochemical Profil Screening and Selected Bioactivity of Myrtus Communis Berries Extracts Obtained from Ultrasound-Assisted and Supercritica Fluid Extraction. *Separations* **2025**, *12*, 8. [CrossRef]
28. Gavril (Rațu), R.N.; Constantin, O.E.; Enachi, E.; Stoica, F.; Lipșa, F.D.; Stănciuc, N.; Aprodu, I.; Râpeanu, G. Optimization of th Parameters Influencing the Antioxidant Activity and Concentration of Carotenoids Extracted from Pumpkin Peel Using a Centra Composite Design. *Plants* **2024**, *13*, 1447. [CrossRef]
29. Bayoï, J.R.; Râpeanu, G.; Stănciuc, N.; Cotărleţ, M.; Constantin, O.E.; Etoa, F.-X. In Vitro Antioxidant Potential and Antimicrobia Activity of Some Cameroonian Plant Extracts. *Ann. Univ. Dunarea de Jos Galati Fascicle VI Food Technol.* **2021**, *45*, 96–116. [CrossRef
30. Bradford, M.M. A Rapid and Sensitive Method for the Quantitation of Microgram Quantities of Protein Utilizing the Principle o Protein-Dye Binding. *Anal. Biochem.* **1976**, *72*, 248–254. [CrossRef] [PubMed]
31. Laemmli, U.K. Cleavage of Structural Proteins during the Assembly of the Head of Bacteriophage T4. *Nature* **1970**, *227*, 680–68 [CrossRef]
32. Mërtiri, I.; Mihalcea, L.; Aprodu, I.; Cioromila, A.C.; Turturică, M.; Râpeanu, G.; Stănciuc, N.; Grigore-Gurgu, L. Phytochemica Profile and Antilisterial Activity of Salvia Officinalis Supercritical Fluid Extract. *Appl. Food Res.* **2025**, *5*, 101379. [CrossRef]
33. Owheruo, J.O.; Akpoghelie, P.O.; Edo, G.I.; Yousif, E.; Jikah, A.N.; Onyibe, P.N.; Ugbune, U.; Isoje, E.F.; Igbuku, U.A.; Ekokotu H.A.; et al. Influence of Spouting Period on Microbiological and Nutritional Attribute of Sesame Seed Flour. *Discov. Food* **202** *5*, 3. [CrossRef]
34. Juárez-Trujillo, N.; González-Avila, C.; Beristain-Guevara, C.I.; Mendoza-López, M.R.; Pascual-Pineda, L.A.; Jiménez-Fernánde M. Nutritional, Physicochemical and Antioxidant Properties of Sprouted and Fermented Beverages Made from *Phalaris canariensi* Seed. *Int. J. Food Sci. Technol.* **2023**, *58*, 5299–5310. [CrossRef]
35. Venturi, M.; Galli, V.; Pini, N.; Guerrini, S.; Granchi, L. Use of Selected Lactobacilli to Increase γ-Aminobutyric Acid (GABA Content in Sourdough Bread Enriched with Amaranth Flour. *Foods* **2019**, *8*, 218. [CrossRef] [PubMed]
36. Atasoy, M.; Álvarez Ordóñez, A.; Cenian, A.; Djukić-Vuković, A.; Lund, P.A.; Ozogul, F.; Trček, J.; Ziv, C.; De Biase, D. Exploitatio of Microbial Activities at Low pH to Enhance Planetary Health. *FEMS Microbiol. Rev.* **2023**, *48*, fuad062. [CrossRef] [PubMed]
37. Haque, M.M.; Islam, M.S.; Parvin, E.; Biswas, P.; Janny, R.J.; Ullah, M.Z.; lal Barua, J. Comparison of the Nutrient Compositions i Red and Green Amaranthus (*Amaranthus hypochondriacus*). *Acad. J. Life Sci.* **2022**, 33–39. [CrossRef]
38. Hur, S.J.; Lee, S.Y.; Kim, Y.-C.; Choi, I.; Kim, G.-B. Effect of Fermentation on the Antioxidant Activity in Plant-Based Foods. *Foo Chem.* **2014**, *160*, 346–356. [CrossRef]
39. Waliat, S.; Arshad, M.S.; Hanif, H.; Ejaz, A.; Khalid, W.; Kauser, S.; Al-Farga, A. A Review on Bioactive Compounds in Sprouts Extraction Techniques, Food Application and Health Functionality. *Int. J. Food Prop.* **2023**, *26*, 647–665. [CrossRef]
40. Maldonado-Alvarado, P.; Pavón-Vargas, D.J.; Abarca-Robles, J.; Valencia-Chamorro, S.; Haros, C.M. Effect of Germination on th Nutritional Properties, Phytic Acid Content, and Phytase Activity of Quinoa (*Chenopodium quinoa* Willd). *Foods* **2023**, *12*, 389 [CrossRef] [PubMed]
41. Leonard, W.; Zhang, P.; Ying, D.; Adhikari, B.; Fang, Z. Fermentation Transforms the Phenolic Profiles and Bioactivities o Plant-Based Foods. *Biotechnol. Adv.* **2021**, *49*, 107763. [CrossRef] [PubMed]
42. Zhang, P.; Zhang, J.; Li, L.; Gu, T.; Chen, S.; Wang, J.; Gao, M. The Release of Bound Phenolics to Enhance the Antioxidant Activity of Cornmeal by Liquid Fermentation with *Bacillus subtilis*. *Foods* **2025**, *14*, 499. [CrossRef]
43. Günal-Köroğlu, D.; Esatbeyoglu, T.; Capanoglu, E. Effect of Germination on the Phenolic Compounds: Content, Bioavailability Food Applications, and Health Benefits. *J. Food Meas. Charact.* **2025**, *19*, 8144–8164. [CrossRef]
44. Adebo, O.A.; Gabriela Medina-Meza, I. Impact of Fermentation on the Phenolic Compounds and Antioxidant Activity of Whol Cereal Grains: A Mini Review. *Molecules* **2020**, *25*, 927. [CrossRef]
45. Melini, F.; Melini, V. Impact of Fermentation on Phenolic Compounds and Antioxidant Capacity of Quinoa. *Fermentation* **2021** *7*, 20. [CrossRef]
46. Rodríguez, M.A.; Fernández, L.A.; Daisley, B.A.; Reynaldi, F.J.; Allen-Vercoe, E.; Thompson, G.J. Probiotics and In-Hive Fermentation as a Source of Beneficial Microbes to Support the Gut Microbial Health of Honey Bees. *J. Insect Sci.* **2023**, *23*, 19 [CrossRef] [PubMed]
47. Matejčeková, Z.; Liptáková, D.; Valík, L'. Evaluation of The Potential of Amaranth Flour for Lactic Acid Fermentation. *J. Pharm Nutr. Sci.* **2016**, *6*, 1–6. [CrossRef]
48. Ziarno, M.; Cichońska, P. Lactic Acid Bacteria-Fermentable Cereal- and Pseudocereal-Based Beverages. *Microorganisms* **2021** *9*, 2532. [CrossRef]
49. Nkhata, S.G.; Ayua, E.; Kamau, E.H.; Shingiro, J. Fermentation and Germination Improve Nutritional Value of Cereals and Legumes through Activation of Endogenous Enzymes. *Food Sci. Nutr.* **2018**, *6*, 2446–2458. [CrossRef]
50. Petrova, P.; Petrov, K. Lactic Acid Fermentation of Cereals and Pseudocereals: Ancient Nutritional Biotechnologies with Modern Applications. *Nutrients* **2020**, *12*, 1118. [CrossRef]

1. Sarker, U.; Oba, S. Antioxidant Constituents of Three Selected Red and Green Color Amaranthus Leafy Vegetable. *Sci. Rep.* **2019**, *9*, 18233. [CrossRef]
2. Khanam, U.K.S.; Oba, S.; Yanase, E.; Murakami, Y. Phenolic Acids, Flavonoids and Total Antioxidant Capacity of Selected Leafy Vegetables. *J. Funct. Foods* **2012**, *4*, 979–987. [CrossRef]
3. Li, H.; Deng, Z.; Liu, R.; Zhu, H.; Draves, J.; Marcone, M.; Sun, Y.; Tsao, R. Characterization of Phenolics, Betacyanins and Antioxidant Activities of the Seed, Leaf, Sprout, Flower and Stalk Extracts of Three *Amaranthus* Species. *J. Food Compos. Anal.* **2015**, *37*, 75–81. [CrossRef]
4. Popoola, O.O. Phenolic Compounds Composition and in Vitro Antioxidant Activity of Nigerian Amaranthus Viridis Seed as Affected by Autoclaving and Germination. *Meas. Food* **2022**, *6*, 100028. [CrossRef]
5. Aguilar-Padilla, J.; Centeno-Leija, S.; Bojórquez-Velázquez, E.; Elizalde-Contreras, J.M.; Ruiz-May, E.; Serrano-Posada, H.; Osuna-Castro, J.A. Characterization of the Technofunctional Properties and Three-Dimensional Structure Prediction of 11S Globulins from Amaranth (*Amaranthus hypochondriacus* L.) Seeds. *Foods* **2023**, *12*, 461. [CrossRef]
6. Bojórquez-Velázquez, E.; Zamora-Briseño, J.A.; Barrera-Pacheco, A.; Espitia-Rangel, E.; Herrera-Estrella, A.; Barba De La Rosa, A.P. Comparative Proteomic Analysis of Wild and Cultivated Amaranth Species Seeds by 2-DE and ESI-MS/MS. *Plants* **2024**, *13*, 2728. [CrossRef]
7. Bojórquez-Velázquez, E.; Barrera-Pacheco, A.; Espitia-Rangel, E.; Herrera-Estrella, A.; Barba De La Rosa, A.P. Protein Analysis Reveals Differential Accumulation of Late Embryogenesis Abundant and Storage Proteins in Seeds of Wild and Cultivated Amaranth Species. *BMC Plant Biol.* **2019**, *19*, 59. [CrossRef]

 https://doi.org/10.3390/microorganisms14020340

microorganisms

MDPI

Article

Comprehensive Safety Assessment of *Lentilactobacillus buchneri* KU200793 as a Potential Probiotic

Suin Kim [1], Huijin Jeong [1], Na-Kyoung Lee [2], Dae-Kyung Kang [3], Hyun-Dong Paik [2], Young-Seo Park [1,*] and Jong Hun Lee [1,*]

[1] Department of Food Science and Biotechnology, Gachon University, Seongnam 13120, Republic of Korea; sue030465@gachon.ac.kr (S.K.); huijin0218@gmail.com (H.J.)
[2] Department of Food Science and Biotechnology of Animal Resource, Konkuk University, Seoul 05029, Republic of Korea; nakyoung_lee@nate.com (N.-K.L.); hdpaik@konkuk.ac.kr (H.-D.P.)
[3] Department of Animal Biotechnology, Dankook University, Cheonan 31116, Republic of Korea; dkkang@dankook.ac.kr
* Correspondence: ypark@gachon.ac.kr (Y.-S.P.); foodguy@gachon.ac.kr (J.H.L.); Tel.: +82-31-750-5378 (Y.-S.P.); +82-31-750-5872 (J.H.L.)

Abstract: The safety profile of *Lentilactobacillus buchneri* KU200793, which has neuroprotective effects, was comprehensively evaluated through both phenotypic and genotypic analyses. Phenotypically, the strain exhibited no β-hemolysis, mucin degradation, indole production, gelatin liquefaction, urease activity, or β-glucuronidase activity. Additionally, it did not produce D-lactate, and only trace amounts of spermidine were detected among the biogenic amines. Furthermore, *L. buchneri* KU200793 did not exhibit bile salt deconjugation, further supporting its safety profile. However, its tetracycline resistance exceeded the threshold set by the European Food Safety Authority. Genotypic analysis using the HGTree program identified *tetA(58)* and *nalD* genes with sequence similarities of 33.64% and 30.17%, respectively, indicating a low level of homology. These findings suggest that tetracycline resistance in *L. buchneri* KU200793 is unlikely to have been acquired through horizontal gene transfer, thereby minimizing the risk of resistance gene dissemination. This study underscores the importance of comprehensive safety assessments to evaluate the suitability of *L. buchneri* KU200793 for probiotic applications.

Keywords: genotypic analysis; *Lentilactobacillus buchneri*; neuroprotective effects; phenotypic analysis; probiotics; safety evaluation

1. Introduction

Lactic acid bacteria (LAB) have long been utilized as probiotics in dietary supplements and food products [1]. LAB serve not only as natural preservatives by enhancing the flavor and acidity of food but also offer various health benefits, including improving mucosal integrity, exhibiting anti-pathogenic properties, aiding in the management of lactose intolerance and food allergies, preventing diarrhea, and alleviating inflammation [2]. The Food and Agriculture Organization (FAO) and the World Health Organization (WHO) defined probiotics as "live microorganisms that, when administered in adequate amounts, confer a health benefit on the host" [3]. Traditionally, probiotics have been studied for their role in immune modulation and gut health. Recently, growing research into the causal links between microbiota and various diseases has increased interest in the therapeutic potential of probiotics for disease management [4,5].

Despite their potential health benefits, the expanding diversity of LAB and their increasing use as probiotics, particularly strains belonging to the genus *Lactobacillus*,

 https://doi.org/10.3390/microorganisms13092067

have raised concerns regarding their safety [6,7]. A report by FAO and WHO highlights four potential adverse effects of probiotics: invasive infections, harmful metabolic processes, excessive immune activation in vulnerable individuals, and horizontal gene transfer (HGT) [8]. Furthermore, certain LAB are opportunistic pathogens linked to bacteremia, a rare but increasingly observed condition in patients with immune disorders or underlying health conditions. These concerns underscore the need for comprehensive safety assessments to ensure the safe use of LAB in food and dietary supplements [9].

Lentilactobacillus buchneri is an obligate heterofermentative, facultative anaerobic bacterium naturally found in diverse ecological environments and is closely associated with food production and fermentation processes [10]. This species is well-known for its beneficial effects, including the conversion of lactic acid to acetic acid, which inhibits the growth of certain yeasts and molds [11]. *L. buchneri* KU200793 exhibits various potential probiotic properties, including high resistance to artificial gastric juice and bile salts, cholesterol-lowering effects, and antimicrobial activity. Additionally, this strain demonstrates beneficial β-galactosidase activity and the absence of β-glucuronidase, indicating its safety. It also demonstrates remarkable antioxidant activity under heat-killed conditions. Furthermore, *L. buchneri* KU200793 has been shown to protect SH-SY5Y neuronal cells from MPP^+-induced neurotoxicity by significantly enhancing cell viability and upregulating brain-derived neurotrophic factor (BDNF) mRNA expression. It also reduces the Bax/Bcl-2 ratio, thereby exerting an anti-apoptotic effect. These findings suggest that *L. buchneri* KU200793 possesses neuroprotective potential through the gut–brain axis and may serve as a promising probiotic candidate for the prevention of neurodegenerative diseases [12].

Although this strain was originally isolated from kimchi, a traditional fermented food, *L. buchneri* KU200793 has not been intentionally applied or evaluated for use in commercial food products or dietary supplements. Considering its potential application as a dietary supplement for humans, a comprehensive safety assessment is essential [13]. Accordingly, this study aimed to evaluate the safety of *L. buchneri* KU200793 based on the guidelines provided by the FAO/WHO (2002) and European Food Safety Authority (EFSA) [14].

2. Materials and Methods

2.1. Bacterial Strains and Culture Conditions

L. buchneri KU200793 (KCCM 12593P) was obtained from Konkuk University (Seoul, Republic of Korea) and grown anaerobically in de Man, Rogosa, and Sharpe broth (MRS; BD, Franklin Lakes, NJ, USA) at 37 °C for 24 h. The MRS broth contained (per liter): 10 g of protease peptone No. 3, 10 g of beef extract, 5 g of yeast extract, 20 g of dextrose, 1 g of polysorbate 80, 2 g of ammonium citrate, 5 g of sodium acetate, 0.1 g of magnesium sulfate, 0.05 g of manganese sulfate, and 2 g of dipotassium phosphate.

Klebsiella pneumoniae subsp. *pneumoniae* KCCM 60022, *Proteus vulgaris* KCCM 40221, *Escherichia coli* KCCM 11234, *Streptococcus pneumoniae* KCCM 41570, *S. pyogenes* KCCM 40411, *S. agalactiae* KCCM 40417, and *Brevibacillus* (*Br.*) *parabrevis* KCCM 41421 were purchased from the Korean Culture Center of Microorganisms (Seoul, Republic of Korea). *Lactiplantibacillus plantarum* KU15122 was obtained from Konkuk University (Seoul, Republic of Korea). *Bacillus cereus* KCCM 11204, *K. pneumoniae* subsp. pneumoniae KCCM 60022, *P. vulgaris* KCCM 40221, *E. coli* KCCM 11234, and *Br. parabrevis* KCCM 41421 were cultured in nutrient broth (BD) (3 g of beef extract and 5 g of peptone, per liter) at 30 °C for 24 h. *Lp. plantarum* KU15122 was cultured in MRS medium at 37 °C for 24 h.

2.2. Cell Culture Conditions

The Caco-2 cell line (ATCC® HTB-37™) was obtained from the American Type Culture Collection (ATCC, Manassas, VA, USA). Caco-2 cells are widely used as an in vitro model of intestinal epithelial cells in probiotic safety assessments, as their viability reflects the absence of cytotoxic effects. The cells were cultured in an incubator (Thermo Fisher Scientific, Waltham, MA, USA) maintained at 37 °C with 5% CO_2. Dulbecco's modified Eagle's medium (DMEM; Gibco, Grand Island, NY, USA) supplemented with 1% (*v*/*v*) penicillin–streptomycin (Gibco), and 10% (*v*/*v*) fetal bovine serum (FBS; Corning Inc., Corning, NY, USA) was used as the culture medium.

2.3. Assessment of Hemolytic Activity

Hemolytic activity was evaluated by streaking bacteria on blood agar base No. 2 (Sigma-Aldrich, St. Louis, MO, USA) supplemented with 5% sheep blood (KisanBio, Seoul, Republic of Korea). The plates were inoculated with *L. buchneri* KU200793, *S. pneumoniae* KCCM 41570, *S. pyogenes* KCCM 40411 and *S. agalactiae* KCCM 40417 and then incubated under anaerobic conditions at 37 °C for 48 h. Strains that formed a green area around the colony were classified as α-hemolytic, those that formed a clear zone were classified as β-hemolytic, and those with no observable color change around the colony were classified as γ-hemolytic.

2.4. Evaluation of Bile Salt Deconjugation

Bile salt deconjugation was assessed using MRS containing 0.5% taurodeoxycholic acid (bile acid; Sigma-Aldrich). Plates inoculated with *L. buchneri* KU200793 and *B. cereus* KCCM 11204 were anaerobically cultured at 37 °C for 48 h. Strains with sedimentation around the bacterial colonies were classified as exhibiting bile salt hydrolase (BSH) activity.

2.5. Cytotoxicity Assay Using Caco-2 Cells

Caco-2 cells were seeded at a density of 5 × 104 cells/well in a 96-well plate in DMEM supplemented with 10% FBS and 1% penicillin–streptomycin. The cells were incubated at 37 °C in a 5% CO_2 incubator for 20 h. *L. buchneri* KU200793 was cultured in MRS broth at 37 °C for 18 h, while *K. pneumoniae* subsp. *pneumoniae* KCCM 60022, used as a positive control, was grown in nutrient broth under the same conditions. Overnight cultures were inoculated into 10 mL of MRS medium at 37 °C for 7 h. Cells were harvested by centrifugation at 16,000× *g* for 1 min. The harvested cells were washed thrice with Dulbecco's phosphate-buffered saline (DPBS; Welgene, Gyeongsan, Republic of Korea) and resuspended in 1 mL of DMEM containing 10% (*v*/*v*) FBS and 1% penicillin–streptomycin [15]. This suspension was added to Caco-2 cells at a multiplicity of infection (MOI; the ratio of LAB cells to Caco-2 cells) of 31.3, 62.5, and 125, and the cultures were incubated at 37 °C in a 5% CO_2 incubator for 24 h. After 24 h, plates were washed twice with DPBS. Subsequently, 200 μL of DMEM containing 10% FBS and 1% penicillin–streptomycin was added to each well, followed by 20 μL of EZ-Cytox solution (DoGenBio Co., Ltd., Seoul, Republic of Korea). The mixture was incubated at 37 °C in a 5% CO_2 atmosphere for 30 min. The EZ-Cytox assay utilizes water-soluble tetrazolium-8, which is reduced by intracellular dehydrogenases in viable cells to produce a water-soluble formazan dye. The amount of formazan formed is proportional to the metabolic activity of viable cells and thus serves as an indirect indicator of cytotoxicity [16]. Absorbance at 450 nm, reflecting cell viability through the amount of formazan produced, was measured using Take3 Micro-Volume Plates (Epoch Microplate Reader; BioTek Instruments, Inc., Winooski, VT, USA).

2.6. Determination of Mucin Degradation

L. buchneri KU200793 was inoculated into four types of media: a basal medium without any carbon source, a basal medium with 0.3% (*w*/*v*) mucin as the sole carbon source, a basal medium supplemented with 1% (*w*/*v*) glucose, and a basal medium containing both 0.3% (*w*/*v*) mucin and 1% (*w*/*v*) glucose. Bacterial growth was assessed after 8 and 24 h of incubation by measuring the optical density (OD) at 600 nm using a spectrophotometer (BioSpec-mini, Shimadzu Co., Kyoto, Japan) and by monitoring pH changes. Growth in mucin-containing media was considered to indicate the strain's ability to utilize mucin as a carbon source, thereby reflecting potential mucin-degrading activity.

2.7. Quantification of D-Lactic Acid Production

The LAB culture was centrifuged at 16,000× *g* for 10 min, and the supernatant was collected. The obtained supernatant was analyzed using an Enzytec™ Liquid D-Lactic acid assay kit (R-Biopharm, Darmstadt, Germany) following the manufacturer's instructions. The D-lactic acid concentration was determined by measuring absorbance at 340 nm using a BioSpec-mini spectrophotometer (Shimadzu Co.). This assay specifically detects D-lactic acid without cross-reactivity to L-lactic acid.

2.8. Analysis of Biogenic Amine Production

L. buchneri KU200793 was cultured in MRS medium supplemented with 400 ppm each of amino acid precursors, including ornithine, histidine, tryptophan, tyrosine, lysine, arginine, and phenylalanine (Sigma-Aldrich) at 37 °C for 20 h. One milliliter of the culture supernatant, 500 μL of saturated sodium carbonate solution, and 800 μL of 1% (*w*/*v*) dansyl chloride acetone solution were mixed and reacted at 45 °C for 1 h. Subsequently, 500 μL of 10% (*w*/*v*) proline and 5 mL of diethyl ether were added to the derivatized mixture, which was shaken for 10 min. The supernatant was collected and evaporated under nitrogen gas, then dissolved in 1 mL of acetonitrile and filtered through a 0.22 μm polyvinylidene difluoride filter. The derivatized biogenic amine (BA) mixture was quantitatively analyzed by high-performance liquid chromatography using a DIONEX UltiMate 3000 system (Thermo Fisher Scientific). The analysis was performed using a 250 × 4.6 mm, 5 μm, Agilent 5 TC-C18 column (Agilent Technologies, Amstelveen, The Netherlands) and a UV detector set at 254 nm, at a column oven temperature of 40 °C. Twenty microliters of the sample was injected into the column at 1 mL/min in a mobile phase consisting of 55:45 (acetonitrile–water) after equilibration for 30 min. The standard BAs evaluated in this study were agmatine, β-phenylethylamine, histamine, putrescine, serotonin, spermidine, tryptamine, and tyramine. To complement the phenotypic data, genes related to the polyamine transport system (*potA–E*) were identified by keyword searching annotated coding sequences in the *L. buchneri* KU200793 genome.

2.9. Antibiotic Susceptibility Test

The antibiotic susceptibility of *L. buchneri* KU200793 was assessed using E-test strips (ETEST, bioMérieux, Marcy-l'Étoile, France). The minimum inhibitory concentration (MIC) of each antibiotic was determined based on the cut-off values established by the EFSA for probiotics [14]. A suspension of bacterial colonies was prepared in 0.88% (*w*/*v*) NaCl solution and adjusted to match the 0.5 McFarland standard. This suspension was uniformly spread onto the surface of agar plates using sterile cotton swabs. E-test strips for antibiotics were placed on the agar surface using sterile forceps, and the plates were incubated at 37 °C for 24 h before evaluating the MIC.

2.10. Genomic Analysis of Antibiotic Resistance and Virulence Factors

The whole-genome sequence of *L. buchneri* KU200793 (NCBI Bioproject accession number: PRJNA1237560) was provided by Dankook University and was used to identify antibiotic resistance. The analysis was performed using Resistance Gene Identifier (RGI) software version 6.0.3 (https://card.mcmaster.ca/analyze/rgi, accessed on 18 February 2025) based on the Comprehensive Antibiotic Resistance Database version 3.2.8 and ResFinder version 4.4.1 (http://genepi.food.dtu.dk/resfinder, accessed on 4 March 2025) based on the ResFinder database. These tools were used to examine the presence of antibiotic resistance genes (ARGs) and assess the resistance levels associated with mutations. The criteria for RGI analysis included the following: Selection criteria: Perfect, Strict, and Loose hits; Nudge $\geq$ 95% identity for Loose hits to Strict (excluding nudge and low-confidence hits); and Sequence Quality: high quality/coverage. Loose hits were considered indicative of ARGs if they exhibited $\geq$70% sequence identity. For ResFinder analysis, the threshold criteria were set at 90% sequence similarity and a minimum length of 60%. Additionally, HGT potential was analyzed using the HGTree2 program, whereas the presence of ARGs within the plasmids and their potential for horizontal transfer were assessed using PlasmidFinder 2.0. PHASTEST was used to identify prophage regions within the genome to assess whether transduction was involved in the horizontal transfer of resistance genes. The involvement of transposons in HGT was assessed using MobileElementFinder.

2.11. Gelatin Liquefaction Assay

The gelatin liquefaction assay was conducted in MRS medium supplemented with 12% (*w*/*v*) gelatin. A single colony of *L. buchneri* KU200793 was introduced into the medium and incubated at 37 °C for 72 h. *Br. parabrevis* KCCM 41421 was used as a positive control and was incubated under identical conditions. After incubation, the test tubes were placed in an ice bath for 30 min. Strains capable of liquefying gelatin were considered positive for gelatinase activity.

2.12. Urease Activity Assay

The urease activity of bacterial cells was assessed by streaking them onto Christensen's urea agar (0.1 g of peptone, 0.1 g of dextrose, 0.5 g of sodium chloride, 0.2 g of monopotassium phosphate, 2 g of urea, 0.012 g of phenol red) slants. *L. buchneri* KU200793 and *P. vulgaris* KCCM 40221, which was used as a positive control, were incubated at 37 °C for six days. Urease activity was determined by observing the changes in pH, as indicated by a color change in the slants from orange to pink. Pink coloration indicated ammonia production, confirming a positive urease reaction.

2.13. Indole Production Assay

Indole production was assessed using tryptophan as a substrate. *L. buchneri* KU200793 and *E. coli* KCCM 11234, used as a positive control, were cultured in tryptophan broth (10.0 g of casein enzyme hydrolysate, 5.0 g of NaCl, 1.0 g of DL-tryptophan, per liter) at 37 °C for 24 h. After incubation, five drops of Kovac reagent (Sigma-Aldrich) were added to the medium. A red ring on the surface of the medium indicated a positive result for indole production.

2.14. Statistical Analysis

All experiments were conducted in triplicate, with the results expressed as the mean $\pm$ standard deviation (SD). Statistical analyses were performed using SPSS 28 (SPSS Inc., Chicago, IL, USA) software. One-way analysis of variance (ANOVA) was used to evaluate the statistical significance of the results, and Duncan's multiple range test was used to identify significant differences between groups. A p-value of <0.05 was considered statistically significant.

3. Results

3.1. Hemolytic Profile of L. buchneri *KU200793*

S. pneumoniae KCCM 41570 exhibited α-hemolysis, characterized by a greenish discoloration around the colonies on blood agar (Figure 1a). In contrast, the positive control,

S. pyogenes KCCM 40411, displayed β-hemolysis, forming a clear hemolytic zone around the colonies (Figure 1b). *S. agalactiae* KCCM 40417 showed no hemolysis on blood agar, and the medium retained its red color (Figure 1c). Similarly, *L. buchneri* KU200793 exhibited no color change around its colonies, indicating the absence of hemolytic activity, and was thus classified as γ-hemolytic (Figure 1d).

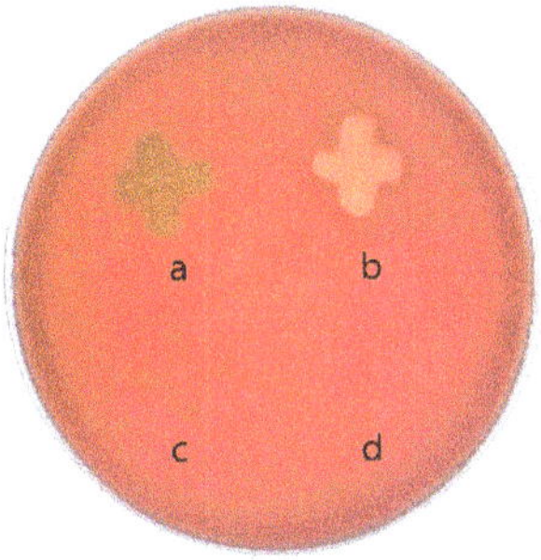

Figure 1. Hemolytic activity of bacterial strains. (a) *S. pneumoniae* KCCM 41570 (α-hemolysis); (b) *S. pyogenes* KCCM 40411 (β-hemolysis); (c) *S. agalactiae* KCCM 40417; and (d) *L. buchneri* KU200793 (γ-hemolysis).

3.2. Bile Salt Deconjugation Activity

Lp. plantarum KU15122 exhibited positive BSH activity, as indicated by the formation of a precipitation halo around the colonies due to bile salt deconjugation. Additionally, the agar appeared cloudy, and the colonies displayed a whitish, granular texture (Figure 2a). In contrast, *L. buchneri* KU200793 did not form any precipitate, confirming the absence of BSH activity (Figure 2b).

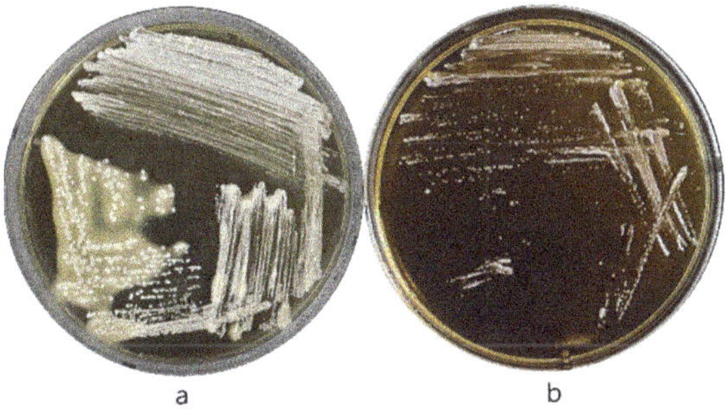

Figure 2. Bile salt deconjugation test. (**a**) *Lp. plantarum* KU15122 (positive control) exhibited a precipitation halo, whereas (**b**) *L. buchneri* KU200793 showed no precipitate.

3.3. Cytotoxic Effects on Caco-2 Cells

L. buchneri KU200793 did not exhibit cytotoxicity under any of the tested MOI conditions. The positive control strain, *K. pneumoniae* subsp. *pneumoniae* KCCM 60022, showed cell viabilities of 64%, 58%, and 48% at MOI values of 31.3, 62.5, and 125, respectively. In contrast, *L. buchneri* KU200793 exhibited significantly higher cell viabilities of 146%, 133%, and 130%, respectively, under the same conditions, surpassing those of the untreated control group (Figure 3). These results clearly indicate that *L. buchneri* KU200793 is not cytotoxic and may even enhance cell viability under all tested MOI conditions.

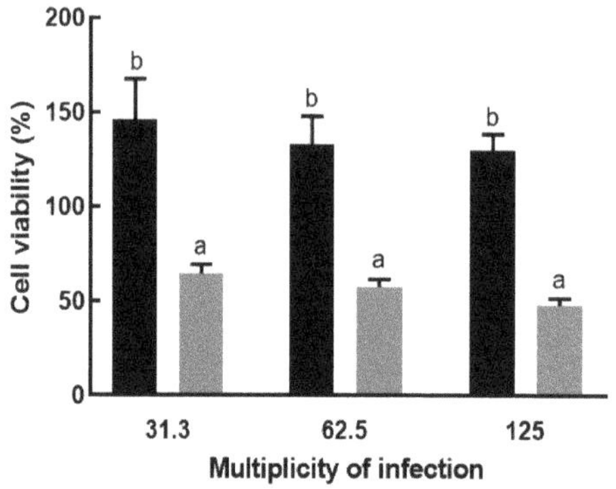

Figure 3. Effect of *L. buchneri* KU200793 on the cell viability of Caco-2 cells. Cells were treated with *L. buchneri* KU200793 (black bar) or *K. pneumoniae* subsp. *pneumoniae* KCCM 60022 (grey bar). Different letters indicate significant differences ($p < 0.05$).

3.4. Mucin Degradation Potential

L. buchneri KU200793 did not exhibit any increase in OD or pH changes in the medium containing 0.3% (*w*/*v*) mucin as the sole carbon source. In contrast, OD values increased in media containing glucose, which is commonly utilized as an optimal carbon source by most LAB (Figure 4). These results demonstrate that *L. buchneri* KU200793 does not exhibit mucin-degrading activity.

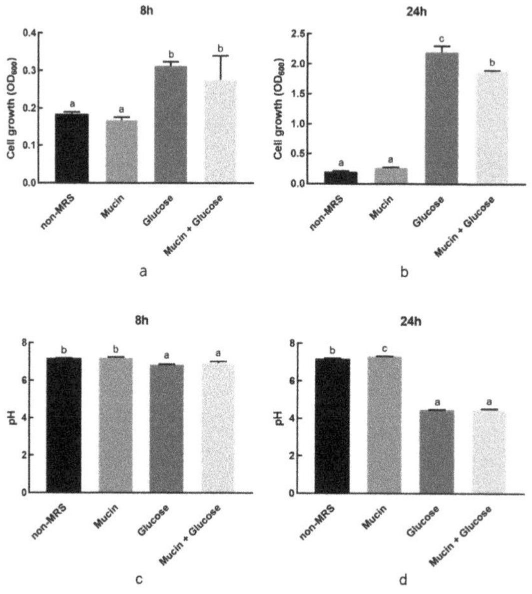

Figure 4. Mucin degradation activity of *L. buchneri* KU200793. (**a**,**b**) Cell growth (OD600) at 8 h and 24 h; (**c**,**d**) culture pH at 8 h and 24 h. Different letters indicate significant differences ($p < 0.05$).

3.5. D-Lactic Acid Production Levels

The D-lactic acid production by *L. buchneri* KU200793 was quantified. The assay results showed that *L. buchneri* KU200793 produced 2 mM D-lactic acid (Table 1).

Table 1. D-Lactic acid production of *Lentilactobacillus buchneri* KU200793.

Strain	D-Lactic Acid (mM)
L. buchneri KU200793	2.0 ± 0.3

The data are expressed as mean ± SD (n = 3).

3.6. Biogenic Amine Production Profile

Under our experimental conditions, the analysis of BA production in *L. buchneri* KU200793 revealed the presence of spermidine at a concentration of 15.27 ppm in the supernatant. However, no detectable levels of other Bas, such as agmatine, histamine, β-phenylethylamine, putrescine, serotonin, tyramine, and tryptamine were observed (Table 2).

Table 2. Biogenic amine production of *Lentilactobacillus buchneri* KU200793.

Biogenic Amines	Concentration (ppm)
Agmatine	n.d. *
β-phenylethylamine	n.d.
Putrescine	n.d.
Histamine	n.d.
Serotonin	n.d.
Spermidine	15.3 ± 7.0
Tryptamine	n.d.
Tyramine	n.d.

* n.d.: Not detected. The data are expressed as mean ± SD (n = 3).

3.7. Antibiotic Resistance Pattern

The antibiotic resistance of *L. buchneri* KU200793 was evaluated based on MIC values (μg/mL) exceeding the thresholds recommended by the EFSA. The results demonstrated that *L. buchneri* KU200793 was susceptible to all tested antibiotics, except for tetracycline (Figure 5, Table 3).

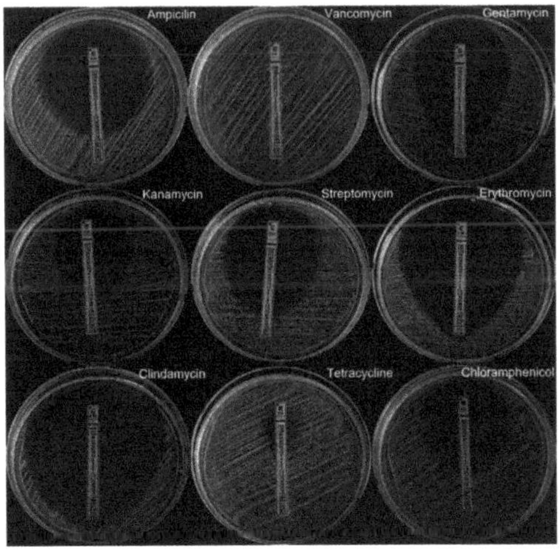

Figure 5. E-test for the determination of antibiotic resistance.

Table 3. Minimum inhibitory concentration and antibiotic susceptibility of *Lentilactobacillus buchneri* KU200793.

Antibiotics	Cut-Off Value (μg/mL) *	MIC (μg/mL)	Assessment
Ampicillin	2	0.38	S ***
Vancomycin	n.r. **	-	-
Gentamycin	16	0.047	S
Kanamycin	32	1	S
Streptomycin	64	0.5	S
Erythromycin	1	0.016	S
Clindamycin	1	0.016	S
Tetracycline	8	48	R ****
Chloramphenicol	4	1	S

* Cut-off value based on European Food Safety Authority guidelines; ** n.r., not required; *** S, Susceptible; **** R, Resistant.

3.8. Genome-Based Identification of Resistance and Virulence Determinants

The complete genome sequence of *L. buchneri* KU200793 was analyzed to identify ARGs and assess the HGT potential. A total of 29 genes were predicted to have undergone HGT, of which 16 were identified as acquired genes and 13 as donated genes. Further analysis using HGTree2 and RGI confirmed that none of these predicted HGT-acquired genes contained ARGs. Additionally, PlasmidFinder 2.0 confirmed the absence of plasmids in *L. buchneri* KU200793. PHASTEST, which was used to evaluate the possibility of transduction, identified one intact prophage region. However, further analyses using RGI and ResFinder confirmed that no ARGs were present in the identified prophage region. The MobileElementFinder program identified eight mobile genetic elements, but ResFinder analysis did not detect any ARGs within these elements. Furthermore, RGI analysis identified *tetA(58)* with 33.64% homology and *nalD* with 30.17% homology (Table 4). The low-sequence homology suggested that these genes were unlikely to function as intact ARGs. Based on these results, we concluded that no functional ARGs were present within the intact prophage regions.

Table 4. Results of antibiotic resistance gene analysis based on genome sequence data.

RGI Criteria	ARO Term	AMR Gene Family	Drug Class	% Identity Matching Region
Loose	*tetA(58)*	Major facilitator superfamily antibiotic efflux pump	tetracycline antibiotic	33.64
	nalD	Resistance-nodulation-cell division antibiotic efflux pump	macrolide antibiotic, fluoroquinolone antibiotic, monobactam, carbapenem, cephalosporin, cephamycin, penam, tetracycline antibiotic, peptide antibiotic, aminocoumarin antibiotic, diaminopyrimidine antibiotic, sulfonamide antibiotic, phenicol antibiotic, penem	30.17

3.9. Gelatin Liquefaction Activity

No gelatin liquefaction was observed in the medium inoculated with *L. buchneri* KU200793. The medium remained solid, indicating that the strain does not produce enzymes capable of hydrolyzing gelatin. In contrast, the positive control strain, *Br. parabrevis* KCCM 41421, exhibited gelatin hydrolysis, resulting in liquefaction of the medium.

3.10. Urease Activity

The urease activity of *L. buchneri* KU200793 was assessed, and no color change was observed in the medium, indicating that the strain does not produce urease. In contrast, the positive control, *P. vulgaris* KCCM 40221, induced a color shift to pink, confirming urease activity and ammonia production.

3.11. Indole Production

L. buchneri KU200793 showed no observable changes in the surface layer after the addition of Kovac's reagent, and no pink ring formation was detected, indicating that this strain did not produce indole. In contrast, the positive control strain, *E. coli* KCCM 11234, exhibited a distinct pink ring on the surface after the addition of Kovac's reagent, suggesting tryptophanase activity, which enables the conversion of tryptophan to indole.

4. Discussion

Safety evaluation of probiotic strains is a crucial prerequisite for their commercial application and potential health benefits. Unverified strains present health risks, including opportunistic infections, HGT of antibiotic resistance, and harmful metabolic activities. Therefore, a comprehensive safety assessment—encompassing genome-based screening for ARGs, metabolic profiling, and both in vivo and in vitro toxicity evaluations—is essential. Only strains that meet rigorous safety standards should be considered for probiotic use. *L. buchneri* is listed in the EFSA's Qualified Presumption of Safety (QPS) list, indicating its safety status for use in food and feed applications in the EU. This species-level recognition provides a basis for investigating the safety characteristics of strain *L. buchneri* KU200793.

Hemolytic activity is a key parameter in probiotic safety evaluation, as β-hemolysis is associated with pathogenic potential. Using blood agar assays, *L. buchneri* KU200793 exhibited no hemolytic activity, consistent with the γ-hemolytic phenotype. The absence of hemolytic activity indicates non-pathogenic properties in probiotic strains and serves as a positive indicator in their safety assessment [17]. Consistent with our findings, a previous study reported no β-hemolytic activity in four other *Lactobacillus* strains, including *L. fermentum* IDCC 3901 [18].

BSH is widely present in the gut microbiota and contributes to microbial survival, but its activity may generate harmful secondary bile acids linked to adverse health effects [19,20]. Owing to the potential risks associated with BSH activity, the Ministry of Food and Drug Safety of Korea recommends cautionary labeling for probiotics with positive BSH activity to prevent excessive consumption. The findings of this study confirm that *L. buchneri* KU200793 is a safe strain with no BSH activity. These results align with those of a previous study reporting the absence of BSH activity in *L. paracasei* subsp. *paracasei* NTU101 [8].

L. buchneri KU200793 enhanced the survival of Caco-2 cells, which are widely used as an intestinal epithelial model for probiotic safety assessment. Cell viability was assessed using a colorimetric assay with water-soluble tetrazolium-8. The amount of formazan produced correlates with the metabolic activity of viable cells, thereby indirectly reflecting cell viability. These findings provide scientific evidence supporting the safety of *L. buchneri* KU200793 and reinforce previous reports that probiotic strains are not expected to be cytotoxic to intestinal epithelial cells [21].

The intestinal mucus layer provides both a favorable niche for microbial survival and a primary barrier against pathogen and toxin entry. Mucin degradation, a known virulence trait of pathogens such as *Vibrio cholerae*, *Helicobacter pylori*, *Salmonella*, *Shigella*, and *Staphylococcus aureus*, facilitates host invasion and increases susceptibility to harmful compounds [22,23]. This study aligns with previous research reporting that *L. rhamnosus* HN001, *L. acidophilus* HN017, and *Bifidobacterium lactis* HN019 do not exhibit mucin-degradation activity [24]. Similarly, an earlier study found that none of the 75 tested *Lactococcus lactis* strains exhibited mucin-degrading activity [25].

In contrast to L-lactic acid, which is efficiently metabolized in humans, D-lactic acid is produced in a strain- and condition-dependent manner and undergoes only limited degradation by intestinal enzymes [26]. Consequently, the excessive consumption of probiotics that produce D-lactic acid can lead to its accumulation in the bloodstream, potentially resulting in D-lactic acidosis. In healthy adults, blood D-lactic acid levels typically range from 5 to 20 μmol/L, which is considered safe [27]. However, when the concentration exceeds 3 mmol/L, D-lactic acidosis may occur, causing various neurological symptoms [28]. The production levels of D-lactic acid vary depending on the type of probiotic strain, and *L. buchneri*, which belongs to the *Lentilactobacillus* genus, is known to produce both D- and L-lactic acid, classifying it as a D, L-lactate-producing strain [29]. *L. buchneri* KU200793 demonstrated a relatively low level of D-lactic acid production, highlighting its safety attributes as a probiotic strain. The amount of D-lactic acid produced by this strain was significantly lower than that produced by *L. lactis* DSM20072, which produces 7.47 mM [30], and *L. rhamnosus* GG, a widely used commercial probiotic strain [26].

BAs produced via amino acid decarboxylation are well known to act as neurotransmitters in humans, and spermidine, an endogenous metabolite, additionally plays a crucial role in cell growth, anti-aging, and autophagy [31]. Fermented and long-stored foods often show elevated BA levels due to microbial contamination, and some Lactobacillus species are major BA producers that generate them under acidic conditions as a survival strategy [32,33]. However, excessive BA accumulation may cause adverse effects such as headaches, neurological disorders, or hypotension, and high intake of histamine and tyramine has been linked to histamine poisoning and hypertensive crisis [34].

The spermidine production in *L. buchneri* KU200793 was relatively low compared with that in previously reported probiotic strains, such as *L. brevis* NZ.Lbr.082 and *L. paracasei* NZ.Lbp.105, which produced 23.06 and 35.81 ppm of spermidine, respectively [35]. Although these strains are not phylogenetically closest to *L. buchneri*, they are commonly used probiotic lactobacilli with established safety profiles and functional similarities, particularly in fermented food contexts [36]. Therefore, the comparison provides a relevant benchmark indicating that the BA production of *L. buchneri* KU200793 remains within a safe and acceptable range. Furthermore, fermented soybean-based foods contain high concentrations of BAs and BA-producing microorganisms. According to the literature, the maximum BA concentrations reported in fermented products such as soy sauce, natto, and *cheonggukjang* were 53.1 ppm, 478.1 ppm, and 79.1 ppm, respectively, indicating that the BA levels present in typical dietary intake are higher than those produced by *L. buchneri* KU200793. In addition, no specific dietary intake recommendations or toxicity thresholds have been established for putrescine, spermidine, or tryptamine in healthy individuals, suggesting that their consumption is not strictly regulated. Moreover, the regulatory authorities in various countries have not set specific permissible limits for spermidine, further supporting its safety [37].

In nature, antibiotics function as competitive tools among microorganisms, but their indiscriminate use has accelerated the emergence of antimicrobial resistance in both pathogenic and commensal species [38]. LAB, including *Lactobacillus* species, are generally considered safe. However, concerns regarding antibiotic resistance in LAB have emerged following reports of antibiotic-resistant *Lactobacillus* strains isolated from cheese [39]. The acquisition of antibiotic resistance reduces treatment efficacy, and the potential transfer of ARGs from probiotics to pathogens represents a critical public health concern. Therefore, assessing the antibiotic resistance profiles of probiotic strains before they can be used for food applications is essential [40]. In the present study, *L. buchneri* KU200793 exhibited tetracycline resistance; however, its overall antibiotic resistance levels were lower than those previously reported for other probiotic strains. *L. salivarius* CGMCC2070 showed resistance to both tetracycline and erythromycin as well as intermediate resistance to gentamicin [41]. In addition, five *L. lactis* strains demonstrated resistance to both tetracycline and streptomycin [42]. Compared with previously reported strains that exhibit resistance to multiple antibiotics, *L. buchneri* KU200793 appears to be a relatively safer probiotic candidate regarding antibiotic resistance. However, further investigations are required to clarify the mechanisms and genomic context of tetracycline resistance, for which we conducted in silico analysis based on whole-genome sequencing data.

Antibiotic resistance can arise either intrinsically, through mutations stably inherited within a strain, or via HGT, which poses a greater risk due to the possibility of ARG dissemination to pathogens [43,44]. Plasmid conjugation, whereby a donor strain transfers a plasmid to a recipient through pilus-mediated contact, represents a major HGT mechanism [45]. Transduction by bacteriophages and mobilization of transposons further contribute to ARG transfer, and integration of resistance-associated transposons into plasmids can accelerate ARG dissemination among bacterial populations [46,47]. Genome analysis confirmed the absence of exogenous resistance determinants commonly associated with HGT, indicating that the tetracycline resistance in *L. buchneri* KU200793 is intrinsic rather than acquired. This result suggests that the strain has a minimal potential for ARG dissemination via horizontal gene transfer.

Gelatinase is a biologically active protease that degrades host proteins such as collagen, casein, and hemoglobin, and its activity has been implicated in bacterial pathogenesis, tissue damage, and the progression of inflammatory diseases such as endocarditis [24,48]. In addition, gelatinase can modulate host factors including cytokines and growth factors, functioning as a virulence determinant in pathogenic bacteria, while bacterial metalloproteinases have also been linked to infection processes and tumor progression [49]. In this study, *L. buchneri* KU200793 showed no gelatin liquefaction, indicating the absence of gelatinase activity and supporting its safety by excluding this virulence-associated trait.

Urease is a nickel-dependent metalloenzyme that hydrolyzes urea into carbon dioxide and ammonia, and its activity has been implicated in urinary tract infections, hepatic encephalopathy, and other pathological conditions [50,51]. Moreover, in *H. pylori*, urease contributes to acid resistance but also causes cytotoxic damage to gastric epithelial cells through ammonia production. In this study, *L. buchneri* KU200793 showed no detectable urease activity, supporting its safety by excluding this virulence-associated trait.

Indole is a nitrogen-containing metabolite generated by microbial tryptophan metabolism via tryptophanase and has been associated with neurotoxic and metabolic effects, including impaired motor function, anxiety, and depression [52]. Once absorbed, indole is converted to indoxyl sulfate in the liver, which induces nephrotoxicity and has been reported at elevated levels in patients with Parkinson's disease [52,53]. In this study, *L. buchneri* KU200793 did not produce indole, consistent with previous findings that none of the tested *Lactobacillus* strains exhibited this activity [54].

5. Conclusions

A comprehensive safety assessment of *L. buchneri* KU200793 demonstrated its potential as a safe probiotic strain. The absence of harmful enzymatic activities, the low production of detrimental metabolites, and the lack of cytotoxicity in Caco-2 cells collectively support its safety. Although tetracycline resistance was detected, genome analysis indicated that this trait is intrinsic and unlikely to be transferred horizontally, minimizing concerns about ARG dissemination. In addition to the previously reported neuroprotective properties of this strain, the present study established its safety, thereby reinforcing the feasibility of its use in food applications and probiotic supplementation. Overall, these findings highlight the importance of rigorous safety evaluations to ensure the responsible development of probiotics and support the potential utilization of *L. buchneri* KU200793 as a safe candidate for incorporation into dietary products.

Author Contributions: Conceptualization, D.-K.K., H.-D.P., Y.-S.P. and J.H.L.; methodology, S.K., N.-K.L., Y.-S.P. and J.H.L.; software, S.K., H.J. and N.-K.L.; validation, S.K., H.J. and N.-K.L.; formal analysis, S.K.; investigation, S.K., H.J., N.-K.L., Y.-S.P. and J.H.L.; resources, D.-K.K., H.-D.P., Y.-S.P. and J.H.L.; data curation, S.K. and H.J.; writing—original draft preparation, S.K. and H.J.; writing—review and editing, Y.-S.P. and J.H.L.; visualization, S.K.; supervision, Y.-S.P. and J.H.L.; project administration, Y.-S.P. and J.H.L.; funding acquisition, H.-D.P. and J.H.L. All authors have read and agreed to the published version of the manuscript.

Funding: This research was funded by the Gachon University research fund of 2024 (GCU-202406430001) and the Korea Institute of Planning and Evaluation for Technology in Food, Agriculture, and Forestry (IPET) through the High Value-Added Food Technology Development Program funded by the Ministry of Agriculture, Food, and Rural Affairs (MAFRA) (grant number 321035052HD020).

Institutional Review Board Statement: Not applicable.

Informed Consent Statement: Not applicable.

Data Availability Statement: The original contributions presented in this study are included in the article. Further inquiries can be directed to the corresponding authors.

Acknowledgments: The authors acknowledge the support provided by the Gachon University Research Fund of 2024 (GCU-202406430001) and the Korea Institute of Planning and Evaluation for Technology in Food, Agriculture, and Forestry (IPET) through the High Value-Added Food Technology Development Program funded by the Ministry of Agriculture, Food, and Rural Affairs (MAFRA) (grant number 321035052HD020).

Conflicts of Interest: The authors declare no conflicts of interest.

References

1. Koll, P.; Mändar, R.; Smidt, I.; Hütt, P.; Truusalu, K.; Mikelsaar, R.H.; Shchepetova, J.; Krogh-Andersen, K.; Marcotte, H. Hammarström, L.; et al. Screening and evaluation of human intestinal lactobacilli for the development of novel gastrointestina probiotics. *Curr. Microbiol.* **2010**, *61*, 560–566. [CrossRef]
2. Gemechu, T. Review on lactic acid bacteria function in milk fermentation and preservation. *Afr. J. Food Sci.* **2015**, *9*, 170–175 [CrossRef]
3. FAO/WHO. *Guidelines for the Evaluation of Probiotics in Food*; FAO/WHO Working Group; FAO: Rome, Italy; WHO: Geneva Switzerland, 2002; pp. 1–11.
4. Barrea, L.; Verde, L.; Auriemma, R.S.; Vetrani, C.; Cataldi, M.; Frias-Toral, E.; Pugliese, G.; Camajani, E.; Savastano, S.; Colao, A. et al. Probiotics and prebiotics: Any role in menopause-related diseases? *Curr. Nutr. Rep.* **2023**, *12*, 83–97. [CrossRef]
5. Chon, J.W.; Youn, H.Y.; Kim, H.J.; Oh, H.; Kang, S.H.; Hwang, W.U.; Jeong, H.; Kim, H.J.; Seo, K.H.; Song, K.Y. Anti-viral activities of probiotics against viral gastroenteritis: A review. *J. Dairy Sci. Biotechnol.* **2023**, *41*, 87–102. [CrossRef]
6. Lee, B.S.; Ban, O.H.; Bang, W.Y.; Chae, S.A.; Oh, S.; Park, C.; Lee, M.; Kim, S.J.; Yang, J.; Jung, Y.H. Safety assessment of *Lactobacillus reuteri* IDCC 3701 based on phenotypic and genomic analysis. *Ann. Microbiol.* **2021**, *71*, 10. [CrossRef]
7. Sarita, B.; Samadhan, D.; Hassan, M.Z.; Kovaleva, E.G. A comprehensive review of probiotics and human health—Current prospective and applications. *Front. Microbiol.* **2025**, *15*, 1487641. [CrossRef] [PubMed]
8. Chen, C.T.; Chao, W.Y.; Lin, C.H.; Shih, T.W.; Pan, T.M. Comprehensive safety assessment of *Lacticaseibacillus paracasei* subsp. *paracasei* NTU 101 through integrated genotypic and phenotypic analysis. *Curr. Issues Mol. Biol.* **2024**, *46*, 12354–12374. [CrossRef] [PubMed]

. Lee, M.; Bang, W.Y.; Lee, H.B.; Yang, S.Y.; Lee, K.S.; Kang, H.J.; Hong, S.M.; Yang, J. Safety assessment and evaluation of probiotic potential of *Lactobacillus bulgaricus* IDCC 3601 for human use. *Microorganisms* **2024**, *12*, 2063. [CrossRef]

0. Nethery, M.A.; Henriksen, E.D.; Daughtry, K.V.; Johanningsmeier, S.D.; Barrangou, R. Comparative genomics of eight *Lactobacillus buchneri* strains isolated from food spoilage. *BMC Genom.* **2019**, *20*, 902. [CrossRef]

1. Heinl, S.; Wibberg, D.; Eikmeyer, F.; Szczepanowski, R.; Blom, J.; Linke, B.; Goesmann, A.; Grabherr, R.; Schwab, H.; Pühler, A.; et al. Insights into the completely annotated genome of *Lactobacillus buchneri* CD034, a strain isolated from stable grass silage. *J. Biotechnol.* **2012**, *161*, 153–166. [CrossRef]

2. Cheon, M.J.; Lim, S.M.; Lee, N.K.; Paik, H.D. Probiotic properties and neuroprotective effects of *Lactobacillus buchneri* KU200793 isolated from Korean fermented foods. *Int. J. Mol. Sci.* **2020**, *21*, 1227. [CrossRef]

3. Shokryazdan, P.; Faseleh Jahromi, M.; Liang, J.B.; Kalavathy, R.; Sieo, C.C.; Ho, Y.W. Safety assessment of two new *Lactobacillus* strains as probiotic for human using a rat model. *PLoS ONE* **2016**, *11*, e0159851. [CrossRef]

4. EFSA Panel on Additives and Products or Substances used in Animal Feed (FEEDAP); Rychen, G.; Aquilina, G.; Azimonti, G.; Bampidis, V.; Bastos, M.L.; Bories, G.; Chesson, A.; Cocconcelli, P.S.; Flachowsky, G.; et al. Guidance on the characterisation of microorganisms used as feed additives or as production organisms. *EFSA J.* **2018**, *16*, e05206. [CrossRef]

5. Lee, M.G.; Kang, M.J.; Kim, S.; Jeong, H.; Kang, D.K.; Paik, H.D.; Park, Y.S. Safety assessment of *Levilactobacillus brevis* KU15006: A comprehensive analysis of its phenotypic and genotypic properties. *Probiotics Antimicrob. Proteins* **2024**, *17*, 1117–1131. [CrossRef]

6. Tail, M.; Zhang, H.; Zheng, G.; Hatami, M.; Skutella, T.; Unterberg, A.; Zweckberger, K.; Younsi, A. The sonic hedgehog pathway modulates survival, proliferation, and differentiation of neural progenitor cells under inflammatory stress in vitro. *Cells* **2022**, *11*, 736. [CrossRef]

7. Kim, T.; Mondal, S.C.; Jeong, C.R.; Kim, S.R.; Ban, O.H.; Jung, Y.H.; Yang, J.; Kim, S.J. Safety evaluation of *Lactococcus lactis* IDCC 2301 isolated from homemade cheese. *Food Sci. Nutr.* **2022**, *10*, 67–74. [CrossRef]

8. Lee, Y.R.; Bang, W.Y.; Baek, K.R.; Kim, G.H.; Kang, M.J.; Yang, J.; Seo, S.O. Safety evaluation by phenotypic and genomic characterization of four Lactobacilli strains with probiotic properties. *Microorganisms* **2022**, *10*, 2218. [CrossRef]

9. Bourgin, M.; Kriaa, A.; Mkaouar, H.; Mariaule, V.; Jablaoui, A.; Maguin, E.; Rhimi, M. Bile salt hydrolases: At the crossroads of microbiota and human health. *Microorganisms* **2021**, *9*, 1122. [CrossRef]

0. Costarelli, V.; Key, T.; Appleby, P.; Allen, D.; Fentiman, I.; Sanders, T. A prospective study of serum bile acid concentrations and colorectal cancer risk in post-menopausal women on the island of Guernsey. *Br. J. Cancer* **2002**, *86*, 1741–1744. [CrossRef]

1. Joung, H.; Chu, J.; Kwon, Y.J.; Kim, K.H.; Shin, C.H.; Ha, J.H. Assessment of the safety and hepatic lipid-lowering effects of *Lactobacillus delbrueckii* subsp. lactis CKDB001. *Appl. Biol. Chem.* **2024**, *67*, 101. [CrossRef]

2. Pradhan, D.; Singh, R.; Tyagi, A.; Rashmi, H.; Batish, V.; Grover, S. Assessing safety of *Lactobacillus plantarum* MTCC 5690 and *Lactobacillus fermentum* MTCC 5689 using in vitro approaches and an in vivo murine model. *Regul. Toxicol. Pharmacol.* **2019**, *101*, 1–11. [CrossRef]

3. Varada, V.V.; Panneerselvam, D.; Pushpadass, H.A.; Mallapa, R.H.; Ram, C.; Kumar, S. In vitro safety assessment of electrohydrodynamically encapsulated *Lactiplantibacillus plantarum* CRD7 and *Lacticaseibacillus rhamnosus* CRD11 for probiotics use. *Curr. Res. Food Sci.* **2023**, *6*, 100507. [CrossRef]

4. Zhou, J.; Gopal, P.; Gill, H. Potential probiotic lactic acid bacteria *Lactobacillus rhamnosus* (HN001), *Lactobacillus acidophilus* (HN017) and *Bifidobacterium lactis* (HN019) do not degrade gastric mucin in vitro. *Int. J. Food Microbiol.* **2001**, *63*, 81–90. [CrossRef]

5. Araújo, C.; Munoz-Atienza, E.; Ramírez, M.; Poeta, P.; Igrejas, G.; Hernández, P.E.; Herranz, C.; Cintas, L.M. Safety assessment, genetic relatedness and bacteriocin activity of potential probiotic *Lactococcus lactis* strains from rainbow trout (*Oncorhynchus mykiss*, Walbaum) and rearing environment. *Eur. Food Res. Technol.* **2015**, *241*, 647–662. [CrossRef]

6. Song, E.J.; Lee, E.S.; Kim, Y.I.; Shin, D.U.; Eom, J.E.; Shin, H.S.; Lee, S.Y.; Nam, Y.D. Gut microbial change after administration of *Lacticaseibacillus paracasei* AO356 is associated with anti-obesity in a mouse model. *Front. Endocrinol.* **2023**, *14*, 1224636. [CrossRef]

7. Łukasik, J.; Salminen, S.; Szajewska, H. Rapid review shows that probiotics and fermented infant formulas do not cause d-lactic acidosis in healthy children. *Acta Paediatr.* **2018**, *107*, 1322–1326. [CrossRef]

8. Uchida, H.; Yamamoto, H.; Kisaki, Y.; Fujino, J.; Ishimaru, Y.; Ikeda, H. D-lactic acidosis in short-bowel syndrome managed with antibiotics and probiotics. *J. Pediatr. Surg.* **2004**, *39*, 634–636. [CrossRef]

9. Jang, Y.J.; Gwon, H.M.; Jeong, W.S.; Yeo, S.H.; Kim, S.Y. Safety evaluation of *Weissella cibaria* JW15 by phenotypic and genotypic property analysis. *Microorganisms* **2021**, *9*, 2450. [CrossRef]

0. Stuivenberg, G.; Daisley, B.; Akouris, P.; Reid, G. In vitro assessment of histamine and lactate production by a multi-strain synbiotic. *J. Food Sci. Technol.* **2021**, *59*, 3419–3427. [CrossRef]

1. Kim, M.J.; Ku, S.; Kim, S.Y.; Lee, H.H.; Jin, H.; Kang, S.; Li, R.; Johnston, T.V.; Park, M.S.; Ji, G.E. Safety evaluations of *Bifidobacterium bifidum* BGN4 and *Bifidobacterium longum* BORI. *Int. J. Mol. Sci.* **2018**, *19*, 1422. [CrossRef]

2. Capozzi, V.; Russo, P.; Ladero, V.; Fernández, M.; Fiocco, D.; Alvarez, M.A.; Grieco, F.; Spano, G. Biogenic amines degradation by *Lactobacillus plantarum*: Toward a potential application in wine. *Front. Microbiol.* **2012**, *3*, 122. [CrossRef]

33. Bin Hafeez, A.; Pełka, K.; Worobo, R.; Szweda, P. In silico safety assessment of *Bacillus* isolated from polish bee pollen and bee bread as novel probiotic candidates. *Int. J. Mol. Sci.* **2024**, *25*, 666. [CrossRef]
34. Ku, S.; Yang, S.; Lee, H.H.; Choe, D.; Johnston, T.V.; Ji, G.E.; Park, M.S. Biosafety assessment of *Bifidobacterium animalis* subsp. *lactis* AD011 used for human consumption as a probiotic microorganism. *Food Control* **2020**, *117*, 106985. [CrossRef]
35. Yilmaz, N. Quantitative analysis of biogenic amine production of different lactic acid bacteria isolated from ready-to-eat packaged fish products. *Vet. Res. Forum* **2024**, *15*, 537–543. [CrossRef]
36. Chen, T.; Zhao, Y.; Fan, Y.; Dong, Y.; Gai, Z. Genome sequence and evaluation of safety and probiotic potential of *Lacticaseibacillus paracasei* LC86 and *Lacticaseibacillus casei* LC89. *Front. Microbiol.* **2025**, *15*, 1501502. [CrossRef]
37. Turna, N.S.; Chung, R.; McIntyre, L. A review of biogenic amines in fermented foods: Occurrence and health effects. *Heliyon* **2024**, *10*, e24501. [CrossRef]
38. Rajasekaran, J.J.; Krishnamurthy, H.K.; Bosco, J.; Jayaraman, V.; Krishna, K.; Wang, T.; Bei, K. Oral microbiome: A review of its impact on oral and systemic health. *Microorganisms* **2024**, *12*, 1797. [CrossRef]
39. Coppola, R.; Succi, M.; Tremonte, P.; Reale, A.; Salzano, G.; Sorrentino, E. Antibiotic susceptibility of *Lactobacillus rhamnosus* strains isolated from parmigiano reggiano cheese. *Lait* **2005**, *85*, 193–204. [CrossRef]
40. Imperial, I.C.; Ibana, J.A. Addressing the antibiotic resistance problem with probiotics: Reducing the risk of its double-edged sword effect. *Front. Microbiol.* **2016**, *7*, 1983. [CrossRef] [PubMed]
41. Jiang, Y.H.; Yang, R.S.; Lin, Y.C.; Xin, W.G.; Zhou, H.Y.; Wang, F.; Zhang, Q.L.; Lin, L.B. Assessment of the safety and probiotic characteristics of *Lactobacillus salivarius* CGMCC20700 based on whole-genome sequencing and phenotypic analysis. *Front. Microbiol.* **2023**, *14*, 1120263. [CrossRef]
42. Kondrotiene, K.; Lauciene, L.; Andruleviciute, V.; Kasetiene, N.; Serniene, L.; Sekmokiene, D.; Malakauskas, M. Safety assessment and preliminary in vitro evaluation of probiotic potential of *Lactococcus lactis* strains naturally present in raw and fermented milk. *Curr. Microbiol.* **2020**, *77*, 3013–3023. [CrossRef] [PubMed]
43. Colautti, A.; Arnoldi, M.; Comi, G.; Iacumin, L. Antibiotic resistance and virulence factors in lactobacilli: Something to carefully consider. *Food Microbiol.* **2022**, *103*, 103934. [CrossRef]
44. de Niederhäusern, S.; Bondi, M.; Messi, P.; Iseppi, R.; Sabia, C.; Manicardi, G.; Anacarso, I. Vancomycin-resistance transferability from VanA enterococci to *Staphylococcus aureus*. *Curr. Microbiol.* **2011**, *62*, 1363–1367. [CrossRef]
45. van Reenen, C.A.; Dicks, L.M. Horizontal gene transfer amongst probiotic lactic acid bacteria and other intestinal microbiota: What are the possibilities? A review. *Arch. Microbiol.* **2011**, *193*, 157–168. [CrossRef]
46. Im, E.J.; Lee, H.H.Y.; Kim, M.; Kim, M.K. Evaluation of enterococcal probiotic usage and review of potential health benefits, safety, and risk of antibiotic-resistant strain emergence. *Antibiotics* **2023**, *12*, 1327. [CrossRef]
47. Mathur, S.; Singh, R. Antibiotic resistance in food lactic acid bacteria—A review. *Int. J. Food Microbiol.* **2005**, *105*, 281–295. [CrossRef]
48. Zhang, F.; Jiang, M.; Wan, C.; Chen, X.; Chen, X.; Tao, X.; Shah, N.P.; Wei, H. Screening probiotic strains for safety: Evaluation of virulence and antimicrobial susceptibility of enterococci from healthy Chinese infants. *J. Dairy Sci.* **2016**, *99*, 4282–4290. [CrossRef] [PubMed]
49. Mortezaei, M.; Dadmehr, M.; Korouzhdehi, B.; Hakimi, M.; Ramshini, H. Colorimetric and label free detection of gelatinase positive bacteria and gelatinase activity based on aggregation and dissolution of gold nanoparticles. *J. Microbiol. Methods* **2021**, *191*, 106349. [CrossRef] [PubMed]
50. Koksharova, O.A.; Butenko, I.O.; Pobeguts, O.V.; Safronova, N.A.; Govorun, V.M. β-N-methylamino-L-alanine (BMAA) causes severe stress in *Nostoc* sp. PCC 7120 cells under diazotrophic conditions: A proteomic study. *Toxins* **2021**, *13*, 325. [CrossRef]
51. Lee, C.S.; Park, M.H.; Kim, S.H. Selection and characterization of probiotic bacteria exhibiting antiadipogenic potential in 3T3-L1 preadipocytes. *Probiotics Antimicrob. Proteins* **2022**, *14*, 72–86. [CrossRef]
52. Jaglin, M.; Rhimi, M.; Philippe, C.; Pons, N.; Bruneau, A.; Goustard, B.; Daugé, V.; Maguin, E.; Naudon, L.; Rabot, S. Indole, a signaling molecule produced by the gut microbiota, negatively impacts emotional behaviors in rats. *Front. Neurosci.* **2018**, *12*, 216. [CrossRef] [PubMed]
53. Li, Q.q.; Zeng, S.p.; Liang, M.h.; Yousaf, M.; Wu, Y.p.; Tang, J.; Xiong, J.; Liu, D.m. Safety and metabolism characteristics of *Lacticaseibacillus rhamnosus* LR-ZB1107-01 based on complete genome and corresponding phenotype. *LWT* **2024**, *204*, 116443. [CrossRef]
54. Lee, D.Y.; Seo, Y.S.; Rayamajhi, N.; Kang, M.L.; Lee, S.I.; Yoo, H.S. Isolation, characterization, and evaluation of wild isolates of *Lactobacillus reuteri* from pig feces. *J. Microbiol.* **2009**, *47*, 663–672. [CrossRef] [PubMed]

microorganisms

MDPI

rticle

)evelopment of a Strain-Specific Detection and Quantification Method for *Bifidobacterium animalis* subsp. *lactis* HN019 Jsing WGS-SNP Analysis and qPCR

)a Mao [1,†], Lei Zhao [2,*,†], Bo Zhao [1], Hongbin Xu [2] and Qinghe Zhang [1]

1 Division of Chemical Metrology and Analytical Science, National Institute of Metrology, Beijing 100029, China; maoda@nim.ac.cn (D.M.); zhaobo@nim.ac.cn (B.Z.); zhangqh@nim.ac.cn (Q.Z.)
2 Key Laboratory of Milk and Dairy Products Detection and Monitoring Technology, State Administration for Market Regulation, Shanghai Institute of Quality Inspection and Technical Research, Shanghai 200233, China; xuhb@sqi.org.cn
* Correspondence: zhaolei19890117@163.com
† These authors contributed equally to this work.

Abstract: Accurate quantification of *Bifidobacterium animalis* subsp. *lactis* HN019, a clinically validated probiotic strain conferring immune modulation, gastrointestinal health, and gut barrier integrity benefits, is essential for diverse applications. To address the critical need for strain-specific detection, we developed a quantitative PCR (qPCR) assay targeting a unique single-nucleotide polymorphism (SNP) within the *gal*K gene, identified through comparative whole-genome sequencing (WGS) analysis of 31 *B. animalis* subsp. *lactis* strains. The assay exhibited exceptional specificity, distinguishing HN019 from 19 other *Bifidobacterium* strains. Sensitivity tests indicated a detection limit of 0.5 pg of DNA and 10^3 CFU/mL of bacterial cells, making it suitable for industrial-scale applications. Additionally, the method exhibited strong repeatability, reproducibility across different qPCR platforms, and resistance to interference from high cell density of *B. animalis* subsp. *lactis* DSMZ 10140. Successful quantification of HN019 in complex multi-strain probiotic powders confirmed its practical reliability. This work establishes a rapid, robust, and scalable tool for precise probiotic strain tracking, addressing critical quality control and regulatory compliance needs within the rapidly expanding probiotic industry.

Keywords: *Bifidobacterium animalis* subsp. *lactis*; strain-specific; identification; quantification; WGS-SNP

1. Introduction

Probiotics are live microorganisms that confer health benefits when administered in adequate quantities and have become integral to modern therapeutic strategies, particularly in the context of precision medicine and microbiome-targeted interventions [1]. Recent insights into host–microbiome interactions, such as the "gut–brain axis" and "gut–liver axis", have expanded their therapeutic potential beyond gastrointestinal health to encompass neurological, metabolic, and immune-related disorders [2–4]. These advancements underscore the need for precise strain-level characterization to ensure consistency in probiotic efficacy and safety.

Among probiotics, *Bifidobacterium animalis* subsp. *lactis* HN019 exemplifies translational promise due to its exceptional gastrointestinal persistence [5] and biological stability [6]. Extensive research has demonstrated its ability to enhance intestinal barrier function, exclude pathogens, and reduce the severity of pediatric infections [7,8]. Systemically, HN019 modulates immune-metabolic pathways by balancing the COX-2/COX-1 ratio [9], suppressing inflammatory cascades [10], and ameliorating conditions like radiation-induced colitis [11]. Clinically, it shortens the duration of respiratory infection-induced fever and improves bowel motility [12,13]. Notably, HN019 has been shown to reduce the risk of viral infections in infants [14] and enhance cellular immune function in elderly populations [15], highlighting its broad applicability across age groups and supporting its use in infant formulas, functional foods, and therapeutics [16].

The efficacy and safety of probiotics are inherently strain-specific, necessitating accurate identification and quantification for product development, quality control, and regulatory compliance. However, existing methods for strain identification and quantification are limited by significant challenges. Colony morphology, while historically used, requires 3–5 days for analysis and lacks resolution for species differentiation, with phenotypes often influenced by environmental factors such as nutrient availability and incubation conditions [17]. Genetic markers like 16S rRNA sequencing enabled species-level identification but failed to resolve subspecies distinctions, such as differentiating *B. animalis* subsp. *lactis* from *B. animalis* [18]. The *atp*D gene (ATP synthase subunit beta) improved subspecies resolution, but it could not distinguish between strains due to high sequence conservation [19]. Molecular fingerprinting techniques like pulsed-field gel electrophoresis (PFGE) and random amplified polymorphic DNA (RAPD) have been employed for strain typing [20]; these techniques are often complex, equipment-intensive, and unsuitable for large-scale industrial applications. To address these challenges, whole-genome sequencing (WGS) has emerged as a transformative tool. WGS-based single-nucleotide polymorphism (SNP) analysis helps identify strain-specific genetic markers with unprecedented precision [21]. When combined with fluorescence quantitative PCR (qPCR), WGS-SNP analysis offers a practical and scalable approach for quality control and regulatory applications in probiotic products.

In this study, we focus on *B. animalis* subsp. *lactis* HN019, employing WGS-SNP analysis to identify strain-specific genetic markers. These markers are integrated with fluorescence quantitative PCR to develop a highly efficient, specific, and stable method for the quantification of HN019. Through the validation of the method's specificity, sensitivity, repeatability, reproducibility, and resistance to interference, we establish a framework for strain-level detection that can be applied both in research settings and across industrial applications. This study provides technical support for the market regulation of probiotic foods and the healthy development of the industry.

2. Materials and Methods

2.1. SNP Analysis

SNPs were analyzed in 31 *B. animalis* subsp. *lactis* strains, including BB-12, Bi07, and others (Table 1), using whole-genome sequences retrieved from the NCBI database. The *B. animalis* subsp. *lactis* HN019 genome served as the reference. Sequence alignments were performed with MUMmer (version 3.23, https://mummer.sourceforge.net/, accessed on 8 February 2025) [22]. Initially, the nucmer utility [23] included with the MUMmer (version 3.23, https://mummer.sourceforge.net/, accessed on 8 February 2025) was used for global pairwise alignment of the genomes from the 30 strains against the HN019 genome, generating a comprehensive map of genomic similarities and differences. To improve alignment accuracy, the delta-filter program (MUMmer version 3.23, https://mummer.sourceforge.net/, accessed on 10 February 2025) was applied to remove low-confidence alignments and redundant genomic regions, retaining only high-quality data [24]. SNPs were identified using the show-snps tool in MUMmer (version 3.23, https://mummer.

sourceforge.net/, accessed on 12 February 2025), extracting variant positions between the HN019 strain and the other 30 strains. The analyses were conducted by Majorbio (Shanghai, China).

Table 1. Strains used for WGS-SNP analysis.

Species	Strain	Assembly Number	Scaffold Number	Genome Size (Mb)
B. animalis subsp. *lactis*	HN019 *	GCA_003606305.1	1	1.935423
	19-D-1	GCA_020309945.1	1	1.963057
	AD011	GCA_000021425.1	1	1.933695
	ATCC 27673	GCA_000471945.1	1	1.963012
	B420	GCA_000277325.1	1	1.938595
	BAMA-B06	GCA_028463865.1	1	1.944252
	BB-12	GCA_000025245.2	1	1.944152
	BF052	GCA_000818055.1	1	1.938624
	BGI-N3	GCA_040084745.1	1	1.944238
	BI040	GCA_038086955.1	1	1.944141
	Bi-07	GCA_000277345.1	1	1.938822
	Bl-04	GCA_000022705.1	1	1.938709
	Bl12	GCA_000414215.2	1	1.944036
	BLC1	GCA_000224965.2	1	1.938583
	CGMCC1.15623	GCA_044866785.1	1	1.944145
	CNCM I-2494	GCA_000220885.1	1	1.943113
	DSM 10140	GCA_000022965.1	1	1.938483
	DSM 15954	GCA_021018785.1	1	1.944152
	GOLDGUT-BB18	GCA_034561975.1	1	1.944144
	H1	GCA_016835115.1	1	1.944352
	H3	GCA_016835135.1	1	1.944351
	HOM2120	GCA_044502405.1	1	1.946900
	i797	GCA_019576095.1	1	1.943538
	IDCC4301	GCA_003428375.1	1	1.944141
	KLDS 2.0603	GCA_000816205.1	1	1.946899
	LPL-RH	GCA_030284625.1	1	1.944223
	MH-02	GCA_029167605.1	1	1.944290
	S7	GCA_003390755.1	1	1.944072
	SF	GCA_029542645.1	1	1.944374
	TCI604	GCA_036923305.1	1	1.944134
	V9	GCA_000092765.1	1	1.944050

*: reference genome.

2.2. Primer and Probe Design

A total of 10,357 SNPs were identified in the *B. animalis* subsp. *lactis* HN019 genome (Table S1). Among these, six variant sites were shared with the other 30 strains, and five of these differential SNPs, located within four single-copy genes, were selected for primer and probe design. Sequences were designed using Primer Express 3.0 (Applied Biosystems, Foster City, CA, USA) [25]. Comparative evaluation revealed that the TaqMan assay targeting the *gal*K SNP exhibited superior specificity for HN019 detection versus assays for other loci, showing no cross-reactivity with phylogenetically related subspecies strains. The sequences of the primers and probe used in this study are listed in Table 2 and were synthesized by Sangon Biotech (Shanghai, China).

Table 2. The sequences of primers and probe.

Gene	The Sequences of Primers and Probe	Product Length (bp)
*gal*K	TGATATGGCGAATGCTTGCA CGGCTTGTGTGTCGTCATG FAM-ACCCTTTTCATTTCCCTGC-MGB	61

2.3. DNA Extraction

Genomic DNA was extracted from the bacterial strains and samples using the Bacterial Genomic DNA Extraction Kit (No. DP302; Tiangen, Beijing, China), following the manufacturer's instructions [26]. The extracted DNA was assessed by electrophoresis on a 1.0% agarose gel, and its concentration, 260/280 ratio, and 260/230 ratio were determined using a DS-11 spectrophotometer (DeNovix Inc., Wilmington, DE, USA) [27]. The DNA samples were then stored at −20 °C for further use.

2.4. Specificity

Primer and probe specificity were assessed using 20 different *Bifidobacterium* strains (Table 3). The qPCR reaction mixture consisted of 10 μL of Takara Prober Taq mix, 1 μL of upstream primer (5 μM), 1 μL of downstream primer (5 μM), 1 μL of probe (10 μM), 1 μL of DNA template, and double-distilled water (Beyotime Biotechnology, Beijing, China) to a final volume of 20 μL. The qPCR amplification was performed on the Roche LightCycler® 480 platform (Roche Diagnostics GmbH, Mannheim, Germany) [28]. The thermal cycling conditions were as follows: Stage 1, 95 °C for 2 min; Stage 2, 95 °C for 5 s, 64 °C for 35 s, for a total of 40 cycles. Sterile double-distilled water (Beyotime Biotechnology, Beijing, China) was used as the blank control.

Table 3. Strains used for specificity test.

No.	Species	Strain
1	*B. animalis* subsp. *lactis*	HN019
2		Bi-07
3		Bl-04
4		BB12
5		DSMZ 10140
6		CICC 21718
7		ATCC 25527
8		ATCC 27673
9	*B. animalis*	ATCC 27672
10		CICC 6165
11		CICC 6250
12	*B. adolescentis*	ATCC 15703
13	*B. longum* subsp. *infantis*	ATCC 15697
14	*B. breve*	ATCC 15700
15		CICC 6185
16		CICC 6181
17	*B. bifidum*	CICC 6173
18	*B. longum*	CICC 6207
19		CICC 6199
20		Bl05

2.5. Stability During Passage

B. animalis subsp. *lactis* HN019 was passaged from 1 to 35 generations. DNA was extracted from bacterial suspensions at the 5th, 15th, 25th, and 35th generations, each at a cell density of 10^8 CFU/mL. The DNA template was adjusted to a concentration of 50 ng/μL for qPCR amplification. Sterile double-distilled water (Beyotime Biotechnology, Beijing, China) was used as the blank control. The qPCR amplification was performed on the Roche LightCycler® 480 platform (Roche Diagnostics GmbH, Mannheim, Germany). The differences in Cq values were analyzed to assess passage stability.

2.6. Sensitivity and Efficiency

Sensitivity was evaluated for both DNA concentration and cell density. For DNA concentration sensitivity, a 50 ng/μL DNA solution was serially diluted in 10-fold steps, followed by qPCR amplification. For bacterial cell density sensitivity, DNA was extracted from bacterial suspensions with concentrations ranging from 10^8 to 10^3 CFU/mL, followed by qPCR amplification. The bacterial cell densities of *B. animalis* subsp. *lactis* HN019 were measured using the Chinese National Standard GB 4789.34-2016 [29]. The qPCR amplification was performed on the Roche LightCycler® 480 platform. Standard curves were constructed by plotting the logarithm of DNA concentration (Log [ng/μL]) or cell density (Log [CFU/mL]) against the Cq value using GraphPad Prism 9 [30]. The slope and R^2 values were calculated, and amplification efficiency was determined using the qPCR Efficiency Calculator (Thermo Fisher Scientific, Waltham, MA, USA, https://www.thermofisher.cn/) [31].

2.7. Repeatability and Reproducibility

Repeatability was evaluated by conducting the experiment three times to test the consistency of the method. Reproducibility was assessed by performing experiments on two different qPCR platforms, Roche LightCycler® 480 and ABI 7500 (Applied Biosystems, Foster City, CA, USA) [32]. The tests were conducted at three different DNA concentrations (50 ng/μL, 5 ng/μL, and 0.5 ng/μL), with each experiment performed in triplicate, using double-distilled water as the blank control.

2.8. Interference Resistance

The anti-interference experiment was designed to validate the specificity and sensitivity of the qPCR assay for quantifying *B. animalis* subsp. *lactis* HN019 in complex microbial matrices containing genetically similar strains. This validation is critical for ensuring accurate quantification in real-world scenarios, such as probiotic products or gut microbiota samples, where closely related subspecies may coexist at high concentrations. Bacterial cell densities of *B. animalis* subsp. *lactis* HN019 at concentrations ranging from 10^8 to 10^3 CFU/mL were mixed with an equal volume of a 10^8 CFU/mL of *B. animalis* subsp. *lactis* DSMZ 10140. DNA was extracted from these mixtures, followed by qPCR amplification. The qPCR amplification was performed on the Roche LightCycler® 480 platform. Each experiment was conducted in triplicate, with double-distilled water as the blank control.

2.9. Actual Sample Test

The developed quantification method for the HN019 strain was applied to analyze actual bacterial powder samples. The powder, containing *B. animalis* subsp. *lactis* HN019, *Lactobacillus acidophilus* NCFM, *Lacticaseibacillus rhamnosus* GG, and oligosaccharides, was prepared according to the Chinese National Standard GB 4789.34-2016 [29]. A 25 g sample was dissolved in 225 mL sterile physiological saline, followed by a 10-fold serial dilution, and the dilution factor was recorded. DNA was extracted from bacterial suspensions (10^3 to 10^8 CFU/mL) following the method described in Section 2.3. The qPCR amplification was performed on the Roche LightCycler® 480 platform, with sterile physiological saline as the blank control. Standard curves were constructed by plotting the logarithm of cell density (Log [CFU/mL]) against the Cq value. In accordance with the Professional Standard of the People's Republic of China for Entry-Exit Inspection and Quarantine SN/T 5642.3-2023 [33], the single-copy gene was used to establish equivalence between genomic copies and CFU, enabling direct reporting of results as CFU/g. The bacterial content of the product was calculated using the following formula:

$$M = 10^{x} \times (A + B)/A \times C$$

where

M = *B. animalis* subsp. *lactis* HN019 content in the bacterial powder (CFU/g);
A = mass of the sample taken for analysis (g);
B = volume of dilution solution (mL);
C = dilution factor of the sample suspension.

2.10. Statistical Analysis

Each experiment was performed thrice. The results were subjected to one-way ANOVA analysis and Tukey's multiple comparisons test by Graphpad Prism (version 9.0), with $p < 0.05$ considered statistically significant.

3. Results

3.1. SNP Analysis and Specificity Test

Using the whole genome sequence of *B. animalis* subsp. *lactis* HN019 as a reference, we compared it with the whole-genome sequences of 30 other *B. animalis* subsp. *lactis* strains from the NCBI database. A total of 6 SNPs specific to the HN019 strain were identified. Among them, the SNP located in the *gal*K gene was selected for the design of primers and probes. After performing qPCR amplification, this SNP-based primer/probe pair was able to effectively differentiate *B. animalis* subsp. *lactis* HN019 from 19 other *Bifidobacterium* strains. As shown in Figure 1, the Cq values for the HN019 strain were 19.25, 19.27, and 19.07, while no significant amplification was observed for the other strains or the blank control. Based on the criteria for interpreting qPCR results outlined by Wang et al. [34], a Cq value below 35 was considered positive.

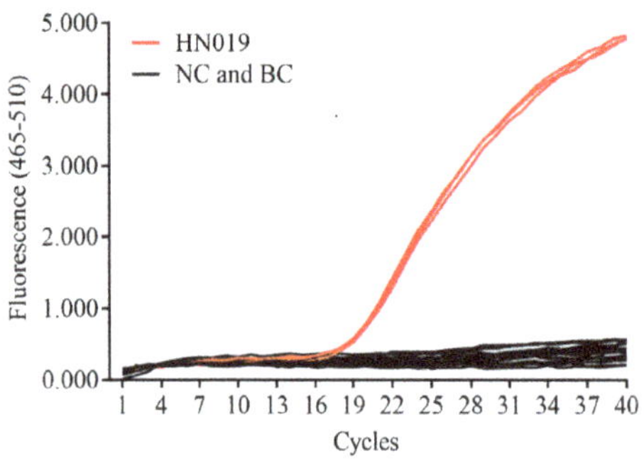

Figure 1. Primer specificity test for the *B. animalis* subsp. *lactis* HN019 strain. NC: Negative Control (strains other than HN019); BC: Blank Control (sterile double-distilled water).

3.2. Passage Stability Test

As shown in Table 4, the Cq value standard deviation for the stability test of the HN019 strain, from the 5th to the 35th generation, was 0.24, with a relative standard deviation of 1.31%. In addition, the investigation revealed no statistically significant disparities in the outcomes of the evaluation between the 5th, 15th, 25th, and 35th generations. This indicates that the SNP site located in the *gal*K gene exhibits good stability throughout the passage process and can be used for the detection of *B. animalis* subsp. *lactis* HN019 in limited generations.

Table 4. Stability of specific amplification during serial passages.

Number of Generations	Cq Value			Mean Cq Value	SD	RSD (%)
5	18.88	18.78	18.98	18.88	0.24	1.31
15	18.92	18.86	18.27	18.68		
25	18.73	18.38	18.89	18.67		
35	17.98	18.4	18.5	18.29		

3.3. Sensitivity Test

To determine the sensitivity of this detection method, a DNA solution with a concentration of 50 ng/μL was serially diluted in 10-fold steps and subjected to qPCR amplification, as shown in Figure 2a. A standard curve was constructed with the logarithmic value of DNA concentration on the *x*-axis and the mean Cq value on the *y*-axis. From Figure 2b, the slope of the standard curve was −3.311, with an amplification efficiency of 100.46% and an R^2 value of 0.9933 (Figure 2b). Based on the data in Figure 2, the detection limit for DNA concentration in this amplification system is approximately 0.5 pg, indicating that the detection method is highly sensitive.

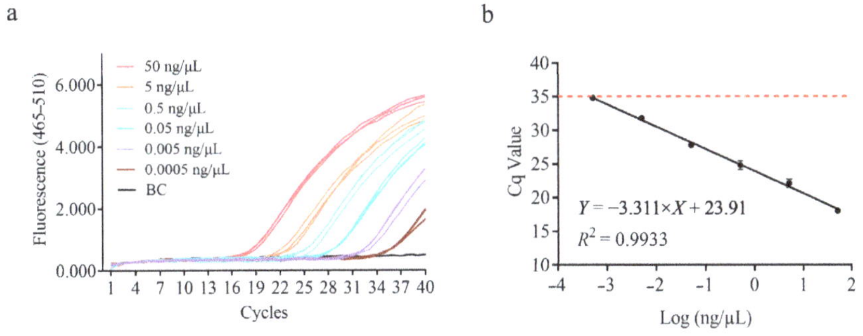

Figure 2. Sensitivity of the specific detection method for *B. animalis* subsp. *lactis* HN019 strain at varying DNA concentrations. (**a**) Amplification curve obtained using the Roche LightCycler® 480 platform for the DNA concentration sensitivity test. (**b**) Standard curve constructed by plotting Cq values against DNA concentration.

To meet the needs of dairy companies for detecting lactic acid bacterial colony counts, quantitative results were correlated with bacterial colony-forming units (CFUs). A gradient dilution of *B. animalis* subsp. *lactis* HN019 bacterial suspension was prepared, DNA was extracted as the template, and qPCR amplification was performed. The amplification results showed that when the bacterial cell density was $\geq 10^3$ CFU/mL, the detection was positive, as shown in Figure 3. A standard curve was constructed with the logarithmic value of bacterial concentration on the x-axis and the average Cq value on the y-axis. The slope was −3.370, with an amplification efficiency of 98.03% and an R^2 value of 0.9944. These results demonstrate that the sensitivity of this method for bacterial suspension concentration is 10^3 CFU/mL, which meets the daily detection needs of dairy companies for 10^6 CFU/mL [35].

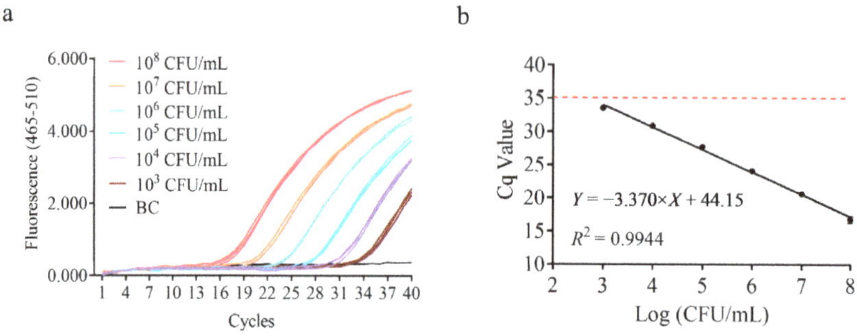

Figure 3. Cell density sensitivity test for specific detection method of *B. animalis* subsp. *lactis* HN019 strain. (**a**) Amplification curve obtained using the Roche LightCycler® 480 platform for the bacterial cell density sensitivity test, BC: Blank Control (sterile double-distilled water). (**b**) Standard curve constructed by plotting Cq values against cell density.

3.4. Repeatability and Reproducibility Test

The repeatability and reproducibility of the detection method were evaluated using DNA samples at three different concentrations (50, 5, and 0.5 ng/µL). As shown in Figure 4a, the RSD for repeatability over time ranged from 1.30% to 2.38%. For reproducibility testing across two different qPCR platforms (Roche LightCycler® 480 and ABI 7500), the RSD ranged from 1.55% to 2.65% (Figure 4b), which is well below the requirement that qPCR detection results should have an RSD value of less than 25% [36]. In the context of repeatability testing, no statistically significant variations were observed among the three test results at 50 ng/µL, 5 ng/µL, and 0.5 ng/µL. In the context of reproducibility testing, no statistically significant discrepancies were observed between the results obtained from two distinct platforms at concentrations of 50 ng/µL, 5 ng/µL, and 0.5 ng/µL. Therefore, the method demonstrated good repeatability and reproducibility and is suitable for use across different qPCR platforms.

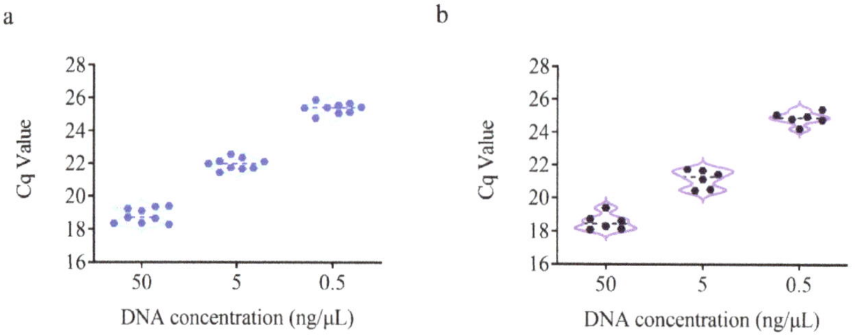

Figure 4. Repeatability (**a**) and reproducibility (**b**) test for specific detection method of *B. animalis* subsp. *lactis* HN019 strain. Violin plots were generated using GraphPad Prism 9, Median (dashed line), kernel density distribution (blue line or purple line) are shown. (**a**) Repeatability test, showing three trials with three replicates (dark blue dots) in each trial, for a total of 9 measurements. (**b**) Reproducibility test, showing experiments conducted on two different qPCR platforms (Roche LightCycler® 480 and ABI 7500), for a total of 6 measurements (dark purple dots).

3.5. Interference Resistance Test

To enhance the efficacy of probiotic products, a combination of target strains and other strains is often added to the products. Therefore, it is essential to verify the interference resistance of the HN019 strain detection method in the presence of closely related strains within the same subspecies. As shown in Figure 5, under the condition of adding 10^8 CFU/mL of DSMZ 10140, the established detection method successfully quantified the HN019 strain, maintaining a sensitivity of 10^3 CFU/mL. These results demonstrate that the method has strong resistance to interference, providing valuable data support for subsequent actual sample testing.

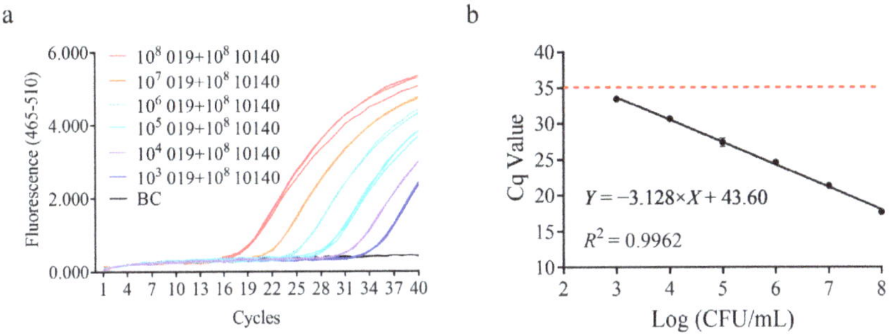

Figure 5. Interference resistance test for specific detection method of *B. animalis* subsp. *lactis* HN019 strain. (**a**) Amplification curve obtained using the Roche LightCycler® 480 platform for the interference resistance test, BC: Blank Control (sterile double-distilled water). (**b**) Standard curve constructed by plotting Cq values against cell density.

3.6. Quantification of HN019 Strain in Probiotic Powder Samples

The strain-specific detection method was used to precisely quantify the HN019 strain in five batches of actual composite probiotic powder samples. Using the standard curve equation $Y = -3.347 \times X + 44.17$, $R^2 = 0.9944$, and the formula 2.9, the HN019 strain content in the five batches of samples was calculated to range from (11.91 ± 0.03) to (12.11 ± 0.12) Log (CFU/g), as shown in Figure 6. A subsequent investigation revealed that there was no significant difference in the content of HN019 in the five batches of probiotic powder samples. The results indicate that the method is equally applicable to composite probiotic powder samples containing large amounts of oligosaccharides and excipients.

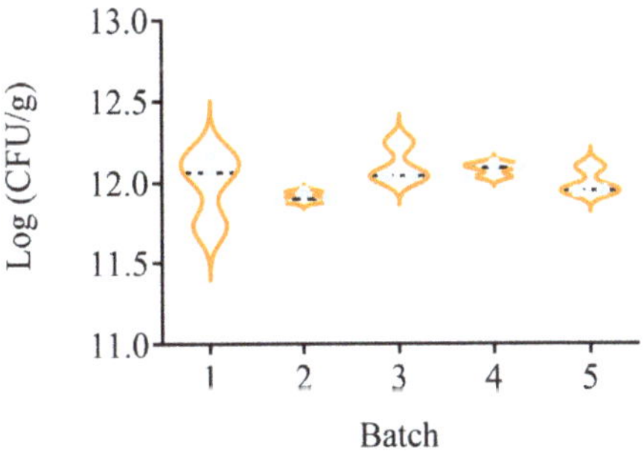

Figure 6. Validation of the specific detection method using five batches of multi-strain bacterial powder samples, including *B. animalis* subsp. *lactis* HN019, *L. acidophilus* NCFM, *L. rhamnosus* GG, and oligosaccharides. Violin plots were generated using GraphPad Prism 9, Median (dashed line), kernel density distribution (orange line) are shown. Each batch was tested three times, with the results of *B. animalis* subsp. *lactis* HN019 represented by light blue dots.

4. Discussion

The selection of *B. animalis* subsp. *lactis* HN019 as the focal strain for this study stems from its well-established yet uniquely complex clinical profile and its prominence in commercially significant probiotic formulations. Originally isolated from dairy sources [37], HN019 has transcended its industrial origins to become a benchmark probiotic strain with documented, dose-dependent health benefits in humans. Critically, robust clinical trials have demonstrated their efficacy in enhancing innate immune responses, such as natural killer cell activity, and reducing the incidence of antibiotic-associated diarrhea. These specific immunomodulatory properties, coupled with their documented stability and persistence in the gut, have positioned HN019 as a cornerstone strain in precision probiotic interventions, particularly for vulnerable populations including infants and the elderly experiencing immunosenescence [7]. However, the very characteristic that makes HN019 valuable also presents a major challenge: its clinical outcomes are demonstrably formulation-dependent and inconsistent across studies, such as null effects observed in constipation trials [38]. This inconsistency underscores a fundamental limitation in the field: the inability to reliably confirm and quantify the presence and dose of the specific, clinically validated HN019 strain within complex probiotic products. Consequently, rigorous strain-specific quantification is not merely beneficial but essential for elucidating HN019's true characteristics in vivo, ensuring product fidelity, validating clinical efficacy claims, and ultimately advancing evidence-based probiotic development.

While traditional culture-based methods provide total viable counts, they fundamentally lack the specificity to distinguish HN019 from closely related strains within the *B. animalis* subsp. lactis subspecies, especially in multi-strain products or those containing low target populations or complex matrices. This gap highlights a persistent problem: the lack of a truly strain-specific, sensitive, and robust quantification tool for HN019. Molecular approaches targeting genes like *tuf*, *Bal-23S*, or *Tal* have been successfully applied to quantify the related BB-12 strain [39–41] but critically fail to differentiate between HN019 and BB-12 due to insufficient genetic resolution. This gap highlights a persistent problem: the lack of a truly strain-specific, sensitive, and robust quantification tool for HN019.

With the rapid advancement of high-throughput sequencing technologies, the complete genome sequences of more *B. animalis* subsp. *lactis* strains have been completed, and genomic data are available through public databases, providing a solid foundation for the application of WGS-SNP analysis. By designing primers and probes for SNPs in the *gal*K gene, we successfully achieved specific detection of the HN019 strain. The selected SNP in *gal*K exhibited robust genetic stability across 35 serial passages, a critical validation for industrial scalability that ensures consistent detectability throughout product lifecycles. According to the criterion of Cq values below 35 for positive results in qPCR, the DNA sensitivity and cell density sensitivity of our quantitative detection method are 0.5 pg and 10^3 CFU/mL, respectively. Additionally, the amplification efficiency exceeds 98%, falling within the optimal range (80–120%) [36], with an R^2 value greater than 0.99, indicating excellent linearity. Evaluation of the method's repeatability and reproducibility showed good consistency across different time intervals and platforms, with RSD values below 3%, confirming its high reliability and adaptability. To further assess the method's resistance to interference, we conducted interference experiments by adding high concentrations of homologous strains. The results indicated that even with interference from other homologous strains and excipients, the method accurately quantifies the HN019 strain at 10^3 CFU/mL, while traditional plate counting methods fail to achieve such specific quantification in multi-strain samples.

The accuracy and reliability of quality control for the HN019 strain, a key probiotic, directly affect consumer health. The quantitative analysis method proposed in this study, based on WGS-SNP analysis and qPCR, offers high specificity, strong repeatability, genetic stability, resistance to interference, and good equipment adaptability, providing regulators and manufacturers with a gold-standard tool to verify HN019 strain identity, viability, and dose claims in commercial products, safeguarding against mislabeling and subpotency. Moreover, the SNP-based design paradigm establishes a blueprint for developing strain-specific assays for other high-value probiotics, advancing the field toward truly personalized microbial therapeutics.

5. Conclusions

As the probiotic market rapidly expands, consumers' demands for the quality and safety of probiotic products are increasing. Accurate and reliable strain detection methods have become one of the key technologies driving the sustainable development of the probiotic industry. This study proposes a high-precision HN019 strain quantification method based on WGS-SNP analysis and quantitative PCR technology, providing scientific and reliable technical support for quality control of probiotic products. Our WGS-SNP-qPCR platform not only unlocks rigorous research into HN019's bioactivity but also sets a new standard for quality control in the rapidly evolving probiotic industry, ensuring that evidence-based health claims translate into reliable consumer benefits.

Supplementary Materials: The following supporting information can be downloaded at: https://www.mdpi.com/article/10.3390/microorganisms13071596/s1. Table S1: Strains used for genomic comparison and specific loci information.

Author Contributions: Conceptualization, B.Z.; methodology, H.X.; software, L.Z.; validation, L.Z. and D.M.; formal analysis, L.Z. and D.M.; investigation, Q.Z.; resources, H.X. and B.Z.; data curation, L.Z.; writing—original draft preparation, L.Z.; writing—review and editing, H.X. and B.Z.; visualization, L.Z. and D.M.; supervision, H.X. and B.Z.; project administration, D.M. and L.Z.; funding acquisition, D.M. and L.Z. All authors have read and agreed to the published version of the manuscript.

Funding: This research was funded by the National Key Research and Development Program of China (grant number 2019YFC1604803) and the Science and Technology Project of the Shanghai Institute of Quality Inspection and Technical Research (grant number KY-2023-6-SP).

Institutional Review Board Statement: Not applicable.

Informed Consent Statement: Not applicable.

Data Availability Statement: The full genome data of 31 *B. animalis* subsp. *lactis* can be found in NCBI GenBank under the assembly number listed in Table 1.

Acknowledgments: We thank Shanghai majorbio Bio-pharm technology Co., Ltd. for helping in high-throughput data analysis.

Conflicts of Interest: The authors declare no conflicts of interest.

References

1. Das, T.K.; Pradhan, S.; Chakrabarti, S.; Mondal, K.C.; Ghosh, K. Current status of probiotic and related health benefits. *Appl. Food Res.* **2022**, *2*, 100185. [CrossRef]
2. Yan, M.; Man, S.; Sun, B.; Ma, L.; Guo, L.; Huang, L.; Gao, W. Gut liver brain axis in diseases: The implications for therapeutic interventions. *Signal Transduct. Target. Ther.* **2023**, *8*, 443. [CrossRef]
3. Loh, J.S.; Mak, W.Q.; Tan, L.K.S.; Ng, C.X.; Chan, H.H.; Yeow, S.H.; Foo, J.B.; Ong, Y.S.; How, C.W.; Khaw, K.Y. Microbiota–gut–brain axis and its therapeutic applications in neurodegenerative diseases. *Signal Transduct. Target. Ther.* **2024**, *9*, 37.
4. Kearns, R. Gut–brain axis and neuroinflammation: The role of gut permeability and the kynurenine pathway in neurological disorders. *Cell. Mol. Neurobiol.* **2024**, *44*, 64. [CrossRef] [PubMed]

5. Lugli, G.A.; Argentini, C.; Tarracchini, C.; Mancabelli, L.; Viappiani, A.; Anzalone, R.; Angelini, L.; Alessandri, G.; Longhi, G.; Bianchi, M.G.; et al. Characterization of a *Bifidobacterium animalis* subsp. *lactis* reference strain based on ecology and transcriptomics. *Appl. Environ. Microbiol.* **2024**, *90*, e01080-24. [CrossRef]
6. D'Ambrosio, S.; Dabous, A.; Sadiq, S.; Casillo, A.; Schiraldi, C.; Cassese, E.; Bedini, E.; Corsaro, M.M.; Cimini, D. *Bifidobacterium animalis* subsp. *lactis* HN019 live probiotics and postbiotics: Production strategies and bioactivity evaluation for potential therapeutic properties. *Front. Bioeng. Biotechnol.* **2024**, *12*, 1379574. [CrossRef] [PubMed]
7. Cheng, J.; Laitila, A.; Ouwehand, A.C. *Bifidobacterium animalis* subsp. *lactis* HN019 effects on gut health: A review. *Front. Nutr.* **2021**, *8*, 790561. [CrossRef]
8. Sharma, M.; Wasan, A.; Sharma, R.K. Recent developments in probiotics: An emphasis on *Bifidobacterium*. *Food Biosci.* **2021**, *41*, 100993. [CrossRef]
9. Putaala, H.; Salusjärvi, T.; Nordström, M.; Saarinen, M.; Ouwehand, A.C.; Bech Hansen, E.; Rautonen, N. Effect of four probiotic strains and *Escherichia coli* O157:H7 on tight junction integrity and cyclo-oxygenase expression. *Res. Microbiol.* **2008**, *159*, 692–698. [CrossRef] [PubMed]
10. Bernini, L.J.; Simão, A.N.C.; de Souza, C.H.B.; Alfieri, D.F.; Segura, L.G.; Costa, G.N.; Dichi, I. Effect of *Bifidobacterium lactis* HN019 on inflammatory markers and oxidative stress in subjects with and without the metabolic syndrome. *Br. J. Nutr.* **2018**, *120*, 645–652. [CrossRef]
11. Invernici, M.M.; Furlaneto, F.A.C.; Salvador, S.L.; Ouwehand, A.C.; Salminen, S.; Mantziari, A.; Vinderola, G.; Ervolino, E.; Santana, S.I.; Silva, P.H.F.; et al. *Bifidobacterium animalis* subsp *lactis* HN019 presents antimicrobial potential against periodontopathogens and modulates the immunological response of oral mucosa in periodontitis patients. *PLoS ONE* **2020**, *15*, e0238425. [CrossRef] [PubMed]
12. Bettocchi, S.; Comotti, A.; Elli, M.; De Cosmi, V.; Berti, C.; Alberti, I.; Mazzocchi, A.; Rosazza, C.; Agostoni, C.; Milani, G.P. Probiotics and Fever Duration in Children with Upper Respiratory Tract Infections: A Randomized Clinical Trial. *JAMA Netw. Open* **2025**, *8*, e250669. [CrossRef]
13. Ala-Jaakkola, R.; Forssten, S.D.; Cheng, J.; Griffon, F.; Metreau, I.; Sturm, Y.; Lecerf, J.M.; Donazzolo, Y.; Junnila, J.; Nordlund, A.; et al. Effect of an 8-week *Bifidobacterium lactis* HN019 supplementation on functional constipation: A multi-center, triple-blind, randomized, placebo-controlled trial. *Mol. Nutr. Food Res.* **2025**, e70081. [CrossRef]
14. Jena, R.; Choudhury, P.K. *Bifidobacteria* in fermented dairy foods: A health beneficial outlook. *Probiotics Antimicrob. Prot.* **2025**, *17*, 1–22. [CrossRef]
15. Waller, P.A.; Gopal, P.K.; Leyer, G.J.; Ouwehand, A.C.; Reifer, C.; Stewart, M.E.; Miller, L.E. Dose-response effect of *Bifidobacterium lactis* HN019 on whole gut transit time and functional gastrointestinal symptoms in adults. *Scand. J. Gastroenterol.* **2011**, *46*, 1057–1064. [CrossRef] [PubMed]
16. Chen, J.; Chen, X.; Ho, C.L. Recent Development of probiotic *Bifidobacteria* for treating human diseases. *Front. Bioeng. Biotechnol.* **2021**, *9*, 770248. [CrossRef]
17. Huang, C.H.; Li, S.W.; Huang, L.; Watanabe, K. Identification and classification for the *Lactobacillus casei* group. *Front. Microbiol.* **2018**, *9*, 1974. [CrossRef] [PubMed]
18. Li, M.N.; Han, Q.; Wang, N.; Wang, T.; You, X.M.; Zhang, S.; Zhang, C.C.; Shi, Y.Q.; Qiao, P.Z.; Man, C.L.; et al. 16S rRNA gene sequencing for bacterial identification and infectious disease diagnosis. *Biochem. Biophys. Res. Commun.* **2024**, *739*, 150974. [CrossRef]
19. Ruiz-Moyano, S.; Tao, N.; Underwood, M.A.; Mills, D.A. Rapid discrimination of *Bifidobacterium animalis* subspecies by matrix-assisted laser desorption ionization-time of flight mass spectrometry. *Food Microbiol.* **2012**, *30*, 432–437. [CrossRef]
20. Huang, C.H.; Liou, J.S.; Huang, L.; Watanabe, K. Developing novel species-specific DNA markers for PCR-based species identification of the *Lactobacillus sakei* group. *Lett. Appl. Microbiol.* **2018**, *66*, 138–144. [CrossRef]
21. Schürch, A.C.; Arredondo-Alonso, S.; Willems, R.J.L.; Goering, R.V. Whole genome sequencing options for bacterial strain typing and epidemiologic analysis based on single nucleotide polymorphism versus gene by gene—Based approaches. *Clin. Microbiol. Infect.* **2018**, *24*, 350–354. [CrossRef] [PubMed]
22. Deschamps, S.; Zhang, Y.; Llaca, V.; Ye, L.; Sanyal, A.; King, M.; May, G.; Lin, H. A chromosome-scale assembly of the sorghum genome using nanopore sequencing and optical mapping. *Nat. Commun.* **2018**, *9*, 4844. [CrossRef]
23. Lo, K.J.; Lin, S.S.; Lu, C.W.; Kuo, C.H.; Liu, C.T. Whole-genome sequencing and comparative analysis of two plant-associated strains of *Rhodopseudomonas palustris* (PS3 and YSC3). *Sci. Rep.* **2018**, *8*, 12769. [CrossRef] [PubMed]
24. Leekitcharoensophon, P.; Kaas, R.S.; Thomsen, M.C.; Friis, C.; Rasmussen, S.; Aarestrup, F.M. snpTree a web server to identify and construct SNP trees from whole genome sequence data. *BMC Genom.* **2012**, *13*, S6.
25. Shao, N.; Li, F.; Nie, K.; Fu, S.H.; Zhang, W.J.; He, Y.; Lei, W.W.; Wang, Q.Y.; Liang, G.D.; Cao, Y.X.; et al. TaqMan real-time rt-pcr assay for detecting and differentiating Japanese encephalitis virus. *Biomed. Environ. Sci.* **2018**, *31*, 208–214.
26. Huang, S.; Wang, X.; Chen, X.; Liu, X.; Xu, Q.; Zhang, L.; Huang, G.; Wu, J. Rapid and sensitive detection of *Pseudomonas aeruginosa* by isothermal amplification combined with Cas12a-mediated detection. *Sci. Rep.* **2023**, *13*, 19199. [CrossRef] [PubMed]

7. Sound, J.K.; Peters, A.; Bellamy-Carter, J.; Rad-Menéndez, C.; MacKechnie, K.; Green, D.H.; Leney, A.C. Rapid cyanobacteria species identification with high sensitivity using native mass spectrometry. *Anal. Chem.* **2021**, *93*, 14293–14299. [CrossRef]
8. Hsieh, A.Y.Y.; Saberi, S.; Ajaykumar, A.; Hukezalie, K.; Gadawski, I.; Sattha, B.; Côté, H.C.F. Optimization of a relative telomere length assay by monochromatic multiplex real-time quantitative pcr on the LightCycler 480: Sources of variability and quality control considerations. *J. Mol. Diagn.* **2016**, *18*, 425–437. [CrossRef]
9. *GB 4789.34–2016*; National Food Safety Standard–Food Microbiological Examination: Examination of *Bifidobacteria*. China Standards Press: Beijing, China, 2016.
0. Yao, Y.T.; Zhang, X.; Wang, C.Y.; Zhang, Y.H.; Li, D.W.; Yang, W.D.; Li, H.Y.; Zou, L.G. Optimizing longifolene production in *Yarrowia lipolytica* via metabolic and protein engineering. *Synth. Syst. Biotechnol.* 2025, *in press*. [CrossRef]
1. Svec, D.; Tichopad, A.; Novosadova, V.; Pfaffl, M.W.; Kubista, M. How good is a PCR efficiency estimate: Recommendations for precise and robust qPCR efficiency assessments. *Biomol. Detect. Quantif.* **2015**, *3*, 9–16. [CrossRef]
2. Von Hertwig, A.M.; Sant'Ana, A.S.; Sartori, D.; da Silva, J.J.; Nascimento, M.S.; Iamanaka, B.T.; Pelegrinelli Fungaro, M.H.; Taniwaki, M.H. Real-time PCR-based method for rapid detection of *Aspergillus niger* and *Aspergillus welwitschiae* isolated from coffee. *J. Microbiol. Methods* **2018**, *148*, 87–92. [CrossRef] [PubMed]
3. *SN/T 5642.3-2023*; Professional Standard of the People's Republic of China for Entry-Exit Inspection and Quarantine–Detection of lactic acid bacteria in dairy products for export–Enumeration method of digital PCR–Part3: *Bifidobacterium animalis*. China Standards Press: Beijing, China, 2023.
4. Wang, R.; Zhang, W.; Ye, R.; Pan, Z.; Li, G.; Su, S. One-step multiplex TaqMan probe-based method for real-time PCR detection of four canine diarrhea viruses. *Mol. Cell. Probes* **2020**, *53*, 101618. [CrossRef]
5. Herbel, S.R.; Lauzat, B.; Von Nickisch-Rosenegk, M.; Kuhn, M.; Murugaiyan, J.; Wieler, L.H.; Guenther, S. Species-specific quantification of probiotic lactobacilli in yoghurt by quantitative real-time PCR. *J. Appl. Microbiol.* **2013**, *115*, 1402–1410. [CrossRef]
6. Broeders, S.; Huber, I.; Grohmann, L.; Berben, G.; Taverniers, I.; Mazzara, M.; Roosens, N.; Morisset, D. Guidelines for validation of qualitative real-time PCR methods. *Trends Food Sci. Technol.* **2014**, *37*, 115–126. [CrossRef]
7. Dekker, J.W.; Wickens, K.; Black, P.N.; Stanley, T.V.; Mitchell, E.A.; Fitzharris, P. Safety aspects of probiotic bacterial strains *Lactobacillus rhamnosus* HN001 and *Bifidobacterium animalis* subsp. *lactis* HN019 in human infants aged 0–2 years. *Int. Dairy J.* **2009**, *19*, 149–154.
8. Cheng, J.; Gao, C.; Ala-Jaakkola, R. Eight-Week Supplementation with Bifidobacterium lactis HN019 and Functional Constipation: A Randomized Clinical Trial. *JAMA Netw. Open* **2024**, *7*, e2436888. [CrossRef] [PubMed]
9. Solano-Aguilar, G.; Dawson, H.; Restrepo, M.; Andrews, K.; Vinyard, B.; Urban, J.F., Jr. Detection of *Bifidobacterium animalis* subsp. *lactis* (Bb12) in the intestine after feeding of sows and their piglets. *Appl. Environ. Microbiol.* **2008**, *74*, 6338–6347.
0. Taipale, T.; Pieniähäkkänen, K.; Isolauri, E.; Larsen, C.; Brockmann, E.; Alanen, P.; Jokela, J.; Söderling, E. *Bifidobacterium animalis* subsp. *lactis* BB-12 in reducing the risk of infections in infancy. *Br. J. Nutr.* **2011**, *105*, 409–416.
1. Requena, T.; Burton, J.; Matsuki, T.; Munro, K.; Simon, M.A.; Tanaka, R.; Watanabe, K.; Tannock, G.W. Identification, detection, and enumeration of human *bifidobacterium* species by PCR targeting the transaldolase gene. *Appl. Environ. Microbiol.* **2002**, *68*, 2420–2427. [CrossRef]

microorganisms

MDPI

Article

Harnessing Waste Bread: From Potential Use in Microbial Growth and Enzyme Production to Techno-Economic Assessment

Sameh Ben Mabrouk [1,†], Bouthaina Ben Hadj Hmida [1,†], Wejdene Sallami [1], Salma Dhaouadi [1], Theodoros Varzakas [2,*] and Slim Smaoui [3]

1 Laboratory of Biochemistry and Enzymatic Engineering of Lipases, National Engineering School of Sfax, University of Sfax, BP 1173, Sfax 3038, Tunisia; benm_sameh@yahoo.fr (S.B.M.); bouthaina.benhadjhmidaa@gmail.com (B.B.H.H.); sellamiwejden22@gmail.com (W.S.); salmadhaouadidh@gmail.com (S.D.)

2 Department of Food Science and Technology, University of the Peloponnese, Antikalamos, 24100 Kalamata, Greece

3 Laboratory of Microbial Biotechnology and Enzymatic Engineering (LMBEE), Centre of Biotechnology of Sfax (CBS), University of Sfax, Road Sidi Mansour Km 6, P.O. Box 1177, Sfax 3018, Tunisia; slim.smaoui@cbs.rnrt.tn

* Correspondence: t.varzakas@uop.gr; Tel.: +30-2721045279

† These authors contributed equally to this work.

Abstract: This study highlights waste bread (WB) as a novel, cost-effective, and nutrient-rich substrate for microbial growth, offering a sustainable alternative to conventional media. As a renewable resource, WB promotes the circular economy by reducing food waste and encouraging biotechnological innovation. The incorporation of WB into microbial culture media enhanced the growth of various reference strains (*E. coli*, *E. faecalis*, *P. aeruginosa*, and *S. aureus*), with at least a two-fold increase compared to conventional Luria-Bertani (LB) medium. Moreover, combining 2% WB with diluted LB (1/10) reduced medium costs by up to 90%. Furthermore, it was confirmed that 1% WB can effectively replace starch during the screening of amylolytic strains. Applying a fractional factorial design, the production of amylase by *Bacillus* sp. BSS (Amy-BSS) was enhanced 15-fold. An analysis of the Pareto diagram revealed that WB was the most significant factor. Additionally, Amy-BSS was applied to hydrolyze polysaccharides in WB, enabling the generation of high-value-added products in food processing. This hydrolysis process yielded 4.6 g/L of fermentable sugars from 1% WB. Evaluating the economic feasibility of WB valorization into value-added products elucidates potential pathways for cost reduction and enhanced environmental sustainability, thereby positioning WB as a viable tool for sustainable development.

Keywords: waste bread; microorganisms; *Bacillus*; amylase; techno-economic assessment

1. Introduction

In recent years, food systems have faced challenges due to climate change, rapid human population growth, accelerating urbanization, and epidemic surges. As a result, global food security gained significant priority from scientific communities worldwide [1,2]. One of the most critical obstacles to global food security is food waste. In this line, current studies revealed that roughly 33% of total food production, corresponding to 1.3 billion tons, is discarded every year. Notably, 14% of food loss arises during post-harvest and distribution phases, while 17% is wasted through retail and consumption, representing TND 750 billion/year in economic losses [3,4].

Owing to its perishable nature, (3–6 days shelf life), bread is the most frequently thrown-away food. Globally, ≈10% equivalent to 900.000 tons of bread is wasted across the entire supply chain, including production, transport, and consumption [5,6]. Meanwhile, the high demand for newly made bread exacerbates its overproduction, generating daily waste [7]. Bread's rich nutrient composition could contribute to rapid quality loss, induced by chemical, physical, and especially mold deterioration. Additionally, during the baking, the starch is converted into a digestible gelatinized, susceptible to fungal attack [8,9].

In Tunisia, bread is a staple food and consumed daily. According to the National Institute of Consumption (INC), 42 kg of bread is wasted per household, amounting to around 113,000 tons of Tunisian bread being discarded annually [10]. In fact, a study on "Bread consumption by Tunisian families" proved that over 900,000 loaves are thrown away daily, with an estimated cost of TND 50 million, which affects the country's economy [11].

Current bread waste disposal management approaches, such as incineration, landfilling, and anaerobic digestion, could pose diverse environmental issues, like soil contamination and generating greenhouse gases. Further, incineration could provoke air pollution through toxic emissions, odors, and disease vectors [12]. In response to these environmental challenges, nations admit an urgent necessity to convert to a zero-waste strategy by 2025. Therefore, developing waste valorization technologies is essential as a core component of sustainable waste management, while concurrently improving economic rationality [13]. In this context, innovative solutions have been investigated, including transforming stale bread into succinic acid [14], animal feed [15], 2,3-butanediol (BDO) [16], bioethanol [2], lactic acid [17], brewing adjunct for craft beer [18], and compost to enrich soil health [19]. Beyond these applications, efforts have focused on leveraging the nutritional profile of waste bread (WB), which consists of approximately 10% protein and 70% carbohydrates. As an upcycled substrate, these substances perform a valued reserve for microbial processes and enzymatic synthesis, which could enhance the efficiency and eco-friendliness of biotechnological claims [20]. Montemurro et al. [21] used WB as substrate to generate polyhydroxyalkanoates by *Haloferax mediterranei*. Similarly, Benabda et al. [22] employed WB to enhance the microbial proliferation of *Rhizopus oryzae* for the amylase and protease synthesis. Additionally, Jung et al. [23] utilized the glucose derived from the WB enzymatic hydrolysis to the *Euglena gracilis* cultivation of microalga, leading to the production of paramylon, which is valued in medicine and cosmetics sectors. In addition to these diverse applications, the food-processing industry provides openings to repurpose WB within the framework of a circular economy. For bakeries, using cost-effective wheat bran (WB) not only helps maintain competitive product pricing but also eliminates disposal costs. Additionally, they can generate extra revenue by selling leftover loaf waste [4]. For instance, Guerra-Oliveira et al. [24] studied the conversion of WB to make sugar-snap cookies. Furthermore, WB can be transformed into sugar hydrolysate to replace sucrose in the sweeteners and confections, to be used as bulking agents in bakery products, or to be processed into fructose syrup [15,25]. In addition, Immonem et al. [8] revealed that WB slurry increases the viscoelasticity of the dough, decreases the water absorption by 13%, and obtains less hard bread compared to the control.

Aligned with a zero-waste industrial strategy, this study investigates the valorization of Tunisian WB through four distinct applications: (1) as an alternative feedstock to replace conventional culture media for selected bacterial strains; (2) as a substitute for commercial starch in screening amylolytic microorganisms; (3) as a fermentation substrate to enhance amylase production; and (4) for the generation of sugar syrup via enzymatic hydrolysis using amylase (Amy-BSS).

2. Materials and Methods

2.1. Waste Bread Preparation and Characterization

The waste bread used in this study was prepared from white wheat flour, using *Saccharomyces cerevisiae* fermentation, following standard French bread-making protocols involving mixing wheat flour, salt, and yeast (https://www.inbp.com/). WB was collected from household food waste and local bakeries (Sfax government) after its shelf-life had expired or in the case of returns. The discarded bread exhibited no indications of mold, was sliced into roughly 1 cm^3 pieces, and was left to dry at room temperature for four days. Afterwards, the slices were blended into a fine powder, using a laboratory blender, and stored at 4 °C until needed. The nutritional composition of the WB was assessed using analytical techniques. The moisture level was evaluated with a KERN DAB desiccator (Kern & Sohn, Balingen, Germany), while the ash content was performed via combustion in a muffle furnace (FLLI GALLI G/P model (Fratelli Galli, Fizzonasco, Italy)). Water activity was determined using a Rotronic RO-HP23-AW meter (Instrumat AG, Renens, Switzerland). The lipids, starch, and proteins were analyzed using MPA FT-NIR spectroscopy (Bruker, Billerica, MA, USA). The results are expressed as a percentage of dry matter.

2.2. Microorganisms, Medium Preparation, and Growth Conditions

Reference microorganisms, namely *Escherichia coli* DH5α, *Enterococcus faecalis*, *Pseudomonas aeruginosa*, and *Staphylococcus aureus*, were chosen for their ability to grow in conventional Luria-Bertani (LB) medium. Their proliferation was tested and compared in both conventional LB (C.LB) and non-conventional media (NC1 and NC2):

1. C.LB (g/L): 10 soya peptone, 5 yeast extract, 5 NaCl;
2. NC1 (g/L): 1 soya peptone, 0.5 yeast extract 0.5 NaCl, 20 g waste bread;
3. NC2 (g/L): 20 g waste bread with no additional nutrients.

A 14 h pre-culture of each microorganism was prepared in C.LB and used to inoculate the test media (NC1 and NC2) in 100 mL Erlenmeyer flasks, at an initial optical density (OD) of 0.4, for 24 h, at 37 °C and 200 rpm. Bacterial growth was monitored by measuring the OD at 600 nm, using a spectrophotometer. WB supernatants were used as blank for NC1 and NC2 fermentation media.

BSS strain isolated from a Tunisian biotope was used to produce amylase. Preliminary biochemical tests confirmed its classification as *Bacillus* species, hence its designation as *Bacillus* spp. BSS. The pre-culture was performed in basic LBS medium (C.LB + 1% soluble starch) at 30 °C, overnight, at 180 rpm. The bacterial cultures (OD = 0.2) were conducted following the experimental matrix (16 experiments with 1 central point) and incubated for 19 h at 30 °C in a rotary shaker.

Four amylolytic *Bacillus* strains (B1–B4) and *E. coli* DH5α (serving as negative control) were streaked in duplicate on solid WB medium (SWB), forming by C.LB +1% WB+ 1.8% agar), and incubated overnight at 37 °C. Following the incubation period, the plated samples were treated with 1% Lugol iodine solution (1 g I_2 + 2 g KI) for 10 min. The presence of starch hydrolysis zones indicated amylase production [26]. A positive control (plate containing LBS medium) was used to verify the effectiveness of waste bread as a starch substitute for amylase-activity screening.

2.3. Crude Enzyme Preparation

The *Bacillus* spp. BSS strain grew in the optimized medium at 30 °C and 160 rpm for 14 h. The cells were harvested by centrifugation at 8000× *g* for 15 min. The resulting supernatant constituted the crude enzyme extract and exhibited an amylase activity of approximately 8 U/mL, using the optimized medium.

2.4. Amylase Activity Assay

Amylase activity of *Bacillus* spp. BSS was assessed using the 3,5-dinitrosalicylic acid (DNS). The reaction mixture consisted of 0.5 mL of 1% starch (Sigma (St. Louis, MO, USA)), 0.45 mL of phosphate buffer (pH = 7; 0.1 M), and 0.05 mL of crude enzyme, and it was kept at 60 °C for 10 min. After incubation, 1.5 mL of DNS was added to stop the enzymatic reaction. The mixture was boiled for 10 min, and 15 mL of deionized water was introduced. An independent control was prepared by mixing the substrate and DNS solution, followed by the enzyme, to remove all reducing sugars from the fermentation. The production of reduced sugars was determined at 550 nm. One unit of amylase activity is defined as the amount of enzyme that releases 1 µmol of reducing sugar per minute under the specified assay conditions. The α-amylase activity was measured as follows:

$$\alpha\text{-amylase activity } (\mathrm{U/mL/min}) = \frac{K \times OD}{180} \times 10^{-3} \times 10^{6} \times \frac{1}{10} \times \frac{1}{0.05} \quad (1)$$

where the molecular weight of glucose = 180; and K = the conversion coefficient of the DNS, which is 1.8.

2.5. Waste Bread Enzymatic Hydrolysis Assay

Enzymatic hydrolysis of WB was conducted using crude amylase produced from *Bacillus* spp. BSS. A total of 9 experiments were carried out following a designed matrix (Table 1). All experiments were carried out in 100 mL Erlenmeyer flasks that contained 20 mL of the various bread suspensions mixed with the enzymatic solution. After each experimental run, samples were collected and centrifuged at 8000× *g* for 15 min. The reducing sugars present in 1 mL of the hydrolyzed waste bread supernatant were measured by the DNS method, with glucose serving as the standard.

Table 1. Variables and experimental domains in the WB hydrolysis (2_V^{4-1}).

Variables	Unity	Level −1	Level 1
X_1: Amylolytic activity	Unities (U)	5	10
X_2: Incubation temperature	°C	50	60
X_3: Waste bread concentration	%	1	3
X_4: Hydrolysis time	Hours (h)	1	3

2.6. Fractional Factorial Design (FFD)

The 2_V^{k-p} fractional factorial plan designed with Minitab version 2019 (64-bit, LLC Minitab (State College, PA, USA) was chosen due to its simplicity, as it requires fewer experiments compared to the other designs. Thus, it can conserve both time and resources. In this study, a FFD 2_V^{5-1} was investigated for enhancing amylase production by *Bacillus* spp. BSS (y1) and the 2_V^{4-1} plan for evaluating the hydrolysis yield of WB using Amy-BSS (y2). The respective experimental variables for each response (y1 and y2) are detailed in Tables 1 and 2.

The fractional factorial design led to the derivation of the subsequent empirical equation [27]:

$$y = \beta_0 + \sum_{i=1}^{k} \beta_i x_i \sum\sum_{j<i} \beta_{ij} x_i x_j + \varepsilon \quad (2)$$

where β_0, β_i, and β_{ij} are the regression coefficients for the constant, linear, and interaction coefficients; y is the response; x_i and x_j are the independent variables; and ε is the experimental error. In FFD, the interaction effects among three or more variables are presumed to be inconsequential.

Table 2. Levels of factors assessed in the optimization of amylase production (2_V^{5-1}).

Variables (g/L)	Level −1	Level 0 *	Level 1
A: Yeast extract	5	7.5	10
B: Soya peptone	5	7.5	10
C: NaCl	0	2.5	5
D: Waste bread	10	15	20
E: Wheat starch	5	7.5	10

* Level 0 is applied to enhance the prediction variance within the experiment and to offer an extra level for assessing potential lack of fit.

2.7. Statistical Analysis and Model Adequacy

The adequacy of the regression model was assessed through multiple statistical parameters. Analysis of variance (ANOVA) with Fisher's test (F-value) assessed the overall model significance. Individual variable relevance and bi-interactions were evaluated using *p*-value and Student's test (*t*-test) at a 95% confidence range. The coefficient of determination (R^2), along with the predicted and adjusted R^2 values, indicates the model's goodness of fit—where values closer to 100% reflect greater predictive accuracy. The experimental designs were carried out in consistent conditions within a single block of measurements, and the sequence of the experiments was kept non-randomized to reduce the impact of uncontrolled variables. All the experiments were conducted in n = 3 ± standard deviation (SD).

2.8. Material Cost Calculation (Techno-Economic Assessment)

The available price on the Sigma Aldrich website (accessed on 16 April 2025) was used to estimate the cost of additional ingredients for media. Peptone, yeast extract, sodium chloride (NaCl), and starch were used to prepare CLB, NC1, and LBS media. The cost assessment was derived by computing the required material quantity, as detailed below:

$$Cost\ per\ unit\ (€) = \frac{\text{material cost}(€)}{100(\text{g})} \times material\ used\ (\text{g}) \quad (3)$$

$$Total\ cost\ (€) = \Sigma\ cost\ per\ unit\ (€) \quad (4)$$

$$Percentage\ price\ decrease\ (\%) = \frac{(\text{original price} - \text{new price})}{\text{originalprice}} \times 100 \quad (5)$$

2.9. High-Performance Liquid Chromatography (HPLC)

The products of starch hydrolysis from waste bread (1%) after 1 h, 2 h, and 3 h of hydrolysis were analyzed by high-performance liquid chromatography (HPLC Agilent 1100 (Agilent, Santa Clara, CA, USA)). The reactions were stopped by boiling the samples for 5 min, filtered through a 0.22 mm membrane, and applied to the HPLC system carbohydrate column (Shodex sugar KS-802H901093 (Shodex, Tokyo, Japan); 8 mm × 300 mm). Isocratic water was the mobile phase with an elution rate of 1 mL per min throughout the analysis. Detection was performed using the retractive index detector (Waters RI 401, profcontrol.de, Schönwalde-Glien, Germany). The hydrolysis products were assigned by comparison with standard glucose, maltose, and unhydrolyzed WB.

3. Results

3.1. Waste Bread Characterization

The proximate analysis of the waste bread is summarized in Table 3. The WB showed moisture content of 17.17% and water activity of 0.761. Starch is the dominant component (61.41%), with an intermediate protein quantity (8.88%). These characteristics highlight the significant bread's potential as a feedstock for microbial fermentation, enzymatic hydrolysis, and nutrient recovery.

Table 3. Compositional analysis of Tunisian white bread waste.

Property	Quantity (%)
Moisture	17.17
Ash	0.64
Water activity	0.761
Lipids	2
Starch	61.41
Proteins	8.88

3.2. Waste Bread for Bacterial Growth

In this study, two types of bread waste-based media (NC1 and NC2) were formulated to test bacterial proliferation. Microbial growth was investigated in diluted culture media. In Figure 1, it can be observed that *E. coli* displayed a two-fold increase in NC2 medium (3.6) compared to CLB (1.81). *E. faecalis* demonstrated a five-fold enhancement in NC2 (3.53 vs. 0.675 in CLB), while *S. aureus* exhibited the highest growth rate in NC2 (4.63), surpassing its growth in both CLB and NC1. Though *P. aeruginosa* thrived best in CLB (4.085), it also showed considerable growth in WB (OD 3.21). These results imply that WB effectively meets the nutritional needs of various bacterial strains, making it a promising alternative culture medium. Notably, the NC1 formulation resulted in only modest growth rates for all microorganisms tested.

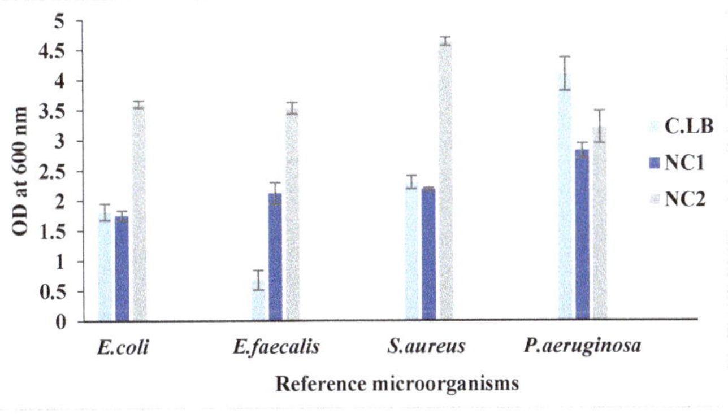

Figure 1. Bacterial growth of reference microorganisms in different culture media (C.LB, NC1, and NC2).

3.3. Waste Bread for Amylolytic Strains Screening

To assess the amylolytic potential of *Bacillus* strains (B1–B4), waste bread plates (SWB) were employed as the only supplier of starch in the screening assay. After an overnight incubation at 37 °C, plates were flooded with iodine solution, revealing various transparent halos around positive bacterial colonies, indicative of starch hydrolysis (Figure 2a,b). However, quantitative analysis demonstrates a relevant difference in enzymatic activity between SWB and LBS plates. Notably, hydrolysis-zone diameters in the LBS plate were larger for all the *Bacillus* strains (B1: 0.55, B2: 0.45, B3: 0.7, and B4: 0.65 cm), respectively, against (0.25, 0.35, 0.6, and 0.55 cm) in the SWB plate.

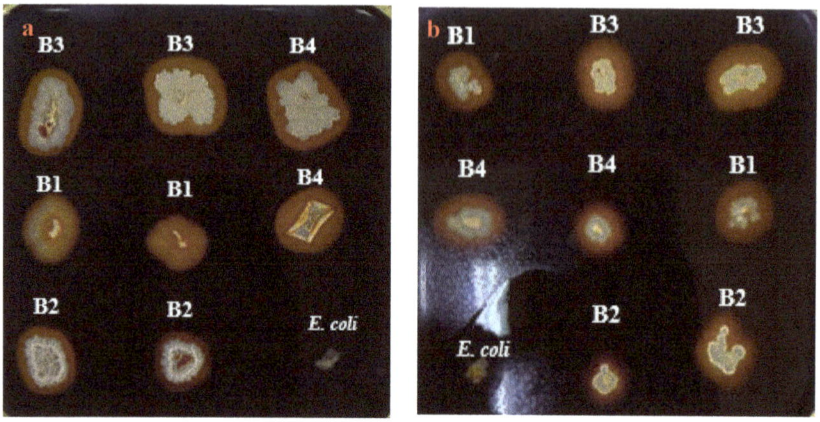

Figure 2. Screening assay of amylolytic strains after incubation at 37 °C for overnight. (**a**) *Bacillus* strains incubated on SWB plate. (**b**) Bacterial colonies on LBS plate.

These variations can be attributed to differences in substrate composition. Commercial LBS medium contains purified starch, whereas the SWB medium relies on the starch content naturally present in the bread waste matrix and the amylaceous fraction of the flour used. Furthermore, the reduction in hydrolysis halo size between LBS and SWB was 54.5%, 22.2%, 14.3%, and 15.4% for B1, B2, B3, and B4, respectively. These reductions ranging from ≈14 to 55% indicate that LBS is a more sensitive medium. However, despite the smaller halo zones observed with SWB, it maintained full detection reliability, showing 100% concordance in identifying positive amylolytic strains. Notably, the cost of preparing SWB was approximately 85% lower than that of LBS, offering a significant economic advantage for large-scale screenings. Despite these variations, using waste bread as a substrate for amylolytic-strain screening preserved detection sensitivity, as the patterns of clearance zones stayed like those seen in commercial starch agar.

3.4. Fractional Factorial Design (FFD)

3.4.1. The 2_{V}^{5-1} for the Optimization of Amylase Production

The measured response of the Amy-BSS activities is indicated in Table 4. Amylase activity varied depending on the fermentation conditions, ranging from 3.76 U/mL to 8.75 U/mL. The highest enzymatic activity was achieved in trials 1 and 6, with 8.75 U/mL and 8.78 U/mL, respectively. It is worth noting that the initial activity of the Amy-BSS in non-optimized medium (conventional LBS) was recorded as 0.69 U/mL. An improvement of ≈13-fold proves the efficiency of this experimental plan.

To determine the significant effect of the main variables and their two-way interactions, ANOVA was performed. The significance of variables was evaluated by the small p-value < 0.05, indicating high significance. From Table 5, all the main factors are significant, having the same p-value < 0.05, except for the NaCl, which had a p-value of 0.46 > 0.05. Conversely, yeast extracts (A), soya peptone (B), waste bread (D), and wheat starch (E) showed different F-values of 46.79, 70.21, 131.72, and 27.03, respectively.

This enables the use of a Pareto chart to visually analyze the impact of the most influential factor. Hence, the Pareto chart (Figure 3) clearly revealed the high influence of WB on the improvement of Amy-BSS production following the addition of soya peptone (A), yeast extract (B), and wheat starch (E). WB had the greatest effect on enzyme production owing to its high starch content, which acts as a highly effective inducer of amylase production, although NaCl (C) itself was not significant. In contrast, its interaction with yeast extract (A × C), soya peptone (B × C), waste bread (D × C), and wheat starch (E × C) significantly affected the enzymatic production. Thus, these interactions were ranked according to their *F*-values, as follows: (C × E), (A × C), (C × D), and (B × C), with *F*-values of 105.83 > 31.29 > 25.86 > 20.82, respectively (Table 5), suggesting their substantial contribution to the response (y_1).

Table 4. Matrix of 2_V^{5-1} design in coded levels for amylase production with experimental values.

Run Order	Yeast Extract (A)	Soya Peptone (B)	NaCl (C)	Waste Bread (D)	Wheat Starch (E)	Experimental Amy-BSS Activity (U/mL)
1	−1	−1	−1	−1	1	8.75 ± 0.32
2	1	−1	−1	−1	−1	7.58 ± 0.53
3	−1	1	−1	−1	−1	5.95 ± 0.57
4	1	1	−1	−1	1	8.35 ± 0.36
5	−1	−1	1	−1	−1	7.53 ± 0.45
6	1	−1	1	−1	1	8.87 ± 0.36
7	−1	1	1	−1	1	3.76 ± 0.20
8	1	1	1	−1	−1	7.42 ± 0.28
9	−1	−1	−1	1	−1	5.85 ± 0.85
10	1	−1	−1	1	1	5.11 ± 0.60
11	−1	1	−1	1	1	5.41 ± 0.53
12	1	1	−1	1	−1	5.57 ± 0.36
13	−1	−1	1	1	1	4.62 ± 0.36
14	1	−1	1	1	−1	8.27 ± 0.50
15	−1	1	1	1	−1	6.73 ± 0.39
16	1	1	1	1	1	4.59 ± 0.42
17	0	0	0	0	0	5.70 ± 0.10

Table 5. Analysis of variance (ANOVA) for the 2_V^{5-1} plan.

Source	DL	F-Value	*p*-Value
Model	16	36.75	<<0.05
Main factors	5	55.26	<<0.05
Yeast extract (A)	1	46.79	<<0.05
Soya peptone (B)	1	70.21	<<0.05
NaCl (C)	1	0.56	0.46
Waste bread (D)	1	131.72	<<0.05
Wheat starch (E)	1	27.03	<<0.05
Interaction 2 factors	10	30.24	<<0.05
A × B	1	0.93	0.342
A × C	1	31.29	<<0.05
A × D	1	25.73	<<0.05
A × E	1	2.32	0.137
B × C	1	20.82	<<0.05
B × D	1	29.60	<<0.05
B × E	1	2.52	0.122
C × D	1	25.86	<<0.05
C × E	1	105.83	<<0.05
D × E	1	57.52	<<0.05

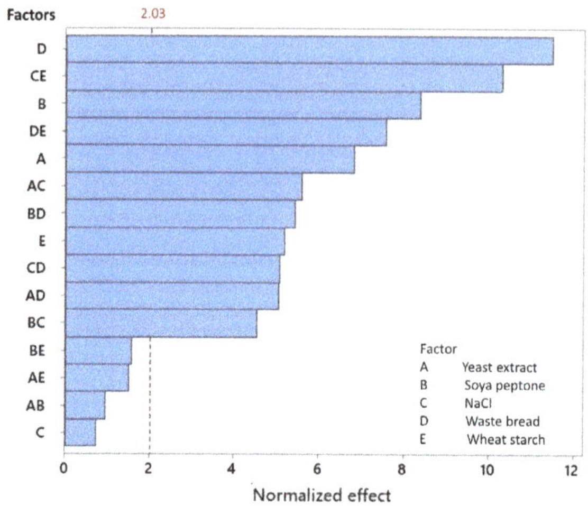

Figure 3. Standard Pareto chart of the normalized effect for the Amy-BSS production.

Additionally, the plot indicates that all data points are normally distributed around a mean of zero and adhere to a straight line.

The mathematical model corresponding to this design is given by the following equation:

$$y1 = 5.61 + 2.36\ A - 4.79\ B + 3.08\ C + 0.666\ D + 7.75\ E + 1.01\ A \times B + 5.86\ B \times C - 2.658\ B \times D + 1.6\ A \times E - 4.78\ B \times C + 2.852\ B \times D - 1.66\ B \times E + 2.665\ C \times D - 10.78\ C \times E - 0.823$$

The goodness of fit was evaluated using ANOVA, which yielded an F-value of 36.75 and a p-value < 0.05. Moreover, the coefficient of determination, R^2, was equal to 0.9453, meaning that 94.53% of the observed variation is attributed to the variable effects. Thus, it can be concluded that the predicted model (R^2pred = 87.7%) fits well with the experimental data (R^2adj = 91.96%). This set of results clearly demonstrates the validity of the model.

Figure 4 shows the optimized levels for the chosen variables. It is observed that the ideal values for yeast extract, soya peptone, NaCl, waste bread, and wheat starch are 10, 5, 0, 10, and 10 g/L, respectively, achieving an amylase activity of 9.64 U/mL with a desirability of d = 1. To ensure the reproducibility of this composition, three independent experiments were executed. An amylase activity of 8.96 ± 0.32 U/mL was obtained, close to the predicted value, confirming the reliability of the method.

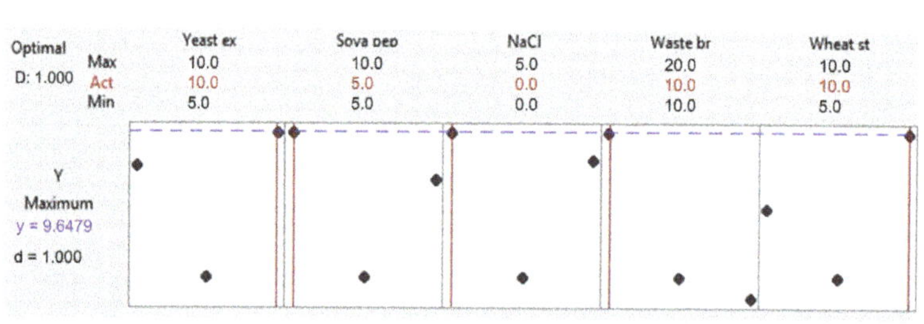

Figure 4. Optimal factors setting for Amy-BSS production. Blue line represents the predicted maximum response. Red vertical lines indicate the optimal levels of each factor used in the solution; Black squares correspond to experimental design points.

3.4.2. The 2_V^{4-1} for the Enzymatic Waste Bread Hydrolysis

An optimal condition for enzymatic hydrolysis of WB by Amy-BSS was defined by 2_V^{4-1} fractional factorial design. Four parameters, amylolytic activity (X_1), incubation temperature (X_2), waste bread concentration (X_3), and hydrolysis time (X_4), were varied at two different levels to investigate their effects and interactions on reducing sugar yield. Table 6 groups the measured experimental responses corresponding to the eight tests runs. The highest amount of released sugar was 4.635 g/L, obtained in run 2 using 10 U of Amy-BSS, a 50 °C incubation temperature, and 1% WB during 3 h of hydrolysis time.

Table 6. Design and data responses resulting in the 2_V^{4-1} plan.

Run	Amylase Activity	Incubation Temperature	Waste Bread	Hydrolysis Time	Experimental Hydrolysis Yield (g/L)
1	−1	−1	−1	−1	1.277 ± 0.099
2	1	−1	−1	1	4.635 ± 0.049
3	−1	1	−1	1	1.877 ± 0.012
4	1	1	−1	−1	1.847 ± 0.015
5	−1	−1	1	1	2.460 ± 0.078
6	1	−1	1	−1	1.857 ± 0.051
7	−1	1	1	−1	1.417 ± 0.006
8	1	1	1	1	3.643 ± 0.023

This result was confirmed to be optimal, as demonstrated by Figure 5, which shows the highest reducing sugar yield under this condition.

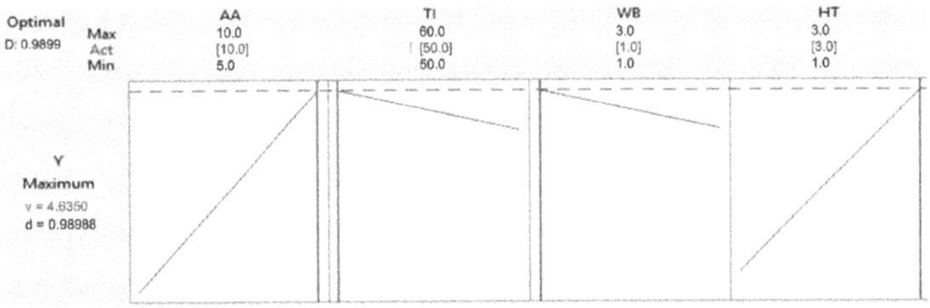

Figure 5. Optimized response design and data responses resulting for the 2_V^{4-1} plan. AA, amylase activity; TI, temperature of incubation; WB, waste bread; HT, hydrolysis time. Blue line represents the predicted maximum response; Red vertical lines indicate the optimal levels of each factor used in the solution.

The statistical significance of each individual factor and their combinations at $\alpha = 5\%$ was evaluated using ANOVA. As shown in Table 7, the main (X_1, X_2, X_3, and X_4) and interaction effects, ($X_1 \times X_2$), ($X_1 \times X_3$), and ($X_1 \times X_4$), are statistically significant ($p << 0.05$).

A first-order polynomial model was used to describe the correlation between the four tested variables and the hydrolysis yield. The fit of the data is described as follows:

$$y2 = -0.367 + 0.4320\ X_1 + 0.00575\ X_2 + 0.60701\ X_3 - 0.3221\ X_4 - 0.005583\ (X_1 \times X_2) - 0.08525\ (X_1 \times X_3) + 0.14658\ (X_1 \times X_4).$$

Table 7. ANOVA for the released sugar by Amy-BSS.

Source Variation	DL	F-Value	*p*-Value
Model	7	1126.51	<<0.05
Main factors	4	1945.07	<<0.05
Amylolytic activity (X_1)	1	3146.82	<<0.05
Incubation temperature (X_2)	1	267.98	<<0.05
Waste bread (X_3)	1	8.57	<<0.05
Hydrolysis time (X_4)	1	4962.70	<<0.05
Interaction 2 factors	3	472.41	<<0.05
$X_1 \times X_2$	1	40.01	<<0.05
$X_1 \times X_3$	1	373.09	<<0.05
$X_1 \times X_4$	1	1103.06	<<0.05

The analysis of the model equation reveals that waste bread (X_3) exerts the strongest positive effect on y_2 (+0.60701), while hydrolysis time (X_4) had a notable negative impact (−0.3221), making them the main drivers of change in y_2. Hence, X_3 acts as a key promoter, and X_4 as an inhibitor. On the contrary, X_2 shows a negligible effect (+0.00575), and X_1 displays a moderate positive influence (+0.4320). Additionally, the antagonistic interactions ($X_1 \times X_2$) and ($X_1 \times X_3$) suggest that high levels of these variables reduce the effectiveness of X_1 due to temperature sensitivity and limited enzyme accessibility. Meanwhile, the synergistic interaction ($X_1 \times X_4$) (+0.146) indicates that extended hydrolysis time enhances the X_1 positive impact on y_2 through improved substrate accessibility and sustained enzyme activity.

The synergistic interaction and its significant effect are further illustrated by the surface plot in Figure 6. The plot reveals that the highest hydrolysis yield occurred on the top-right corner, corresponding to the combination of high Amy-BSS concentration (10 U) and extended hydrolysis time (3 h). This trend aligns with enzymatic action: at low amylase concentrations, the reaction rate is limited by the scarcity of active sites for starch binding, increasing both the enzyme load and duration-maximized substrate conversion. These results experimentally validate the ($X_1 \times X_4$) synergy identified in the model equation, demonstrating how optimized enzyme–time enhanced hydrolysis yield.

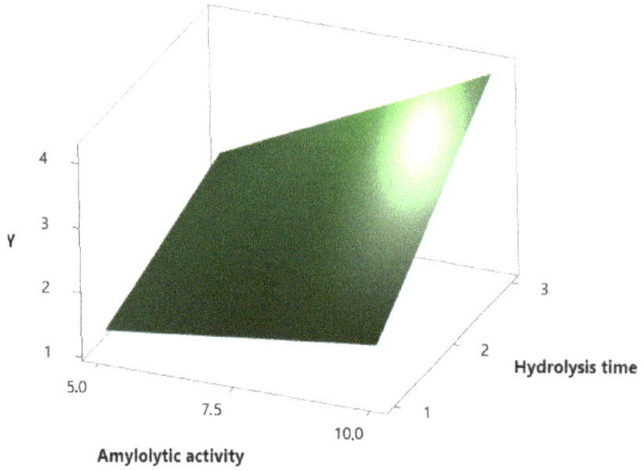

Figure 6. The surface plot of the synergistic interaction (amylolytic activity × hydrolysis time). y is the hydrolysis yield.

The validity of the adjusted model was assessed based on the results in Table 7. The regression analysis indicates that the model is highly significant, with a very low *p*-value ($p < 0.05$) and a high F value (1226.51).

Meanwhile, the coefficient of determination (R^2 = 99.83%) indicates that only 0.17% of the total variability was not explained by the model, suggesting that the considered variables were sufficient to make reliable predictions. Additionally, the small difference between the adjusted R^2 (99.74%) and the estimated R^2 (99.59%) demonstrates a strong correlation between the estimated and experimental data.

3.5. High-Performance Liquid Chromatography (HPLC)

HPLC analysis revealed that glucose, and other sugars with a degree of polymerization (DP) ≥ 3 were the final products of WB starch (Figure 7). It appears that the amylase generated by *Bacillus* spp. BSS is an endo-amylase. Indeed, its action on WB starch primarily produces oligosaccharides with a DP ≥ 3, along with a fraction of glucose and maltose. Conversely, exoamylases act in a progressive manner from the non-reducing ends of the polysaccharide chain, resulting in the sequential release of only maltose or glucose.

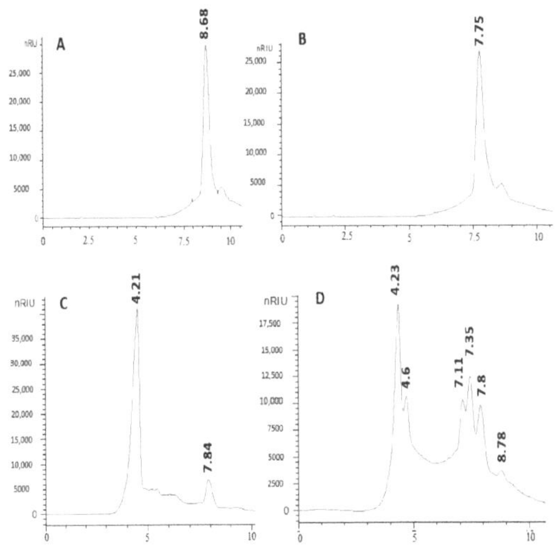

Figure 7. Hydrolysis products of 3% WB using 10U of Amy-BSS: (**A**) glucose solution of 5 mg/mL, (**B**) maltose solution of 5 mg/mL, (**C**) 3% unhydrolyzed WB, and (**D**) hydrolyzed WB by Amy-BSS.

4. Discussion

The growing global population has led to an increase in food quantities, resulting in a surge in food waste (FW). This is part of the aftermath of an absolute linear economy model. Recently, the circular economy has gained traction as a sustainable alternative. It enables FW recycling in the industrial process via new technologies. This approach aligns with the concept of "cradle to cradle", where the recycled materials retain their features and can be re-introduced to the environment without harm [28]. Bread ranks as the most discarded food waste globally due to its rapid spoilage. Its valorization through multiple biological approaches offers a promising pathway for promoting a circular economy.

To develop effective valorization approaches, a comprehensive waste stream analysis is required. Hence, the characterization study (Table 3) showed that WB contained 61.41% starch, which is in accordance with Kumar et al. [4], who reported a content of 50–70%. Additionally, the moisture content (17.17%) and water activity (0.761) are critical parameters affecting bread stability. In food processing, moisture content influences preservation, storage, packaging, and transportation. Thus, the high level of moisture leads to mold growth within 3–7 days. Our outcomes are in close agreement with the findings of Gobbetti et al. [29], who reported a water activity of about 0.97. Notably, the ash quantity, 0.64%, was significantly lower than that observed by Abidin et al., i.e., 6.74% [30].

Given its rich starch content and nutrient profile, WB serves as an attractive low-cost substrate for fermentation [31]. In general, bacterial cultivations are achieved successfully using conventional medium, such as Luria-Bertani broth (LB). However, LB is expensive due to the elevated-cost components, such as yeast extract, soya peptone, and beef extract, which are the basic nitrogen additives [13]. The development of an alternative growth media using WB aims to strike a balance between where we can obtain a cost-effective medium with good biomass yields. The used strains (*E. coli*, *E. faecalis*, *S. aureus*, and *P. aeruginosa*) were routinely cultured in LB. Interestingly, our results have shown superior microbial growth in the novel formulated media (NC1 and NC2) compared to LB.

The variations observed in the growth of the pathogenic microorganisms using the waste bread media (NC1 and NC2) can be explained by the difference in the metabolic process and nutrient preferences of each strain. In other words, the improved growth of *E. coli*, *E. faecalis*, and *S. aureus* in NC2 could be attributed to their ability to consume waste bread and satisfy their need to achieve better growth, which likely explains their superior growth in NC2 compared to the control. Meanwhile, *P. aeruginosa* may prefer more complex and pure nitrogen and carbon sources (commercial), even its demonstrated viable proliferation. Hence, the differential growth patterns could be influenced by the ability of each microorganism to metabolize a specific nutrient released from bread waste. This result validates WB as a cheaper nitrogen source for microbial cultivation.

Our findings are in concordance with Verni et al. [5], who confirm WB's efficacy as a nutrient rich growth medium for starter yeasts and fungi, surpassing classical media. Meanwhile, lactic acid bacteria [13] and fungi [32] showed significant microbial growth on the WB-based medium. In addition, the effective cultivation of these strains on WB-based media (NC1 and NC2) confirms the ability of the media to support microbial growth under conditions relevant to food microbiology. This indicates that WB is suitable for potential applications in food safety monitoring and environmental pathogen screening. Conversely, some researchers have used enzymatic pretreatment of WB before it was incorporated into the fermentation media. For instance, Carsanba et al. [33] treated WB with amylase, glucoamylase, and protease before its use to cultivate *Yarrowia lipolytica* strain K57. However, such pretreatments increase the overall cost of the fermentation process, counteracting the objective of waste valorization for cost reduction. Above all, when comparing CLB with NC1, we showed that the costs of the fermentation medium could be reduced by 90%, and with NC2, the expenses could be eliminated completely. Additionally, expired bread can successfully substitute pure starch for screening amylolytic bacteria, as was previously shown. Furthermore, using SWB instead of LBS reduced costs by 46.68% (Table 8).

Traditionally, α-amylase production has relied on submerged fermentation with starch as the sole carbon and energy source. However, this conventional source faces significant economic challenges, as the cost of the carbon source constitutes a major portion of amylase production expenses. This restriction often makes the process unfeasible economically [31]. To address this challenge, exploring agro-waste as a substitute substrate is one of the solutions. These wastes naturally occurred with carbon storage, acting as an energy source throughout microbial fermentation [34]. This results in a promising option for economical industrial amylase production. The use of agricultural waste, such as moong husk, soybean cake, rice bran, and groundnut shell, as media supplements in α-amylase production for *Bacillus amyloliquefaciens* KCP2, *Bacillus tequilensis* TB5, *Aspergillus*, respectively, has been reported [35–37]. Recently, WB has sparked interest as a promising fermentation feedstock for enzymatic synthesis due to its richness in carbohydrates, calories, proteins, and essential vitamins [38]. Its nutritional profile has been investigated in this study as an ideal energy and carbon source for Amy-BSS production.

Table 8. Cost comparison of different culture media used.

Media	Media Components	Component Cost (EUR/100 g)	Component Used (g/L)	Cost per Unit (EUR)	Total Cost (EUR)
CLB	Peptone soya	28	10	2.8	5.14
	Yeast extract	33.68	5	1.68	
	NaCl	13.2	5	0.66	
NC_1	Peptone soya	28	1	0.28	0.514
	Yeast extract	33.68	0.5	0.168	
	NaCl	13.2	0.5	0.066	
	Waste bread	0	10	0	
NC_2	Waste bread	0	0	0	0
LBS	Peptone soya	28	10	2.8	9.64
	Yeast extract	33.68	5	1.68	
	NaCl	13.2	5	0.66	
	Soluble starch	45	10	4.5	

Take into consideration that EUR 1 = TND 3.34 (exchange rate as of 16 April 2025).

To further optimize and reduce the cost of amylase production, a fractional factorial design (FFD) was employed in this study to systematically evaluate the effects of key process variables. Unlike full factorial or response surface designs, FFD requires fewer experimental runs while still revealing valuable insights into key variables. When executing 17 runs (Table 4), it has been found that WB significantly improved Amy-BSS production from 0.69 U/mL to 8.96 U/mL. This enhancement confirms that WB is a valuable carbon/substrate source. Furthermore, based on the optimized response (Figure 5), it was demonstrated that a low level of soy peptone was sufficient, reducing the required amount of nitrogen sources thanks to the compensatory effect of WB. Additionally, the absence of NaCl needed for Amy-BSS production suggests that *Bacillus* BSS's sodium requirement was met by the inherent salt quantity present in the WB. Moreover, bread contains the key amino acids (nicotinic acid, thiamine, biotin, and pantothenic acids), as reported by Abd-Elhalim et al. [39]. These elements support *Bacillus* spp. BSS growth while minimizing dependence on synthetic amino acid supplementation. The re-purpose of WB lowers Amy-BSS production cost, further preserving bacterial biomass yield. A major advantage of WB is its non-seasonal availability, unlike conventional agro-residues, facilitating consistent integration into circular economy systems.

Alternatively, in order to apply waste bread in food processing, its modification through hydrolysis is crucial. Hydrolysis breaks the complex configuration of WB starch into fermentable sugars that can be easily converted into valuable food supplements. As opposed to lignocellulosic waste, WB is a more efficient source of fermentative sugars due to its high starch quantity [33]. Notably, sugar generated from WB is without inhibitors, commonly found in lignocellulosic compounds such as furans, organic acids, and phenols, which may reduce the activity of hydrolytic enzymes [2,40]. Nevertheless, the starch WB can be hydrolyzed using enzymatic or acid hydrolysis. Commonly, dilute acid hydrolysis through either HCl or H_2SO_4 has been applied to cleave glycosidic bonds, yielding lower-molecular-weight carbohydrates [41]. However, chemical hydrolysis demands excessive energy input, corrosive substances, and severe operating settings, which can generate harmful byproducts. In contrast, enzymatic pretreatment offers a more sustainable and eco-friendly approach, leveraging the high specificity and efficiency of enzymes to cleave starch under mild conditions (40–50 °C). Therefore, enzymatic hydrolysis consumes and hydrolyzes with no toxic substance compared to the chemical approach [17,42].

Commercial enzymes, namely α-amylase (Grindamyl A14000), amyloglucosidase (Grindamyl PlusSweet), maltogenic amylase (MALT), and protease (Corolase 7089), were investigated for the enzymatic hydrolysis of WB, as reported by Rosa-Sibakov et al. [15]. Yet, to make an economical process, substituting commercial enzymes with house laboratory enzymes is crucial. Therefore, we have prioritized the use of Amy-BSS for this purpose. As evidenced by FFD, the reducing sugar yield through WB hydrolysis was significantly dependent on hydrolysis time, Amy-BSS unities, WB ration, and incubation temperature. Our results proved that fermentable sugar production increased with longer time and higher Amy-BSS unities, in accordance with Abidin et al. [30], who found that the maximum reducing sugar levels are obtained at an enzyme concentration of 6% (w/v), as opposed to 2% and 4%. Additionally, when hydrolysis periods increase, the concentration of reducing sugars also increase, peaking at 180 min. In addition, Mihajlovski et al. [43] reported that hydrolysis yield was influenced by fermentation time and WB concentration. In contrast, in our findings, 1% of WB was optimal because higher substrate increased viscosity, thereby restricting Amy-BSS action. Further, incubation temperature (50°) showed a positive effect, which contradicts the Kahlouche et al. [44] results. They found that changing the temperature from 20 °C to 100 °C had no discernible impact on the hydrolysis of waste bread (p-value of $0.069 > 0.05$). The main differences between the two studies could account for this discrepancy. Interestingly, our investigation used a bacterial α-amylase, while Kahlouche et al. [44] used a commercial α-amylase from a fungus source, *Aspergillus oryzae* (A8220-50ML, Sigma-Aldrich, Søborg, Denmark). There are notable physiological and biochemical distinctions between these two enzyme sources, which are from different kingdoms. Furthermore, to lower the economic cost of industrial-scale applications, we utilized a crude extract from a laboratory strain, non-commercial, even though their enzyme was refined and commercially produced.

As illustrated in Figure 7, the fermentable sugars produced through Amy-BSS activity were primarily glucose, along with sugars having a DP ≥ 3, demonstrating a promising potential for reuse in new bread-making applications. It has been demonstrated that certain strains, including *Bacillus* species and *Leuconostoc mesentroides*, may convert waste bread into mannitol, a sugar alcohol used in the food sector [45]. Mihajlovski et al. [43] identified maltose, glucose, and maltotriose as main sugars products generated during WB hydrolysis using enzymes from *Hymenobacter* sp. CKS3. In a nutshell, by replacing conventional sugar with enzymatically hydrolyzed WB syrup, the food industry, particularly bakeries, can significantly reduce production costs. These syrups can completely replace standard sugars in the making of bread and boost shelf-life through improved hygroscopicity and crumb tenderization, as well as flavor and crust color via Maillard reactions [15]. Such practical incorporation of bread waste-derived components offers bakeries a scalable solution to align production with both financial and environmental goals.

5. Conclusions

This study highlights the feasibility of valorizing Tunisian waste bread (WB) as a sustainable and cost-effective substrate for various biotechnological applications, supporting its role in enhancing circular economy. The formulation of WB-based culture media not only reduced production costs by up to 90% compared to conventional LB media but also significantly enhanced microbial growth, confirming WB's economic and functional advantages. The use of just 1% WB as a replacement for commercial starch in amylolytic-strain screening further validated its applicability as a sustainable alternative. Moreover, the application of FFD led to a 15-fold increase in amylase production by the Amy-BSS strain, with WB identified as the most influential variable, directly supporting the aim of optimizing bioprocess efficiency. In addition, enzymatic hydrolysis of WB under mild conditions yielded 4.6 g/L of fermentable sugars, demonstrating its viability as a raw material for bio-based product development. Altogether, these outcomes confirm that Tunisian WB is a robust and efficient resource, aligning with the study's goal of advancing sustainable waste valorization within a circular bioeconomy framework.

Author Contributions: S.B.M., conceptualization, investigation, methodology, and writing—original draft; B.B.H.H., methodology, software, formal analysis, and writing—original draft; W.S. and S.D., methodology; T.V. and S.S., writing—review and editing, visualization, supervision, funding. All authors have read and agreed to the published version of the manuscript.

Funding: This research received no external funding.

Institutional Review Board Statement: Not applicable.

Informed Consent Statement: Not applicable.

Data Availability Statement: The original contributions presented in this study are included in the article. Further inquiries can be directed to the corresponding author.

Acknowledgments: Authors are grateful to Bouassida Hichem from Tunisian food production company (STPA) for his help on the WB composition analysis.

Conflicts of Interest: The authors declare no conflicts of interest.

References

Tamasiga, P.; Ouassou, E.H.; Onyeaka, H.; Bakwena, M.; Happonen, A.; Molala, M. Forecasting disruptions in global food value chains to tackle food insecurity: The role of AI and big data analytics—A bibliometric and scientometric analysis. *J. Agric. Food Res.* **2023**, *14*, 100819. [CrossRef]

Hafyan, R.H.; Mohanarajan, J.; Uppal, M.; Kumar, V.; Narisetty, V.; Maity, S.K.; Gadkari, S. Waste bread valorization: A review of sustainability aspects and challenges. *Bioresour. Technol.* **2024**, *8*, 1334801. [CrossRef]

Aït-Kaddour, A.; Hassoun, A.; Tarchi, I.; Loudiyi, M.; Boukria, O.; Cahyana, Y.; Ozogul, F.; Khwaldia, K. Transforming plant-based waste and by-products into valuable products using various "Food Industry 4.0" enabling technologies: A literature review. *Sci. Total Environ.* **2024**, *955*, 176872. [CrossRef] [PubMed]

Kumar, V.; Brancoli, P.; Narisetty, V.; Wallace, S.; Charalampopoulos, D.; Dubey, B.K.; Kumar, G.; Bhatnagar, A.; Bhatia, S.K.; Taherzadeh, M.J. Waste bread—A potential feedstock for sustainable circular biorefineries. *Bioresour. Technol.* **2023**, *369*, 128449. [CrossRef]

Verni, M.; Minisci, A.; Convertino, S.; Nionelli, L.; Rizzello, C.G. Wasted bread as substrate for the cultivation of starters for the food industry. *Front. Microbiol.* **2020**, *11*, 293. [CrossRef]

Bhardwaj, A.; Soni, R.; Singh, L.P.; Mor, R.S. A simulation approach for waste reduction in the bread supply chain. *Logistics* **2023**, *7*, 2. [CrossRef]

Asqardokht-Aliabadi, A.; Sarabi-Aghdam, V.; Homayouni-Rad, A.; Hosseinzadeh, N. Postbiotics in the Bakery Products: Applications and Nutritional Values. *Probiot. Antimicrob. Proteins* **2025**, *17*, 292–314. [CrossRef]

Immonen, M.; Maina, N.H.; Wang, Y.; Coda, R.; Katina, K. Waste bread recycling as a baking ingredient by tailored lactic acid fermentation. *Int. J. Food Microbiol.* **2020**, *327*, 108652. [CrossRef]

Narisetty, V.; Cox, R.; Willoughby, N.; Aktas, E.; Tiwari, B.; Matharu, A.S.; Salonitis, K.; Kumar, V. Recycling bread waste into chemical building blocks using a circular biorefining approach. *Sustain. Energy Fuels* **2021**, *5*, 4842–4849. [CrossRef]

10. African Manager. Available online: https://en.africanmanager.com/113000-tons-of-bread-is-thrown-away-in-tunisia-every-year/ (accessed on 19 April 2025).

11. Ben Rejeb, I.; Baraketi, S.; Charfi, I.; Khwaldia, K.; Gargouri, M. Valorization of bread waste, a nonconventional feedstock for starch extraction using different methods: A comparative study. *Euro-Mediterr. J. Envion. Integr.* **2024**, *9*, 1485–1498. [CrossRef]

12. Sahu, P.K.; Singh, R.; Shrivastava, M.; Darjee, S.; Mageshwaran, V.; Phurailtpam, L.; Rohtagi, B. Microbial production of α-amylase from agro-waste: An approach towards biorefinery and bio-economy. *Energy Nexus* **2024**, *14*, 100293. [CrossRef]

13. Iosca, G.; Turetta, M.; De Vero, L.; Bang-Berthelsen, C.H.; Gullo, M.; Pulvirenti, A. Valorization of wheat bread waste and cheese whey through cultivation of lactic acid bacteria for bio-preservation of bakery products. *LWT* **2023**, *176*, 114524. [CrossRef]
14. Gadkari, S.; Kumar, D.; Qin, Z.H.; Lin, C.S.K.; Kumar, V. Life cycle analysis of fermentative production of succinic acid from bread waste. *Waste Manag.* **2021**, *126*, 861–871. [CrossRef] [PubMed]
15. Rosa-Sibakov, N.; Sorsamäki, L.; Immonen, M.; Nihtilä, H.; Maina, N.H.; Siika-aho, M.; Katina, K.; Nordlund, E. Functionality and economic feasibility of enzymatically hydrolyzed waste bread as a sugar replacer in wheat bread making. *J. Food Process Preserv.* **2022**, *46*, e16378. [CrossRef]
16. Narisetty, V.; Zhang, L.; Zhang, J.; Lin, C.S.K.; Tong, Y.W.; Show, P.L.; Bhatia, S.K.; Misra, A.; Kumar, V. Fermentative production of 2,3-Butanediol using bread waste—A green approach for sustainable management of food waste. *Bioresour. Technol.* **2022**, *358*, 127381. [CrossRef]
17. Cox, R.; Narisetty, V.; Nagarajan, S.; Agrawal, D.; Ranade, V.V.; Salonitis, K.; Venus, J.; Kumar, V. High-level fermentative production of lactic acid from bread waste under non-sterile conditions with a circular biorefining approach and zero waste discharge. *Fuel* **2022**, *313*, 122976. [CrossRef]
18. Dall'Acqua, K.; Klein, M.P.; Tech, B.I.; Fontana, A.; Crepalde, L.T.; Wagner, R.; de Oliveira, F.D.C.; Sant'Anna, V. Understanding the utilization of wasted bread as a brewing adjunct for producing a sustainable wheat craft beer. *Microorganisms* **2025**, *13*, 66. [CrossRef] [PubMed]
19. Cacace, C.; Rizzello, C.G.; Brunetti, G.; Verni, M.; Cocozza, C. Reuse of wasted bread as soil amendment: Bioprocessing, effects on alkaline soil and escarole (*Cichorium endivia*) production. *Foods* **2022**, *11*, 189. [CrossRef]
20. Sigüenza-Andrés, T.; Pando, V.; Gómez, M.; Rodríguez-Nogales, J.M. Optimization of a simultaneous enzymatic hydrolysis to obtain a high-glucose slurry from bread waste. *Food Chem.* **2022**, *396*, 1793. [CrossRef]
21. Montemurro, M.; Salvatori, G.; Alfano, S.; Martinelli, A.; Verni, M.; Pontonio, E.; Villano, M.; Rizzello, C.G. Exploitation of wasted bread as substrate for polyhydroxyalkanoates production through the use of *Haloferax mediterranei* and seawater. *Front. Microbiol.* **2022**, *13*, 1000962. [CrossRef]
22. Benabda, O.; M'hir, S.; Kasmi, M.; Mnif, W.; Hamdi, M. Optimization of protease and amylase production by Rhizopus oryzae cultivated on bread waste using solid-state fermentation. *J. Chem.* **2019**, *2019*, 3738181. [CrossRef]
23. Jung, J.M.; Kim, J.Y.; Kim, J.H.; Kim, S.M.; Jung, S.; Song, H.; Kwon, E.E.; Choi, Y.E. Zero-waste strategy by means of valorization of bread waste. *J. Clean. Prod.* **2022**, *365*, 132795. [CrossRef]
24. Guerra-Oliveira, P.; Belorio, M.; Gómez, M. Waste bread as main ingredient for cookie elaboration. *Foods* **2021**, *10*, 1759. [CrossRef]
25. Riaukaite, J.; Basinskiene, L.; Syrpas, M. Bioconversion of waste bread to glucose fructose syrup as a value-added product. In *Foodbalt Conference Proceedings*; Foodbalt: Kaunas, Lithuania, 2019; pp. 120–124.
26. Andualem, B. Isolation and screening of amylase producing thermophilic spore forming Bacilli from starch rich soil and characterization of their amylase activities using submerged fermentation. *Int. Food Res. J.* **2014**, *21*, 831–837.
27. Beldjoudi, S.; Kouachi, K.; Bourouina-Bacha, S.; Lafaye, G.; Soualah, A. Kinetic study of methyl orange decolorization by the Fenton process based on fractional factorial design. *React. Kinet. Mech. Catal.* **2020**, *130*, 1123–1140. [CrossRef]
28. Dymchenko, A.; Geršl, M.; Gregor, T. Trends in waste bread utilisation. *Trends Food Sci. Technol.* **2023**, *132*, 93–102. [CrossRef]
29. Gobbetti, M.; De Angelis, M.; Di Cagno, R.; Calasso, M.; Archetti, G.; Rizzello, C.G. Novel insights on the functional/nutritional features of the sourdough fermentation. *Int. J. Food Microbiol.* **2019**, *302*, 103–113. [CrossRef] [PubMed]
30. Abidin, Z.Z.; Hassan, S.R. Kinetic studies of enzymatic hydrolysis pre-treatment of waste bread under preliminary conditions. *J. Eng. Sci. Res.* **2023**, *7*, 9–14. [CrossRef]
31. Ibenegbu, C.C.; Leak, D.J. Simultaneous saccharification and ethanologenic fermentation (SSF) of waste bread by an amylolytic Parageobacillus thermoglucosidasius strain TM333. *Microb. Cell Factories* **2022**, *21*, 251. [CrossRef]
32. Svensson, S.E.; Bucuricova, L.; Ferreira, J.A.; Souza Filho, P.F.; Taherzadeh, M.J.; Zamani, A. Valorization of waste bread to a fiber and protein-rich fungal biomass. *Fermentation* **2021**, *7*, 91. [CrossRef]
33. Carsanba, E.; Agirman, B.; Papanikolaou, S.; Fickers, P.; Erten, H. Valorisation of waste bread for the production of yeast biomass by Yarrowia lipolytica bioreactor fermentation. *Fermentation* **2023**, *9*, 687. [CrossRef]
34. Werle, L.B.; Abaide, E.R.; Felin, T.H.; Kuhn, K.R.; Tres, M.V.; Zabot, G.L.; Mazutti, M.A. Gibberellic acid production from Gibberella fujikuroi using agro-industrial residues. *Bioresour. Technol.* **2020**, *25*, 101608. [CrossRef]
35. Bhatt, B.; Prajapati, V.; Patel, K.; Trivedi, U. Kitchen waste for economical amylase production using Bacillus amyloliquefaciens KCP2. *Biocatal. Agric. Biotechnol.* **2020**, *26*, 101654. [CrossRef]
36. Paul, J.S.; Beliya, E.; Tiwari, S.; Patel, K.; Gupta, N.; Jadhav, S.K. Production of biocatalyst α-amylase from agro-waste 'rice bran' by using Bacillus tequilensis TB5 and standardizing its production process. *Biocatal. Agric. Biotechnol.* **2020**, *26*, 101648. [CrossRef]
37. Abdulsalam, M.; Fari, H.I.; Tiamiyu, B.B.; Salam, O.L. Optimizing α-amylase production from locally isolated Aspergillus species using selected agro waste as substrate. *Biosci. Biotechnol. Res. Commun.* **2022**, *15*, 424–430. [CrossRef]
38. Zain, M.Z.M.; Shori, A.B.; Baba, A.S. Potential functional food ingredients in bread and their health benefits. *Biointerface Res. Appl. Chem.* **2022**, *12*, 6533–6542.

39. Abd-Elhalim, B.T.; Gamal, R.F.; El-Sayed, S.M.; Abu-Hussien, S.H. Optimizing alpha-amylase from Bacillus amyloliquefaciens on waste bread for effective industrial wastewater treatment and textile desizing through response surface methodology. *Sci. Rep.* **2023**, *13*, 19216.
40. Guo, H.; Zhao, Y.; Chang, J.S.; Lee, D.J. Inhibitor formation and detoxification during lignocellulose biorefinery: A review. *Bioresour. Technol.* **2022**, *361*, 127666. [CrossRef]
41. Torabi, S.; Satari, B.; Hassan-Beygi, S.R. Process optimization for dilute acid and enzymatic hydrolysis of waste wheat bread and its effect on aflatoxin fate and ethanol production. *Biomass Convers. Biorefin* **2021**, *11*, 2617–2625. [CrossRef]
42. Banu, J.R.; Merrylin, J.; Usman, T.M.; Kannah, R.Y.; Gunasekaran, M.; Kim, S.H.; Kumar, G. Impact of pretreatment on food waste for biohydrogen production: A review. *Int. J. Hydrogen Energy* **2020**, *45*, 18211–18225. [CrossRef]
43. Mihajlovski, K.; Rajilić-Stojanović, M.; Dimitrijević-Branković, S. Enzymatic hydrolysis of waste bread by newly isolated *Hymenobacter* sp. CKS3: Statistical optimization and bioethanol production. *Renew. Energy* **2020**, *152*, 627–633. [CrossRef]
44. Kahlouche, F.Z.; Zerrouki, S.; Bouhelassa, M.; Rihani, R. Experimental optimization of enzymatic and thermochemical pretreatments of waste bread by central composite design study for bioethanol production. *Water Sci. Technol.* **2022**, *85*, 3436–3450. [CrossRef] [PubMed]
45. Sadeqi, V.; Astani, A.; Khanafari, A. Investigating the possibility of microbial production of mannitol from waste bread. *Iran. J. Health Saf. Environ.* **2016**, *3*, 587–591.

microorganisms

MDPI

Article

Effects of Fucoidan on the Inhibition of Biofilm Formation of *Salmonella enterica* Subsp. *enterica* Serovar *Typhimurium* on Seafoods and Its Molecular Antibiofilm Mechanisms

Anamika Roy [1], Pantu Kumar Roy [1,*], Sung Rae Cho [2] and Shin Young Park [1,*]

[1] Department of Seafood Science and Technology, Institute of Marine Industry, Gyeongsang National University, Tongyeong 53064, Republic of Korea; anamika32roy@gmail.com

[2] South Sea Fisheries Research Institute, National Institute of Fisheries Science, Yeosu 59780, Republic of Korea; srchoi1113@korea.kr

* Correspondence: vetpantu88@gmail.com (P.K.R.); sypark@gnu.ac.kr (S.Y.P.); Tel.: +82-55-772-9143 (S.Y.P.); Fax: +82-55-772-9149 (S.Y.P.)

Abstract: Foodborne illnesses, particularly those caused by *Salmonella enterica* subsp. *enterica* Serovar *Typhimurium*, present a significant challenge to public health, especially within the seafood industry due to biofilm formation on foods. This study investigated the antibiofilm potential of fucoidan, a sulfated polysaccharide, against *Salmonella enterica* subsp. *enterica* Serovar *Typhimurium* biofilm on crab and shrimp surfaces. Fucoidan's minimum inhibitory concentration (MIC) was determined to be 150 µg/mL. Sub-MIC (1/8, 1/4, 1/2, and MIC) were evaluated for their impact on inhibition of biofilm formation. Fucoidan treatment resulted in significant, dose-dependent inhibition in biofilm formation, achieving 2.61 log CFU/cm^2 and 2.45 log CFU/cm^2 reductions on crab and shrimp surfaces, respectively. FE-SEM analysis confirmed biofilm disruption and cell membrane damage. Real-time PCR showed the down-regulation of quorum-sensing (*luxS*) and virulence (*rpoS*, *avrA*, and *hilA*) genes. These results propose that fucoidan has the ability as a natural antibacterial agent for controlling *Salmonella enterica* subsp. *enterica* Serovar *Typhimurium* biofilms in seafood processing, thereby enhancing food safety and minimizing contamination.

Keywords: biofilms; *Salmonella enterica* subsp. *enterica* Serovar *Typhimurium*; crab shells; shrimp shells; fucoidan

1. Introduction

Foodborne infections are caused by eating tainted food, and their increasing prevalence is a serious public health issue [1,2]. Biofilm formation on seafood surfaces compromises food safety and raises the cross-contamination risk during processing, posing a direct threat to public health [3,4]. Bacterial cells permanently attached to food and food contact surfaces and immersed in a self-made matrix of extracellular polymeric substances (EPS) are recognized as biofilms [5]. *Salmonella* spp. is a pathogenic bacterium known to cause foodborne illness [6]. The Centers for Disease Control and Prevention (CDC) estimated that salmonellosis leads to infections of 1.35 million, 26,500 hospitalizations, and 420 fatalities each year in the United States [7]. The rise in salmonellosis cases has been attributed to *Salmonella* spp. biofilms on various food products and food-contact surfaces within food processing facilities [1]. Compared to planktonic cells, bacteria in biofilms can better adapt to various environmental situations. Biofilms are challenging to eliminate, representing a substantial hygienic concern in the seafood industry [8]. These biofilms on biotic and abiotic surfaces impact the safety and quality of food products.

Effectively controlling biofilm formation remains a major challenge in the food sector, as they contribute to food contamination, damage processing surfaces and equipment, and pose serious health risks [3]. Biofilms are responsible for 80% of persistent bacterial infections, making it increasingly difficult to eliminate these pathogens [9]. As many chronic infections stem from biofilms, finding effective treatments has become a significant challenge [5,10]. Recently, intense research has been engaged on the expansion of new pharmaceuticals with improved efficacy, with a particular emphasis on innovative approaches and technological advancements. *Salmonella enterica* subsp. *enterica* Serovar *Typhimurium* is known for its ability to form robust biofilms, which promote its persistence in the environment and its resistance to conventional treatments [6,11]. In the seafood industry, biofilm formation by pathogenic bacteria, such as *Salmonella enterica* subsp. *enterica* Serovar *Typhimurium* on the surfaces of crabs and shrimps presents a significant challenge to food safety and public health.

Quorum sensing (QS) and extracellular polymeric substances (EPS) play critical roles in biofilm formation, stability, and pathogenicity in many bacteria, including *Salmonella Typhimurium* and *Listeria monocytogenes* [4]. QS is a cell-to-cell communication mechanism that regulates bacterial behavior in a population-density-dependent manner through signaling molecules known as autoinducers. In *S. Typhimurium*, the *luxS* gene is involved in the synthesis of autoinducer-2 (AI-2), a key signal in QS [6]. QS modulates the expression of numerous genes associated with virulence, motility, and biofilm development, facilitating bacterial adaptation and survival in diverse environments.

EPS, the major component of the biofilm matrix, is composed of polysaccharides, proteins, lipids, and extracellular DNA. It provides structural integrity to the biofilm, mediates adhesion to surfaces, offers protection against environmental stressors and antimicrobial agents, and hosts immune responses. The production of EPS is tightly regulated and often coordinated by QS systems, making these two processes intricately linked.

Fucoidan is a sulfated polysaccharide, numerous algal species generated in large quantities, and its biological characteristics have been thoroughly studied [12–14]. By altering the permeability of the cell wall or creating a gel-like barrier on the bacterial membrane, fucoidan is thought to have antibacterial properties that prevent bacterial nutrition exchange and suppress bacterial pathogens [15–17]. The search for new antimicrobial agents has been spurred by the growing resistance of microorganisms to traditional antibiotics [18,19]. Because of their alleged less adverse effects and affordability, natural substances like fucoidan hold promise as possible sources of antimicrobial medications [9,11]. Emerging evidence indicates that fucoidan may also disrupt biofilm formation and reduce the stability of pre-formed biofilms in various microbial species, making it a potential agent for combating biofilm-related infections [20,21]. However, its efficacy in the reduction in *Salmonella enterica* subsp. *enterica* Serovar *Typhimurium* biofilms on seafood surfaces, such as crab and shrimp, remains largely unexplored.

This study aims to assess the impacts of fucoidan of sub-minimum inhibitory concentrations (sub-MIC) in preventing *Salmonella enterica* subsp. *enterica* Serovar *Typhimurium* biofilm formation on crab and shrimp and its influence on the expression of key genes at the molecular level. The results will offer effective perceptions of the possible use of fucoidan as an antimicrobial agent in the seafood industry, enhancing food safety and minimizing microbial contamination.

2. Materials and Methods

2.1. Bacterial Strain Culture and Growth

In this study, *Salmonella enterica* subsp. *enterica* Serovar *Typhimurium* (*S. Typhimurium* ATCC 13311) was utilized as the test organism. The bacterial strains, with a cell density of 10^7–10^8 Log CFU/mL, were stored at −80 °C in tryptic soy broth (TSB; BD Difco, Franklin Lakes, NJ, USA) accompanied by 30% glycerol. Initially, 100 µL of the bacterial stock was inoculated and incubated at 37 °C into 10 mL of TSB with continuous shaking at 170 rpm using a shaking incubator (Vision Scientific, VS–8480, Seoul, Republic of Korea). A 100 µL aliquot was subsequently transferred into 10 mL of TSB and incubated for 24 h. After 18 h incubation, the bacterial culture was centrifuged at 5400 rpm for 10 min at 4 °C, washed two times with sterile phosphate-buffered saline (PBS; Sigma-Aldrich, St. Louis, MO, USA), and resuspended in the same buffer. The final suspension was then diluted using peptone water (PW) (Oxoid, Basingstoke, UK) to accomplish 10^5 Log CFU/mL. This prepared inoculum (10^5 Log CFU/mL) was used to facilitate biofilm formation on crab and shrimp.

2.2. Fucoidan and Minimum Inhibitory Concentration (MIC) Determination

In this study, fucoidan was obtained from Sigma-Aldrich (St. Louis, MO, USA). The compound was dissolved in DW at 1 mg/mL concentration. The MIC of fucoidan against *S. Typhimurium* ATCC 13311 was determined by a previous method [22]. A two-fold serial dilution technique was employed to establish the MIC in TSB. In a 96-well plate (Corning Incorporated, Corning, NY, USA), 100 µL of fucoidan (0–300 µg/mL) was diluted serially in TSB and mixed with 100 µL of a bacterial suspension (10^5 Log CFU/mL), resulting in 200 µL per well. The cultures were incubated at 37 °C for 24 h and bacterial growth was assessed by evaluating absorbance at 600 nm using a microplate reader (Spectra Max 190, Sunnyvale, CA, USA). After overnight incubation, 100 µL aliquots from wells showing no visible growth were plated on xylose lysine deoxycholate (XLD) agar (Thermo Scientific, Oxoid, Basingstoke, UK) to enumerate bacterial colonies. The experiment was performed three times, and the MIC was revealed to be 150 µg/mL of fucoidan. For subsequent experiments, the sub-MIC concentrations (0, 1/8, 1/4, 1/2, and MIC) of fucoidan were utilized.

2.3. Crabs and Shrimp Sample Preparation

With minor changes, sample preparation was conducted following methods described in previous studies [23]. Shrimp (*Penaeus monodon*) and crab (*Corystes cassivelaunus*) were purchased from the local market and with a sterile scalpel, they were cut into 2 × 2 cm^2 coupons. Any residual meat was removed, and the shells were thoroughly washed with sterile DW. To eliminate background microflora on each side of the coupons, the coupons were exposed to UV-C light for 15 min before bacterial inoculation. The prepared coupons were then submerged into 10 mL of TSB in a 50 mL Falcon tube, inoculated with *S. Typhimurium* ATCC 13311 (10^5 Log CFU/mL), and incubated at 37 °C for 24 h under static conditions for experimentation.

2.4. Biofilm Formation Process

The technique was conducted with minor modifications from a previous method [24]. Biofilm inhibition was observed at sub-MIC and MIC, which may not have directly eliminated the bacteria but potentially influenced their virulence factors. The experimental setup included a control (without fucoidan) and fucoidan concentrations of 1/8, 1/4, 1/2, and MIC.

Fucoidan at the designated concentrations (0, 1/8, 1/4, 1/2, and MIC) was added to 10 mL of TSB in 50 mL Falcon tube with food surface samples. A 100 µL suspension of bacteria (10^5 Log CFU/mL) was introduced, and the mixture was homogenized by a vortex mixer (Scientific Industries, SI-0256, Bohemia, NY, USA). The samples were incubated at 37 °C for 24 h to allow biofilm development.

Following biofilm formation, loosely attached bacteria were removed by washing the coupons twice with DW. The surfaces were then immersed in 10 mL of PW (containing

10 sterile glass beads) and vortexed for 2 min to detach adherent bacteria. The recovered bacterial suspension was serially diluted and plated on XLD to assess bacterial colony counts. After incubation at 37 °C for 24 h, CFU were enumerated. Biofilm inhibition was quantified by calculating the colony of bacteria in bacterial populations (Log CFU/cm^2) at each fucoidan concentration compared to the control.

2.5. Biofilm Visualization by Field Emission Scanning Electron Microscopy (FE-SEM)

To visually confirm biofilm inhibition by fucoidan (at concentrations of 0, 1/8, and 1/2 MIC) on crab and shrimp coupons, FE-SEM was employed. The sample preparation procedure was adapted with minor adjustments from the previous protocol [25]. In brief, the crab and shrimp coupons were fixed in 2.5% glutaraldehyde in PBS and stored at room temperature for 4 h. Following fixation, the coupons were subjected to ethanol treatments (50%, 60%, 70%, 80%, and 90%) for 15 min each and two times in 100% ethanol immersions.

The samples were then dehydrated by sequential soaking in increasing concentrations of hexamethyldisilazane (33%, 50%, 66%, and 100% in ethanol) for 15 min each. After dehydration, the samples were air-dried in a fume hood for 3 h and coated with platinum using a sputter coater (Q150T Plus, Quorum, UK) and examined using a FE-SEM (Hitachi/Baltec S-4700, Tokyo, Japan) to observe the biofilm morphology.

2.6. Relative Expression of Gene by Real-Time PCR (RT–PCR)

This test was conducted with minor modifications based on a previously described protocol [26,27] to analyze the impact of fucoidan on the expression of quorum-sensing and virulent genes in *S. Typhimurium* ATCC 13311. Fucoidan-treated bacterial suspensions (10^5 log CFU/mL) were inoculated into a Falcon tube (10 mL of TSB) and incubated at 37 °C for 24 h to facilitate biofilm formation. After the incubation, RNA was extracted from the bacterial pellet using the RNeasy Mini kit (Qiagen, Hilden, Germany), following centrifugation.

The RNA concentrations and purity were assessed using a spectrophotometer (NanoDrop, Bio-Tek Instruments, Chicago, IL, USA) at absorbance ratios of 260/280 nm and 260/230 nm. Complementary DNA (cDNA) was synthesized using the Maxime RT PreMix kit (iNtRON Biotechnology Co., Ltd., Seoul, Gyeonggi-do, Republic of Korea). Primers used for the RT-qPCR analysis are listed in Table 1, with 16S rRNA serving as the housekeeping gene.

Table 1. Names of primers related to virulence and QS.

Target Primers	Sequence (5′–3′)	Product Size (bp)
16S rRNA	F: CAGAAGAAGCACCGGCTAAC R: GACTCAAGCCTGCCAGTTTC	167
rpoS	F: GAATCTGACGAACACGCTCA R: CCACGCAAGATGACGATATG	171
avrA	F: GAGCTGCTTTGGTCCTCAAC R: AATGGAAGGCGTTGAATCTG	173
hilA	F: ATTAAGGCGACAGAGCTGGA R: GCAGAAATGGGCGAAAGTAA	134
LuxS	F: CGGGTTGCAAAAACGATGA R: GTTGAGGTGGTCGCGCATA	150

For quantitative PCR (RT-qPCR), a 20 µL reaction mix containing 1 µL of cDNA, appropriate primers, and Power SYBR Green PCR Master Mix (Applied Biosystems, Thermo Fisher Scientific, Warrington, UK) was prepared. RT-qPCR was performed with the following thermal cycling conditions: initial denaturation at 95 °C for 20 s, followed by annealing at 50 °C for 20 s, and extension at 72 °C for 20 s. Melting curve analysis was performed to confirm specificity, and gene expression was analyzed using the $2^{-\Delta\Delta Ct}$ method.

2.7. Statistical Analysis

Each experiment was performed in triplicate. The results are presented as the mean ± standard error of the mean (SEM). One-way analysis of variance (ANOVA), followed by Duncan's multiple range test, was analyzed to assess significance. Statistical significance was considered as *p*-value (<0.05). The data were evaluated using SAS software version 9.2 (SAS Institute Inc., Cary, NC, USA).

3. Results

3.1. MIC Determination

Fucoidan's inhibitory effect was found to be dose-dependent and varies across different species. MIC was defined as the lowest concentration of fucoidan at which no visible bacterial growth was observed. To evaluate its antimicrobial activity against *S. Typhimurium* ATCC 13311, further experiments were conducted to determine the MIC of fucoidan. With fucoidan concentrations (ranging from 0 to 150 μg/mL), a significant effect on bacterial growth was observed up to 150 μg/mL ($p \geq 0.05$) (Figure 1). Based on these results, the sub-MIC range was established as 31.25–125 μg/mL, which was used in all subsequent experiments. The MIC of fucoidan against *S. Typhimurium* ATCC 13311 was revealed at 150 μg/mL. Sub-MIC and MIC concentrations (1/8, 1/4, 1/2, and MIC) were employed for further investigations.

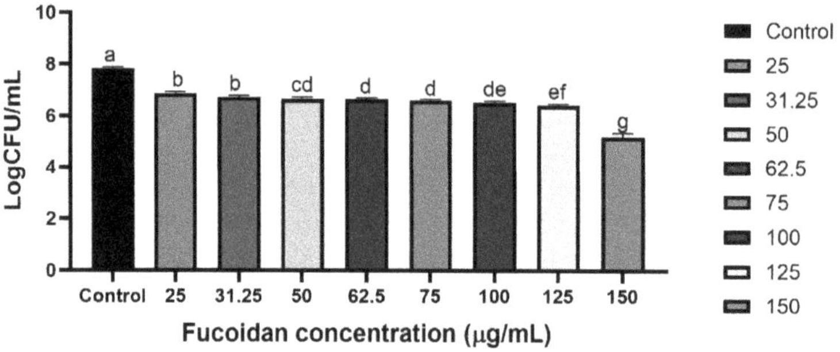

Figure 1. Minimum inhibitory concentrations of Fucoidan against *S. Typhimurium* ATCC 13311. Data are presented as mean ± SEM from three independent experiments. Within each treatment, values with different letters (a–g) indicate significant differences according to Duncan's multiple range test ($p < 0.05$).

3.2. Eradication Impact of Fucoidan on Crab and Shrimp Against S. Typhimurium ATCC 13311

The biofilm formation of *S. Typhimurium* ATCC 13311 on crab and shrimp coupons was effectively inhibited by fucoidan, as demonstrated in Figures 2 and 3, respectively. As the concentration increased, the inhibition of biofilm also showed a corresponding increase. On crab surfaces, the biofilm inhibition values of *S. Typhimurium* ATCC 13311 were 0.12, 1.45, 2.13, and 2.50 log CFU/cm^2 at fucoidan concentrations of 1/8, 1/4, 1/2, and MIC, respectively (Figure 2). Similarly, on shrimp coupons, the biofilm inhibition values were 0.19, 1.36, 2.21, and 2.45 log CFU/cm^2 at fucoidan concentrations of 1/8, 1/4, 1/2, and MIC (Figure 3).

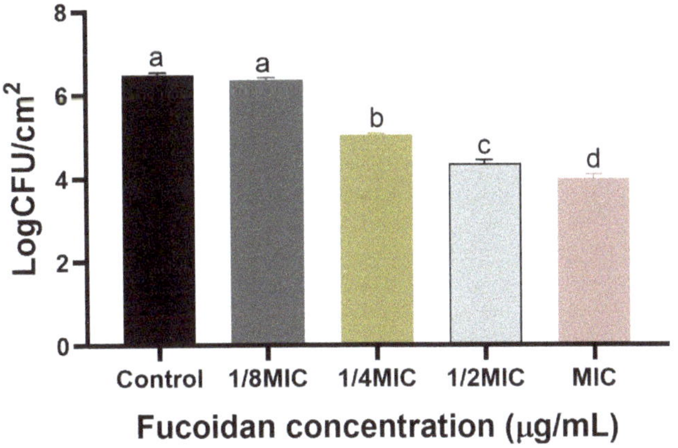

Figure 2. Biofilm formation by *S. Typhimurium* ATCC 13311 on crab coupons in the presence of sub-MIC of fucoidan (0, 1/8, 1/4, 1/2, and MIC) under optimal conditions (37 °C, pH 7.0). Data are expressed as mean ± SEM from three independent replicates. Within each treatment, values with different letters (a–d) indicate significant differences according to Duncan's multiple range test ($p < 0.05$).

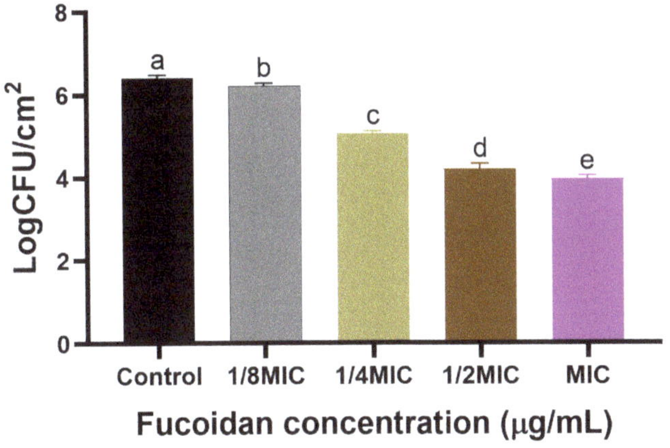

Figure 3. Biofilm formation by *S. Typhimurium* ATCC 13311 on shrimp coupons in the presence of sub-MIC of fucoidan (0, 1/8, 1/4, 1/2, and MIC) under optimal conditions (37 °C, pH 7.0). Data are expressed as mean ± SEM from three independent replicates. Within each treatment, values with different letters (a–e) indicate significant differences according to Duncan's multiple range test ($p < 0.05$).

3.3. Confirmation of Biofilm Reduction Visually by Fucoidan

Figure 4 presents the visual validation of biofilm inhibition by fucoidan. In the control samples, the biofilms exhibited well-organized architectural structures with intact cell-to-cell contacts, with smooth and regular cells having intact cell membranes (Figure 4A,D). In contrast, the fucoidan-treated samples displayed bacterial cells with a rough and irregular appearance, indicating a loss of their normal shape (Figure 4C,F). The red color marks in the control samples (Figure 4A,D) indicate the attachment of biofilm cells, while in the fucoidan-treated groups, the biofilms showed significant lysis, as evidenced by the reduced presence of red-stained cells (Figure 4B,C,E,F).

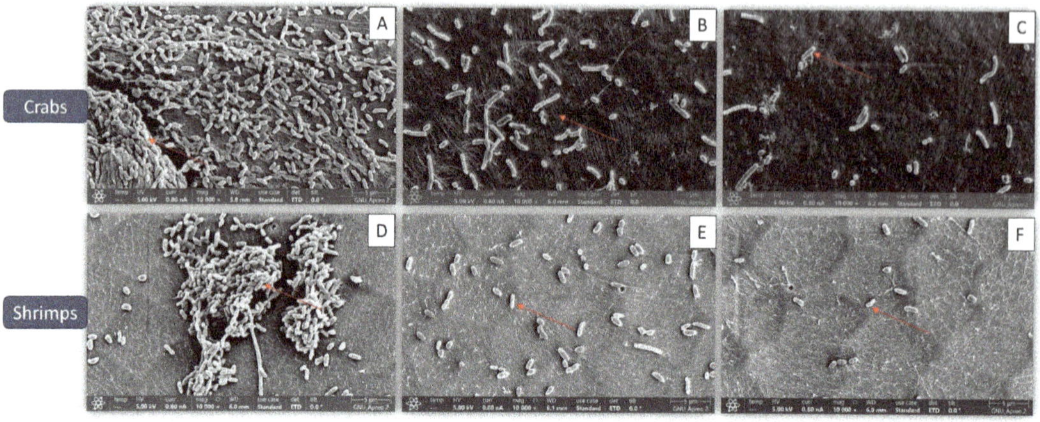

Figure 4. Representative scanning electron micrographs of *S. Typhimurium* ATCC 13311 biofilm formation on crabs and shrimp under optimal conditions in the presence of different sub-MIC concentrations of fucoidan: (**A**) control; (**B**) 1/8 MIC; (**C**) 1/2 MIC for crabs, and (**D**) control; (**E**) 1/8 MIC; (**F**) 1/2 MIC for shrimp.

3.4. Relative Expression Levels

Figure 5 shows the levels of relative expression of virulence (*rpoS*, *avrA*, and *hilA*) as well as the QS gene (*luxS*) in *S. Typhimurium* ATCC 13311, measured by qPCR in the presence of sub-MIC of fucoidan (ranging from 0 to 125 μg/mL). Gene expression was significantly downregulated at all sub-MIC concentrations of fucoidan ($p < 0.05$).

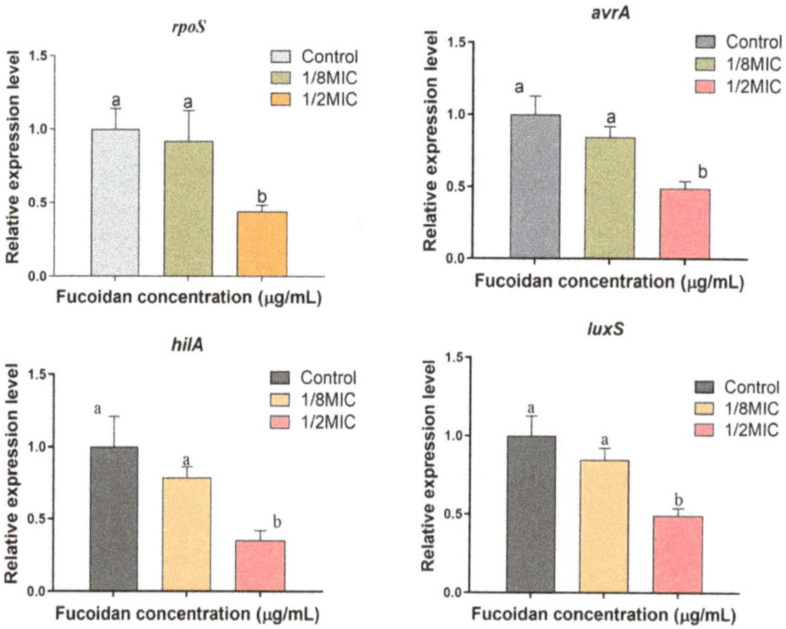

Figure 5. Relative expression levels of *rpoS*, *avrA*, *hilA*, and *luxS* genes in *S. Typhimurium* ATCC 13311 suspension supplemented with different concentrations of fucoidan. Different superscript letters (a, b) indicate significant differences ($p < 0.05$) based on three independent replicates.

4. Discussion

This study explored the efficacy of fucoidan in biofilm inhibition of *S. Typhimurium* ATCC 13311 on crab and shrimp and provided insights into its potential mechanisms of action. The findings demonstrated that fucoidan, even at sub-MICs, significantly reduced biofilm formation and destabilized pre-formed biofilms. Fucoidan, a sulfated polysaccharide derived from *Fucus vesiculosus*, exhibited antimicrobial properties against food-pathogenic bacteria [28,29]. In previous studies, the MICs ranged from 125 to 1000 μg/mL [30]. Fucoidan concentrations exceeding 250 μg/mL were highly effective in inhibiting biofilm formation and planktonic cell growth of *Streptococcus mutans* and *S. sobrinus* [30]. These results underscore the potential application of fucoidan as a natural antimicrobial and antibiofilm agent in the seafood industry.

Biofilm growth contributes to ongoing contamination and resistance to traditional cleaning techniques, it presents serious problems for the seafood sector [31]. According to the current study, fucoidan decreased *S. Typhimurium* ATCC 13311 ability to form biofilms on crabs and shrimp in a concentrate-dependent manner, with significant reductions shown at 1/2 MIC and MIC concentrations (Figure 1). Previous investigations have documented similar antibiofilm activities of fucoidan against various pathogens, such as *Staphylococcus aureus* and *Escherichia coli* [32]. The 150 μg/mL MIC value for *S. Typhimurium* ATCC 13311 found in this investigation is consistent with reports of fucoidan's antibacterial effectiveness against Gram-negative bacteria. For instance, MIC values for *Pseudomonas aeruginosa* and *E. coli* ranged from 100 to 200 μg/mL [33]. Together, these results demonstrate the broad-spectrum potential and dose-dependent nature of fucoidan's antibacterial qualities. Its promise as a food-safe antimicrobial intervention is highlighted by the observed decreases in biofilm-associated bacterial counts, which were as high as 2.50 log CFU/cm^2 on crab coupons and 2.45 log CFU/cm^2 on shrimp coupons. In a previous study, fucoidan F85 was tested at concentrations ranging from 0 to 250 μg/mL. The growth control group in BHI broth increased from approximately 5.2 log CFU/mL (initial cell count) to around 8.6 log CFU/mL after 18 h, maintaining this level with a slight decline of about 1 log cycle after 48 h. In contrast, *S. mutans* treated with fucoidan at 125 μg/mL showed no growth, while *S. mutans* treated with fucoidan F85 at 250 μg/mL exhibited a consistent decrease, reaching approximately 3.3 log CFU/mL after 48 h [30]. FE-SEM imaging provides evidence of fucoidan's mode of action by identifying notable structural anomalies in individual bacterial cells and the biofilm matrix. The smooth and undamaged looks of untreated control cells were sharply contrasted with the rough and uneven surfaces of bacterial cells in fucoidan-treated samples. These results align with earlier research showing that fucoidan breaks down the EPS matrix, which is essential for preserving the structure of biofilms and promoting bacterial adhesion [34]. In addition to this experiment, virulence-associated genes (*rpoS, hilA, avrA*, and *luxS*) were shown to be significantly downregulated in *S. Typhimurium* ATCC 13311 biofilms treated with fucoidan. These genes are important for bacterial communication, biofilm control, and pathogenicity; fucoidan may interfere with these functions to prevent the formation of biofilms [35]. In the present study, the application of sub-inhibitory concentrations of fucoidan significantly reduced biofilm formation on various seafood coupons. This antibiofilm activity may be attributed to the disruption of QS signaling and EPS production. Gene expression analysis revealed significant downregulation of *luxS* in *S. Typhimurium* ATCC 13311, indicating a possible QS-inhibitory effect of fucoidan. Additionally, genes associated with virulence and stress response (*hilA, avrA, rpoS*) were also suppressed, further suggesting that fucoidan interferes with key regulatory pathways involved in biofilm development and maintenance.

By inhibiting QS and reducing EPS synthesis, fucoidan compromises biofilm integrity and diminishes the protective advantages conferred by the biofilm matrix. These findings highlight fucoidan's potential as a natural QS and EPS-targeting agent, offering an effective strategy to combat persistent biofilms in food industry settings.

All of the observed decreases in gene expression point to fucoidan's complex antibiofilm mechanisms. Inhibiting biofilm formation and destabilizing pre-existing biofilms, fucoidan targets quorum-sensing, stress response, and virulence pathways. These results are similar to previous studies that suggested natural polysaccharides might function as potent antibiofilm agents by interfering with molecular pathways [36,37].

The seafood industry faces significant challenges with biofilm-associated contamination, which poses serious food safety and public health concerns. According to this study, fucoidan shows promise as a natural remedy for these problems. Fucoidan, which comes from brown algae, has several benefits over conventional chemical disinfectants, such as being naturally derived, having minimal toxicity, and being environmentally friendly [28,38]. Additionally, its effectiveness at sub-MIC levels increases its applicability by cutting costs in industrial applications and lessening the danger of the development of antibiotic resistance [39]. Sensory evaluation was conducted on butter samples from both control and treatment groups, including varying fucoidan concentrations over different storage durations [40]. The results demonstrated that butter samples containing 0.5% fucoidan exhibited significantly reduced acid and peroxide values, extending shelf life as supported by sensory assessments [41]. No alterations in taste were detected in any samples on day zero, and even at higher fucoidan concentrations, no adverse effects on flavor were observed. An increase in fucoidan concentration resulted in slightly lighter butter color and enhanced texture characteristics, such as improved softness and spreadability. By day 40, a slight increase in rancid flavor was noted across all samples; however, the control group displayed the most pronounced rancidity, whereas the samples treated with the highest fucoidan concentrations maintained acceptable sensory quality without unpleasant organoleptic changes [42]. Shelf life, defined as the duration during which a food product remains acceptable in terms of safety, nutrition, and sensory attributes, was notably prolonged in the fucoidan-treated samples.

However, this study also points to several limitations. The effectiveness of fucoidan has to be confirmed in actual seafood processing settings, as the trials were carried out in a controlled laboratory setting. To guarantee its usefulness in the sector, research should also be carried out on its affordability and any effects on the sensory aspects of fish products. While this study highlights the promising antibiofilm and antimicrobial properties of fucoidan against *S. Typhimurium* ATCC 13311 on seafood coupons, translating these findings into industrial application necessitates careful consideration of several practical factors. Regulatory approval is a foremost concern; although fucoidan is derived from edible brown seaweed and is generally regarded as safe (GRAS) in some jurisdictions, its application as an antimicrobial agent in food processing environments would require compliance with food safety regulations governed by authorities such as the FDA, EFSA, or regional food safety agencies. Comprehensive toxicological assessments and efficacy validations under industrial conditions would be essential for its approval.

Moreover, sensory impacts on seafood products must be thoroughly evaluated. Even though fucoidan is considered low in toxicity and natural in origin, its potential influence on the organoleptic properties—including taste, texture, aroma, and appearance—of treated seafood could affect consumer acceptability. It is imperative that sensory testing be integrated into future studies to ensure the maintenance of product quality post-treatment.

Formulation and application challenges also merit close attention. Fucoidan's stability under various processing conditions—such as temperature, pH, and exposure to other food-grade chemicals—must be addressed. Effective delivery mechanisms, such as incorporation into edible coatings, sprays, or wash solutions, should be optimized for uniform application and maximal biofilm inhibition. Furthermore, large-scale production of standardized fucoidan with consistent bioactivity is necessary to ensure batch-to-batch reproducibility, a requirement for commercial scalability.

In addition, although fucoidan holds significant promise as a natural, eco-friendly antimicrobial in the seafood industry, its practical implementation demands multi-faceted exploration involving regulatory, sensory, and formulation studies. Addressing these challenges through interdisciplinary research will be critical to unlocking fucoidan's full potential in enhancing food safety and reducing microbial contamination.

Fucoidan's synergistic effects with other natural antimicrobials should be investigated in future studies to increase its effectiveness against infections linked with biofilms.

The study's results imply that fucoidan may be used as a safe and efficient natural antibiofilm agent to combat S. *Typhimurium* ATCC 13311 on seafood coupons. It offers a viable method for enhancing food safety and minimizing microbial contamination in the seafood sector because of its capacity to prevent the creation of new biofilms, decrease existing biofilms, and alter quorum-sensing pathways. To maximize its usage and broaden its application to additional foodborne diseases and industrial settings, more research is necessary.

5. Conclusions

Fucoidan demonstrates strong antimicrobial and antibiofilm activity against S. *Typhimurium* ATCC 13311 on seafood coupons, significantly reducing bacterial counts by 2.50 log CFU/cm^2 on crabs and 2.45 log CFU/cm^2 on shrimp at sub-MIC levels. FE-SEM analysis confirmed its ability to disrupt the EPS matrix, likely by interfering with quorum sensing and virulence gene expression. With its natural origin, low toxicity, and effectiveness at sub-MIC levels (31.25–125 μg/mL), fucoidan presents a sustainable alternative to chemical disinfectants in the seafood industry. Further research is needed to assess its industrial application, potential synergistic effects, and impact on product quality, making it a promising tool for enhancing food safety and reducing microbial contamination.

Author Contributions: Conceptualization, A.R.; methodology, A.R. and P.K.R.; software, A.R.; validation, A.R. and P.K.R.; formal analysis, A.R. and P.K.R.; investigation, A.R.; resources, A.R.; data curation, A.R. and P.K.R.; writing—original draft preparation, A.R. and P.K.R.; writing—review and editing, A.R. and P.K.R.; visualization, A.R.; supervision, S.Y.P.; project administration, S.Y.P.; funding acquisition, S.R.C. and S.Y.P. All authors have read and agreed to the published version of the manuscript.

Funding: This study was supported by the Basic Science Research Program through the National Research Foundation of Korea (NRF) funded by the Ministry of Education, grand number 2021R1I1A3A04037468. This research was also supported by the National Institute of Fisheries Science, Ministry of Oceans and Fisheries, Republic of Korea (R2025055).

Institutional Review Board Statement: Not applicable.

Informed Consent Statement: Not applicable.

Data Availability Statement: The original contributions presented in this study are included in the article. Further inquiries can be directed to the corresponding authors.

Conflicts of Interest: The authors declare no conflicts of interest.

References

1. Roy, P.K.; Ha, A.J.; Nahar, S.; Hossain, M.I.; Ashrafudoulla, M.; Toushik, S.H.; Mizan, M.F.R.; Kang, I.; Ha, S.D. Inhibitory effects of vorinostat (SAHA) against food-borne pathogen *Salmonella enterica* serotype Kentucky mixed culture biofilm with virulence and quorum-sensing relative expression. *Biofouling* **2023**, *39*, 617–628. [CrossRef] [PubMed]
2. Dawan, J.; Zhang, S.; Ahn, J. Recent Advances in Biofilm Control Technologies for the Food Industry. *Antibiotics* **2025**, *14*, 254. [CrossRef] [PubMed]
3. Roy, P.K.; Kim, S.H.; Jeon, E.B.; Park, E.H.; Park, S.Y. Inhibition of *Listeria monocytogenes* Cocktail Culture Biofilms on Crab and Shrimp Coupons and the Expression of Biofilm-Related Genes. *Antibiotics* **2023**, *12*, 1008. [CrossRef] [PubMed]
4. Cervantes-Huamán, B.R.H.; Vega-Sánchez, A.; Rolón-Verdún, P.; Gervilla-Cantero, G.; Rodríguez-Jerez, J.J.; Ripolles-Avila, C. Effect of Cinnamomum cassia essential oil combined with enzymes on the elimination and regrowth potential of *Listeria monocytogenes* and *Salmonella enterica* biofilms formed on stainless steel surfaces. *Food Control* **2025**, *172*, 111120. [CrossRef]
5. Yang, P.; Huo, Y.; Yang, Q.; Zhao, F.; Li, C.; Ju, J. Synergistic anti-biofilm strategy based on essential oils and its application in the food industry. *World J. Microbiol. Biotechnol.* **2025**, *41*, 81. [CrossRef]
6. Pelyuntha, W.; Yamik, D.Y.; Vetboocha, N.; Vongkamjan, K. Effect of novel phage cocktail on *Salmonella* recovered from broiler sources and its anti-biofilm effect on food contact surface model. *Food Control* **2025**, *169*, 111000. [CrossRef]
7. CDC. Salmonella Infection (Salmonellosis). Available online: https://www.cdc.gov/salmonella/ (accessed on 15 March 2025).
8. Kim, Y.K.; Roy, P.K.; Ashrafudoulla, M.; Nahar, S.; Toushik, S.H.; Hossain, M.I.; Mizan, M.F.R.; Park, S.H.; Ha, S.-D. Antibiofilm effects of quercetin against *Salmonella enterica* biofilm formation and virulence, stress response, and quorum-sensing gene expression. *Food Control* **2022**, *137*, 108964. [CrossRef]
9. Giaouris, E.; Habimana, O. The Battle Against Biofilms: A Focus on Novel Antimicrobial Strategies and Their Mechanisms of Action. *Antibiotics* **2025**, *14*, 111. [CrossRef]
10. Rohilla, A.; Kumar, V.; Ahire, J.J. Unveiling the persistent threat: Recent insights into *Listeria monocytogenes* adaptation, biofilm formation, and pathogenicity in foodborne infections. *J. Food Sci. Technol.* **2024**, *61*, 1428–1438. [CrossRef]
11. Alsuwat, M.A.; Shah, A.A.; Ullah, S.; Khan, R.U.; Alissa, M.; Khan, M.S. Microbial Biofilm Formation to Mitigate Foodborne Pathogens Strategies and Control Measures. *Indian J. Microbiol.* **2025**, 1–6. [CrossRef]
12. Tabassum, N.; Khan, F.; Kang, M.G.; Jo, D.M.; Cho, K.J.; Kim, Y.M. Inhibition of Polymicrobial Biofilms of *Candida albicans Staphylococcus aureus/Streptococcus mutans* by Fucoidan-Gold Nanoparticles. *Mar. Drugs* **2023**, *21*, 123. [CrossRef] [PubMed]
13. Zhang, S.; Chen, L.; Shang, N.; Wu, K.; Liao, W. Recent Advances in the Structure, Extraction, and Biological Activity of Sargassum fusiforme Polysaccharides. *Mar. Drugs* **2025**, *23*, 98. [CrossRef]
14. Vladkova, T.G.; Staneva, A.D.; Avramova, I.A.; Ivanova, I.A.; Gospodinova, D.N. Fucoidan-Containing, Low-Adhesive Siloxane Coatings for Medical Applications: Inhibition of Bacterial Growth and Biofilm Development. *Materials* **2023**, *16*, 3651. [CrossRef]
15. Nazari, M.; Taheri, M.; Nouri, F.; Bahmanzadeh, M.; Alikhani, M.Y. The antimicrobial and antibiofilm effects of gentamicin, imipenem, and fucoidan combinations against dual-species biofilms of *Staphylococcus aureus* and *Acinetobacter baumannii* isolated from diabetic foot ulcers. *Ann. Clin. Microbiol. Antimicrob.* **2024**, *23*, 101. [CrossRef] [PubMed]
16. Duan, X.; Agar, O.T.; Barrow, C.J.; Dunshea, F.R.; Suleria, H.A.R. Improving potential strategies for biological activities of phlorotannins derived from seaweeds. *Crit. Rev. Food Sci. Nutr.* **2025**, *65*, 833–855. [CrossRef]
17. Azeem, K.; Fatima, S.; Ali, A.; Ubaid, A.; Husain, F.M.; Abid, M. Biochemistry of Bacterial Biofilm: Insights into Antibiotic Resistance Mechanisms and Therapeutic Intervention. *Life* **2025**, *15*, 49. [CrossRef]
18. Jeong, G.J.; Khan, F.; Tabassum, N.; Cho, K.J.; Kim, Y.M. Strategies for controlling polymicrobial biofilms: A focus on antibiofilm agents. *Int. J. Antimicrob. Agents* **2024**, *64*, 107243. [CrossRef] [PubMed]
19. Farrukh, M.; Munawar, A.; Nawaz, Z.; Hussain, N.; Hafeez, A.B.; Szweda, P. Antibiotic resistance and preventive strategies in foodborne pathogenic bacteria: A comprehensive review. *Food Sci. Biotechnol.* **2025**, 1–29. [CrossRef]
20. Elkhalifa, A.M.E.; Nazar, M.; Ali, S.I.; Khursheed, I.; Taifa, S.; Ahmad Mir, M.; Shah, I.H.; Malik, M.; Ramzan, Z.; Ahad, S.; et al. Novel Therapeutic Agents for Management of Diabetes Mellitus: A Hope for Drug Designing against Diabetes Mellitus. *Life* **2024**, *14*, 99. [CrossRef]
21. Anestopoulos, I.; Kiousi, D.E.; Klavaris, A.; Maijo, M.; Serpico, A.; Suarez, A.; Sanchez, G.; Salek, K.; Chasapi, S.A.; Zompra, A.A.; et al. Marine-derived surface active agents: Health-promoting properties and blue biotechnology-based applications. *Biomolecules* **2020**, *10*, 885. [CrossRef]
22. McGurrin, A.; Suchintita Das, R.; Soro, A.B.; Maguire, J.; Flórez Fernández, N.; Dominguez, H.; Torres, M.D.; Tiwari, B.K.; Garcia-Vaquero, M. Antimicrobial Activities of Polysaccharide-Rich Extracts from the Irish Seaweed *Alaria esculenta*, Generated Using Green and Conventional Extraction Technologies, Against Foodborne Pathogens. *Mar. Drugs* **2025**, *23*, 46. [CrossRef]
23. Roy, P.K.; Song, M.G.; Jeon, E.B.; Kim, S.H.; Park, S.Y. Antibiofilm Efficacy of Quercetin against *Vibrio parahaemolyticus* Biofilm on Food-Contact Surfaces in the Food Industry. *Microorganisms* **2022**, *10*, 1902. [CrossRef]
24. Kim, S.H.; Roy, P.K.; Park, S.Y. Synergistic Effects of Combined Flavourzyme and Floating Electrode-Dielectric Barrier Discharge Plasma on Reduction of *Escherichia coli* Biofilms in Squid (*Todarodes pacificus*). *Microorganisms* **2024**, *12*, 1188. [CrossRef]

25. Ashrafudoulla, M.; Park, J.; Toushik, S.H.; Shaila, S.; Ha, A.J.W.; Rahman, M.A.; Park, S.H.; Ha, S.D. Synergistic mechanism of UV-C and postbiotic of *Leuconostoc mesenteroides* (J. 27) combination to eradicate *Salmonella Thompson* biofilm in the poultry industry. *Food Control* **2024**, *164*, 110607. [CrossRef]
26. Roy, P.K.; Kim, G.; Fang, X.; Hassan, B.; Soysa, M.D.; Shin, S.T.; Cho, J.K. Optimization of post-activation systems to improve the embryonic development in porcine parthenogenesis and somatic cell nuclear transfer. *J. Embryo Transf.* **2017**, *32*, 95–104. [CrossRef]
27. Roy, P.K.; Fang, X.; Hassan, B.; Shin, S.T.; Cho, J.K. Effects of roscovitine on in vitro development of porcine oocyte using brilliant cresyl blue. *J. Embryo Transf.* **2017**, *32*, 111–122. [CrossRef]
28. Cao, J.; Qin, L.; Liu, M.; Yao, M.; Wang, K.; Lin, H.; Qu, C.; He, Y.; Xue, C.; Miao, J. Fucoidan from sea cucumber cooking liquid: Structural analysis, physicochemical properties, and anti-Helicobacter pylori potential. *Int. J. Biol. Macromol.* **2025**, *306*, 141593. [CrossRef]
29. Mensah, E.O.; Kanwugu, O.N.; Panda, P.K.; Adadi, P. Marine fucoidans: Structural, extraction, biological activities and their applications in the food industry. *Food Hydrocoll.* **2023**, *142*, 108784. [CrossRef]
30. Jun, J.Y.; Jung, M.J.; Jeong, I.H.; Yamazaki, K.; Kawai, Y.; Kim, B.M. Antimicrobial and Antibiofilm Activities of Sulfated Polysaccharides from Marine Algae against Dental Plaque Bacteria. *Mar. Drugs* **2018**, *16*, 301. [CrossRef]
31. Gemba, M.; Rosiak, E.; Kołożyn-Krajewska, D. Development of predictive models of biofilm formation by *C. sakazakii*, *E. cloacae* on surfaces used in the food industry and medicine. *Int. J. Food Microbiol.* **2025**, *434*, 111131. [CrossRef]
32. Rajasekaran, J.; Viswanathan, P. Anti-bacterial and antibiofilm properties of seaweed polysaccharide-based nanoparticles. *Aquac. Int.* **2023**, *31*, 2799–2823. [CrossRef]
33. Khan, F.; Manivasagan, P.; Lee, J.-W.; Pham, D.T.N.; Oh, J.; Kim, Y.-M. Fucoidan-stabilized gold nanoparticle-mediated biofilm inhibition, attenuation of virulence and motility properties *in Pseudomonas aeruginosa* PAO1. *Mar. Drugs* **2019**, *17*, 208. [CrossRef] [PubMed]
34. Roy, P.K.; Mizan, M.F.R.; Hossain, M.I.; Han, N.; Nahar, S.; Ashrafudoulla, M.; Toushik, S.H.; Shim, W.-B.; Kim, Y.-M.; Ha, S.-D. Elimination of *Vibrio parahaemolyticus* biofilms on crab and shrimp surfaces using ultraviolet C irradiation coupled with sodium hypochlorite and slightly acidic electrolyzed water. *Food Control* **2021**, *128*, 108179. [CrossRef]
35. Han, S.; Choi, M.W.; Byun, K.H.; Kim, B.H.; Song, M.S.; Kang, I.; Ha, S.D. Characterization of *Salmonella* ser. Enteritidis-specific bacteriophages and biocontrol strategy to reduce *S. Enteritidis* on egg products using bacteriophages and essential oil compounds. *Food Control* **2024**, *160*, 110304. [CrossRef]
36. Shivaprasad, D.P.; Kaushik, A.; Taneja, N.K.; Lakra, A.; Bharadwaj, D.K.; Juneja, V.K.; Taneja, P.; Chauhan, K.; Oberoi, H.S. Breaking the biofilm barrier: Vitamin C as a novel strategy against multidrug-resistant *Salmonella Typhimurium*. *Int. J. Food Sci. Technol.* **2025**, *60*, vvae082. [CrossRef]
37. Lawal, H.; Saeed, S.I.; Gaddafi, M.S.; Kamaruzzaman, N.F. Green Nanotechnology: Naturally Sourced Nanoparticles as Antibiofilm and Antivirulence Agents Against Infectious Diseases. *Int. J. Microbiol.* **2025**, *2025*, 8746754. [CrossRef]
38. Galinskaitė, A.; Gruškienė, R.; Kavleiskaja, T.; Stanevičienė, R.; Servienė, E.; Sereikaitė, J. Bioactive Fucoidan-Based Three-Component Colloidal Particles for Food Safety. *Food Bioprocess Technol.* **2025**, 1–13. [CrossRef]
39. Karthikeyan, A.; Thirugnanasambantham, M.K.; Khan, F.; Mani, A.K. Bacteria-Inspired Synthesis of Silver-Doped Zinc Oxide Nanocomposites: A Novel Synergistic Approach in Controlling Biofilm and Quorum-Sensing-Regulated Virulence Factors in *Pseudomonas aeruginosa*. *Antibiotics* **2025**, *14*, 59. [CrossRef]
40. Alipour, A.; Marhamatizadeh, M.H.; Mohammadi, M. Studying the shelf life of butter containing fucoidan, by evaluating sensory and chemical properties. *Food Sci. Nutr.* **2023**, *11*, 2956–2963. [CrossRef]
41. Augustyńska-Prejsnar, A.; Ormian, M.; Sokołowicz, Z.; Kačániová, M. The effect of the addition of hemp seeds, Amaranth, and Golden flaxseed on the nutritional value, physical, sensory characteristics, and safety of poultry Pâté. *Appl. Sci.* **2022**, *12*, 5289. [CrossRef]
42. Pittia, P.; Heer, M. Space food for the future: Nutritional challenges and technological strategies for healthy and high-quality products. In *In-Space Manufacturing and Resources: Earth and Planetary Exploration Applications*; John Wiley & Sons, Inc.: Hoboken, NJ, USA, 2022; Volume 58, pp. 251–268.

microorganisms

MDPI

Review

Probiotic Potential of Traditional and Emerging Microbial Strains in Functional Foods: From Characterization to Applications and Health Benefits

Chijioke Christopher Uhegwu [1,2] and Christian Kosisochukwu Anumudu [1,3,*]

1 Microbiology Unit, Bioscience Department, Federal University Otuoke, Otuoke 562103, Nigeria; chijiokeuhegwu@gmail.com
2 Bioinformatics and Genomics Research Unit, Genomac Institute, Ogbomosho 210213, Nigeria
3 School of Chemical Engineering, University of Birmingham, Birmingham B15 2TT, UK
* Correspondence: cka329@student.bham.ac.uk

Abstract: Global consumer demand for probiotic-enriched functional foods has increased as consumers become increasingly aware of the connection between what they eat and its role in their long-term health. Compared with conventional foods that primarily deliver fundamental nutrients, functional foods include biologically active compounds capable of influencing physiological processes. While traditionally used probiotic strains like *Lactobacillus* and *Bifidobacterium* are still at the center of this trend, there is growing interest in the exploration of emerging and novel microbial candidates that harbor new functional properties. This review addresses the characterization, modes of action, technological limitations, regulatory guidelines, and prospective health benefits of new probiotic strains in functional foods. The review further highlights the need for precise strain selection, novel encapsulation technologies for viability, and strict safety assessments in accordance with EFSA's QPS (Qualified Presumption of Safety) and the United States FDA GRAS (Generally Recognized As Safe) specifications. Current research focuses on the classical benefits of probiotics, including gut microbiota modulation, immunomodulation, antimicrobial activity, lowering of cholesterol, and mental health. However, long-term clinical validation, strain specificity, personalized application, and effective communication to consumers are some areas where gaps remain. Addressing these challenges through the incorporation of omics technologies, synthetic biology, and more detailed microbiome–host interaction studies will be the key to unlocking the full potential of next-generation probiotics and sustaining consumer trust in this emerging market.

Keywords: probiotics; functional foods; emerging microbial strains; fermented foods; gut microbiome

1. Introduction

1.1. Background

Functional foods are foods that have been scientifically established to deliver a positive impact on one or more specific physiological functions in the human body, other than simply providing nutritional benefits. These impacts can be towards considerably enhanced health, welfare, or lower risk of developing certain diseases [1]. With the increasing public awareness of the link between nutrition and health, there is more global momentum toward the use of food in disease prevention and health promotion. Consumers are also becoming more interested in dietary products that promote immune function, gastrointestinal health, and overall well-being. This is especially seen in the growing popularity of functional foods and dietary supplements enriched with bioactive components [2]. The concept of functional foods initially gained traction in Japan in the 1980s and was formally established by the Japanese Ministry of Health and Welfare in 1991 through the Foods for Specified Health Uses (FOSHU) system [3]. The system created an effective legal pathway for foods to carry approved health claims, provided they are backed by robust scientific and clinical evidence. Since then, the concept has expanded globally as consumers increasingly seek food products that actively support their health and lifestyle goals. Since diet is considered to be closely linked to several disease conditions and significant age-related chronic disease states, researchers and the food industry have turned their attention to functional foods that can provide more than calories and nutrients but also bioactive compounds with proven health impacts [1,4]. Similarly, consumers are increasingly becoming aware of the connection between what they eat and their long-term health conditions. Thus, many consumers are shifting from conventional foods that primarily deliver fundamental nutrients such as carbohydrates, proteins, fats, vitamins, and minerals to functional foods which contain biologically active compounds capable of influencing physiological processes such as probiotic yogurts, beverages with supplemented plant sterols, or prebiotic fibers. These foods have been found to contain vitamins, minerals, dietary fiber, antioxidants, phytochemicals, and bioactive peptides which are beneficial to human health [5,6]. This has fueled more interest in the consumption of functional foods as they help in disease prevention, immunity, digestive health, and other targeted health goals [7]. This increasing demand for functional foods has given rise to a rapidly expanding market for these products and this continues to grow. The combined global market value of functional foods and beverages was approximately USD 258 billion in 2021 and is expected to rise to USD 530 billion by the year 2028 [8], indicating the vast potential and business opportunity for developing new products, bridging the gap between nutrition and well-being.

Of all the bioactive compounds used to enhance the health attributes of functional foods, probiotics are of utmost significance. Probiotics are live microorganisms including bacteria and yeast which are incorporated into functional foods for consumption or application on the body to provide some health benefits [9]. According to Hill et al. [10], probiotics must be "safe for their intended use" and have "defined contents, appropriate viable count at end of shelf life, and suitable evidence for health benefits." The International Scientific Association of Probiotics and Prebiotics (ISAPP) reaffirmed these ideas in a policy statement in 2018 [11]. Globally, probiotics rank as the fastest emerging class of dietary functional food supplements, and a growing body of research continues to indicate their extensive benefits to health [12]. Their overall acceptability comes from their recognized role of maintaining digestive and immunological well-being. Currently, probiotics are consumed regularly not only in yogurt but also in cheeses, beverages, cereals, and other emerging food delivery vehicles [13]. These probiotics, especially bacteria have been demonstrated to have potential such as neutralizing toxins [14], enhancing sensorial appeal when microencapsulated [15], and having improved survival for transit through the hostile environment of the gastrointestinal tract [16]. Several approaches are employed in introducing and maintaining probiotics in foods. Newer fortification strategies have been implemented to improve the potential and usage of probiotics. They have been utilized as additives in green tea fortified with soy [17], as added fortifications with vitamin D in products for weight management [18], incorporated in fruit juices to increase their health promotion [19], or as prophylactic treatments for cancer [20]. The global sales of probiotic supplements has continued to rise and was estimated at USD 96 billion in 2020 [1].

While the strains of *Lactobacillus* and *Bifidobacterium* have traditionally been the most widely utilized probiotics based on their established safety and effectiveness, scientific advancements and consumer interest have increased the exploration of new microbial strains as potential probiotics from diverse sources, including plants, fruits, and fermented foods. Hence, this review provides an overview of the current state of emerging probiotics in functional foods, from strain selection and characterization to their mechanisms of action, regulatory considerations, technological challenges, and future perspectives for research and applications.

1.2. Sources of Novel and Conventional Probiotic Strains

Probiotic strains, whether well-known or novel, can be isolated from a wide range of natural or traditional sources. Over the years, several strains of yeasts and LAB with techno-functional properties have been isolated from ethnic beverages, unique sources such as plants, animal hosts, and marine ecosystems [21]. Table 1 summarizes different sources of both conventional and emerging probiotic candidates and their main functional features.

Table 1. Representative conventional and emerging probiotic species, their isolation sources, and key functional properties.

Species/Strains	Isolation Source	Key Functional Traits	References
Weissella confusa MD1 and *Weissella cibaria* MD2	Fermented batter, traditional fermented foods such as horreh	Co-aggregation with pathogens; lysozyme and acid tolerance; cholesterol reduction; antimicrobial potential; exopolysaccharide production; antioxidant activity.	[22,23]
Bacillus coagulans	Milk	Spore-forming abilities, gastrointestinal survivability; enzyme production; heat tolerance.	[24]
Lactobacillus plantarum (e.g., SY11, SY12, AAS3)	Kimchi, dry fish based fermented food	Acid and bile tolerance; adhesion to intestinal cells; cholesterol reducing property; antimicrobial and antioxidant activity.	[25,26]
L. paraplantarum (e.g., SC61)	Jangajii (fermented vegetable)	Antioxidant and immunostimulatory activity; stability in artificial gastric and bile conditions, non-production of β-glucuronidase, suitable antibiotic susceptibility, and attachment to intestinal cells.	[27]
Weissella hellenica BCC 7239	Nham (fermented pork sausage)	Production of bacteriocins, bactericidal effects against both Gram-positive and Gram-negative organisms	[28]
Fructobacillus fructosus MCC 3996	Flower nectar	Resistance to gastric conditions; co-aggregation with pathogens, hydrophobicity, and the absence of hemolytic activity.	[29]
Saccharomyces cerevisiae (e.g., KU200270, KU200280, and KU200284)	Cucumber *jangajji* and other fermented foods	Antioxidative properties; gastric and bile resistance; adhesion to epithelial cells.	[30]
Aureobasidium pullulans (e.g., Y39, Y40, Y41, Y43)	Kalamata table olive	Auto-aggregation ability; hydrophobicity; adhesion to Caco-2 cells; absence of hemolytic activity.	[31]
Lacticaseibacillus casei (e.g., SB71, SB73 and SB93)	Marine ecosystem	Inability to form biogenic amines; adherence to Caco-2 cells; cholesterol assimilation; and tolerance to NaCl, bile and low pH.	[32]
Leuconostoc (*citreum* and *mesenteroides* subsp. *mesenteroides*)	Traditional fermented foods such as Horreh	Exopolysaccharide production; antioxidant activity; acid tolerance	[23]
Pediococcus pentosaceus	Traditional fermented foods	Exopolysaccharide production; antioxidant activity; acid tolerance	[23]
Enterococcus (*faecium* and *faecalis*)	Traditional fermented foods (e.g., Kimchi, Horreh)	Absence of antibiotic resistance or virulence factors; auto aggregation ability; hydrophobicity; resistance to gastrointestinal conditions.	[23,33]
Akkermansia muciniphila	Human intestinal microbiota	Mucin degradation; modulation of host metabolism; gut barrier reinforcement	[34]
Faecalibacterium prausnitzii	Human gut	Butyrate production; anti-inflammatory and gut-protective effects	[35]

1.3. Conventional vs. Emerging Probiotic Strains

For several decades, the global probiotic industry has relied on a relatively small number of well-studied, widely accepted microbial species. Those relevant to the human gut microbiota among them are majorly members of the genera *Lactobacillus* (now reallocated to many new genera such as *Lactiplantibacillus*, *Ligilactobacillus*, *Lacticaseibacillus* and *Limosilactobacillus*) and *Bifidobacterium* [21,36]. These bacteria are naturally found in the human gut and fermented foods and have a long record of safe ingestion backed up by a strong body of scientific evidence [21,37]. Some of the common examples include strains of *Lactobacillus acidophilus*, *Lacticaseibacillus rhamnosus* GG, *Limosilactobacillus reuteri*, *Bifidobacterium bifidum*, *Bifidobacterium longum*, *Bifidobacterium animalis* subsp. *lactis*, etc. These strains have reportedly been characterized by survival under stomach acidic conditions, ability to colonize the intestinal epithelium, ability to compete with and inhibit pathogens thereby keeping the gut healthy, producing essential metabolites such as short-chain fatty acids, and modulating the host's immune system [2,37]. Due to their proven efficacy and safety, they have been used over the years in different products such as in fermented foods, supplements, infant formula, and even as therapeutic treatments of gastrointestinal and immune diseases [21,31,38].

While milk and other dairy products continue to be the major vehicle for probiotics delivery to consumers, interest in their non-dairy probiotic food counterparts including cereals, fruits, and vegetables has expanded over the last few decades [39]. This is primarily due to changing food habits, i.e., the growing rate of vegetarianism and veganism, and the accelerated rate of milk allergy, cholesterol, and lactose intolerance which limits the consumption of dairy products by many [5]. Identification of new strains suitable for these foods and niche markets opens the door to more opportunities for functional foods innovation. However, it is noteworthy that most of the probiotic cultures used in food products commercially available today are derived from human gut microbiota or traditional dairy fermentations [21]. The dairy sector which is strongly associated with probiotics, is the largest functional food market, accounting for nearly 33% of the broad market, while cereal products have just over 22% [13]. For stable supplementation of probiotics into plant or cereal-based foods, strains that come directly from plants or plant fermentations tend to be better adapted to thrive. The use of strains well adapted to these substrates is more effective as non-dairy food matrices have particular technological, environmental and physiological challenges, that typical dairy or gut-derived probiotic strains would not be able to survive in optimally [40]. Thus, the optimization of novel probiotic strains to survive, be potent and viable in non-dairy food product matrixes is critical in the development of new functional foods that cater to different dietary needs. Technological advancements have made it feasible to modify food components in a regulated manner, such as modifying food components, among other methods, to change specific structural aspects of foods such as fruit and vegetables [41], making them conducive for probiotic supplementation. These foods are nutrient-rich and contain essential minerals, vitamins, dietary fibers, and antioxidants which makes them perfect substrates for probiotics culture [21]. The idea of using fewer milk components as vehicles of probiotic agents or even substituting milk with other media, like cereals, fruits, and vegetables, is also encouraged by traditions and financial considerations that restrict the use of dairy products due to religious, raw material unavailability, and other practical reasons [13]. Hence, it has become crucial to leverage the numerous advancements in scientific innovation and enhanced screening technologies like genomics, metagenomics, innovative culturing techniques and metabolomics, to optimize the customization of probiotic products using newer strains and food matrices for specific nutritional needs, disease prevention, and even adjunct therapy for conditions like obesity, diabetes, and inflammatory bowel disease. To provide a clearer perspective of the current frontiers of probiotic research, a conceptual comparison between conventional, well-established probiotics and newly emerging candidates that exhibit probiotic-like properties but are still under safety and regulatory evaluation is presented herein (Table 2).

Table 2. Conceptual Comparison between Conventional Probiotics and Emerging Probiotic Candidates.

Criteria	Conventional Probiotics (e.g., *Lactobacillus*, *Bifidobacterium*)	Emerging Probiotics (e.g., *Akkermansia muciniphila*, *Faecalibacterium prausnitzii*)
Safety status	Well-established (GRAS/QPS) [2,10]	Ongoing safety evaluation and limited regulatory approval [21]
Isolation sources	Traditional fermented foods (yogurt, kefir, sauerkraut), dairy products, and human/animal gut [5,6]	Novel environments (soil, plants, insects, marine microbiota, human gut) [21,32,35]
Health benefits (evidence base)	Well-documented gut health and anti-diarrheal effects, lactose intolerance relief; multiple clinical trials [5,10]	Limited but growing number of clinical studies; promising roles in obesity, diabetes, inflammatory bowel disease, and metabolic syndrome [42]
Functional traits	Antimicrobial activity; acid and bile salt tolerance, epithelial adhesion [25,26]	Immune modulation, gut barrier enhancement, production of short-chain fatty acid (SCFA) [43,44]
Mechanistic understanding	Mechanisms relatively well-characterized (competition with pathogens, production of antimicrobials, adhesion, immune modulation) [2].	Mechanisms still being evaluated (mucin degradation, signaling via metabolites like SCFAs, anti-inflammatory pathways) [42,44]
Industrial application	Widely commercialized in yogurts, cheeses, beverages, infant formula, dietary supplements [38].	Limited commercial applications; potential in next-generation probiotics (capsules, synbiotics, functional beverages) [42]
Challenges	Strain-specific variability; genetic instability in industrial settings [45]	Difficulties in cultivation, safety uncertainties, lack of regulatory approval, stability issues in food matrices [21]
Future perspectives	Continued use in conventional foods and nutraceuticals; exploration of strain engineering for enhanced traits	Potential game-changers in personalized nutrition, microbiome-targeted therapies, and precision probiotics once validated

2. Biological and Functional Characteristics of Probiotic Strains

For new probiotic strains to be validated, a rigorous scientific framework is required to ensure safety, efficacy, and viability [2]. Several international organizations, most notably the Food and Agriculture Organization (FAO) and the World Health Organization (WHO), have provided globally accepted definitions and guidelines for the classification and characterization of specific strains of microorganisms to qualify as a probiotic for use in foods and dietary supplements. These criteria include that probiotic strains must be (i) adequately characterized; (ii) safe for the intended use; (iii) supported by at least one positive human clinical trial; and (iv) alive at an efficacious dose in the product throughout shelf life [46]. As presented in Figure 1, these criteria can serve as a useful tool in determining whether a candidate strain, or combination of strains, can serve as probiotics or not.

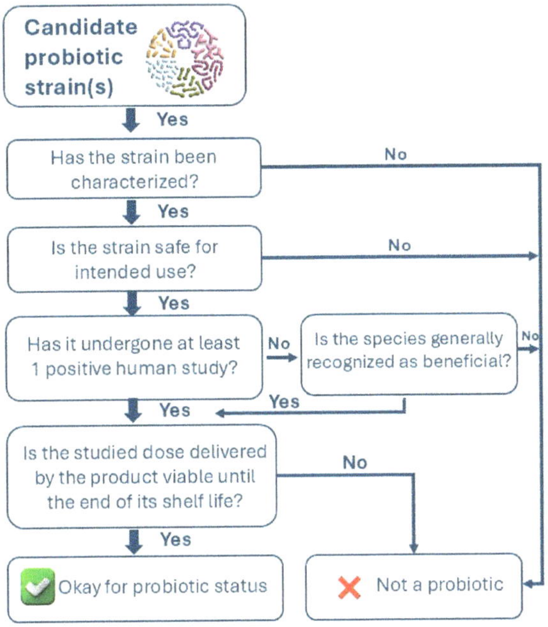

Figure 1. Decision tree used to assess whether a potential probiotic meets the requirements (Adapted from Binda et al. [2]).

The lactic acid bacteria (LAB) that are mostly used as probiotics are mainly species of *Lactobacillus* (*L. acidophilus*), *Lacticaseibacillus* (*L. rhamnosus* and *L. casei*), *Lactiplantibacillus* (*L. plantarum*), *Ligilacto bacillus* (*L. salivarius*), and *Limosilactobacillus* (*L. reuteri* and *L. fermentum*). Furthermore, strains of the *Bifidobacterium* genus (*B. longum*, *B. bifidum*, *B. lactis*, *B. animalis*, *B. breve*, and *B. infantis*) and yeasts (*Saccharomyces cerevisiae*) are commonly used [21]. Probiotic strains are majorly selected, putting into consideration several aspects such as the consumer's safety (hemolytic activity, mucin degradation, and antibiotic susceptibility), physiological functionalities (bile salt and acid tolerance, auto-aggregation, bile salt deconjugation, co-aggregation with pathogens, cell surface hydrophobicity, ability to survive when exposed to gastrointestinal conditions and antagonistic action against pathogens) [21]. Furthermore, technological aspects such as tolerance to sodium chloride (NaCl), production of bioactive compounds, and proteolytic and lipolytic activity are also taken into consideration [47]. The "step-by-step screening approach" as shown in Figure 2 and discussed in the following sub-sections is majorly followed for the selection of potential probiotic microorganisms isolated from traditional and emerging sources. The strains are generally subjected to a series of tests to determine their characteristics. The strains that exhibit the greatest number of functional characteristics and no undesirable traits are chosen at the conclusion of this process. Additionally, proper strain identification and typing are essential because many probiotic attributes are strain specific. Even within the same species, different strains can vary in their ability to survive gastrointestinal stress, adhere to intestinal cells, or exert antipathogenic effects. Hence, strain-level identification using genetic and molecular methods such as 16S rRNA sequencing or whole-genome sequencing (WGS) is crucial for accurate probiotic characterization and safety assurance [2,48].

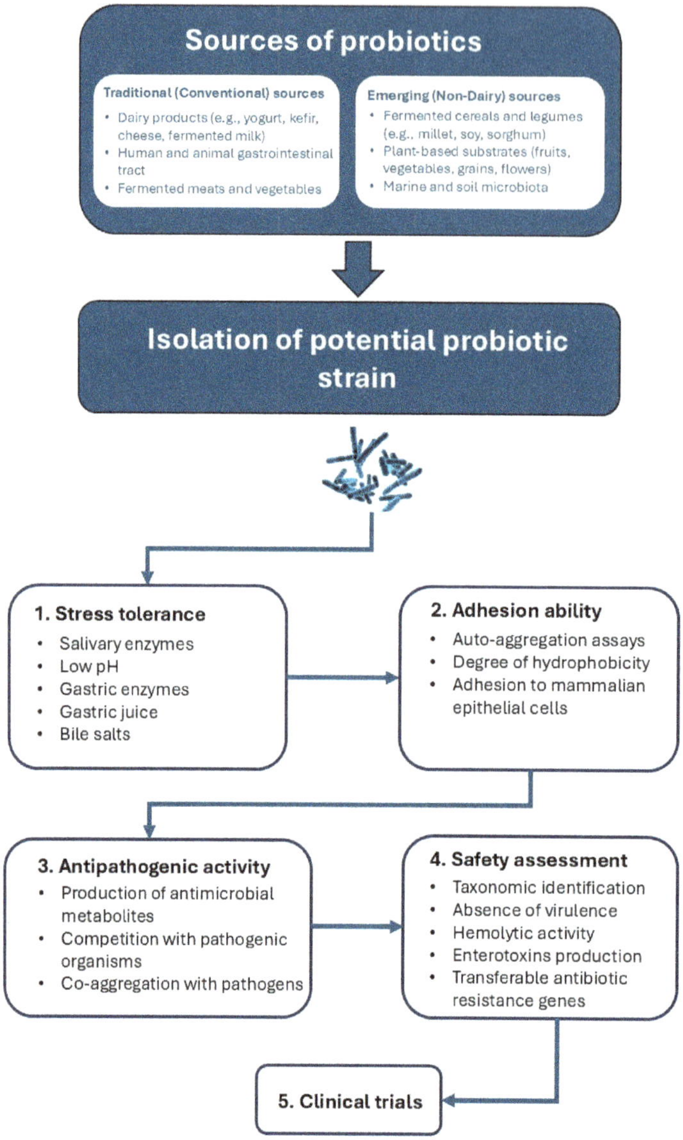

Figure 2. The step-by-step screening approach used for characterization of probiotic strains according to WHO/FAO guidelines.

2.1. Ability to Tolerate Stress

The ability to withstand the variety of stress conditions associated with food processing and the human GIT is a crucial test in the selection of food-grade emerging microbial strains. A potential probiotic strain must demonstrate resilience from the moment it is swallowed until it is effective in the gut before it can be considered acceptable for consumption as a probiotic in foods or dietary supplements. Upon ingestion, probiotics must survive enzymes naturally present in the mouth, such as amylase and lysozyme. While most Gram-positive bacteria are vulnerable to the activity of lysozyme, many strains of LAB have demonstrated significant resistance lysozyme, and can thus, survive as part of the human oral microbiota [49,50]. Once swallowed, the probiotic must withstand harsh conditions in the gastrointestinal tract, which involves survival of the highly acidic stomach environment, digestive enzymes like pepsin, and later, bile salts and pancreatic juices in the small intestine [50]. The probiotic must also withstand mild heat stress from the body temperature. To survive these conditions, candidate strains must possess good acid and bile salt tolerance or mechanisms to neutralize or expel such inhibitory agents [2]. For probiotics, tolerance to bile and acidity is necessary for their survival in the upper digestive system and passage to the large intestine. Thus, a potential probiotic's survival in the gastrointestinal tract is predicted by its ability to survive in simulated gastrointestinal settings [51]. For confirmation of these survival properties, possible strains are cultivated in media containing different pH conditions and stress elements such as pepsin, lysozyme, amylase, bile salts, NaCl, Oxgall, and porcine pancreatic or gastric juices added. Their viability is then determined by quantifying viable cell numbers or growth over time [50].

Studies have demonstrated that different isolated microorganisms with potential probiotic properties are able to survive simulated gastrointestinal conditions. For instance, Garcia et al. [52] evaluated five selected *Lactobacillus* strains (*L. brevis* 59, *L. pentosus* 129, *L. paracasei* 108, *L. plantarum* 49, and *L. fermentum* 111) isolated from byproducts of fruit pulp processing for desirable probiotic-related properties and found that exposure to pH 5.0 and to bile salts (0.15, 0.30, and 1.00%) to simulate gastrointestinal conditions did not decrease the counts of the *Lactobacillus* strains. Results from their study also suggest that acid and bile salt tolerance are generally strain-specific, indicating that some strains may not withstand the harsh conditions of the gastrointestinal tract [52]. However, gastrointestinal stress resistance varies greatly across species and strains and only a few isolated strains pass these rigorous screening tests which have been confirmed by different experimental studies. For example, out of the fifteen potential probiotic yeast strains isolated from pistachio fruits (*Pistacia vera*) in the study by Fernández-Pacheco et al. [53], only 65% of them had a high survival rate in the gastrointestinal conditions they were tested in. In another study by Maragkoudakis et al. [54], they found that out of 29 *Lactobacillus* strains tested, only six survived when pH 1 was applied for an hour, and eight strains could not survive pepsin at pH 2. Generally, pH values ranging from 2 to 5 and bile salts concentrations from 0.3 to 2% are considered critical selection limits for potential probiotic microorganisms [21,55].

2.2. Adhesion Ability

It is essential to determine the potential of emerging probiotic strains to adhere to the gastrointestinal epithelial cells. Effective adhesion is a functionally significant attribute since it allows the probiotic to transiently colonize the gut mucosa, produce biofilms, and exhibit actions like competitive exclusion of harmful microorganisms, modulation of the immune response, and interaction with host tissues [50]. The adhesion is a complex molecular interaction between the contact and binding of two biological membranes (the probiotic cell surface and the host gut epithelial cells). The physicochemical characteristics of microbial cell surfaces play a major role in determining this interaction. Specifically, the extent, strength and selectivity with which a probiotic strain adheres to mucosal surfaces is influenced by surface proteins, exopolysaccharides, and other extracellular structures [56,57]. Microbe adhesion to epithelial cells is further influenced by the cell surface's hydrophobic characteristics as well as its ability to auto-aggregate. While cell surface hydrophobicity enables a better connection between microbes and human epithelial cells, microbial cell auto-aggregation allows for the probiotic cells to clump together in flocs, which allows them to be more locally dense in the gut and adhere and colonize epithelial surfaces better [50]. The process can be measured by a simple test where the strain is added to phosphate-buffered saline (PBS) solution, and absorbance changes are tracked with time to see how well the cells aggregate together [55,58]. Cell surface hydrophobicity is another parameter of a strain's attachment capability. Hydrophobic cell surfaces are readily interlocked with human epithelial cell membranes, consisting of high lipid concentrations, for superior contact and firm attachment. This is measured on a routine basis by the Microbial Adhesion to Hydrocarbon (MATH) test which involves mixing the microbial suspension with a hydrocarbon phase (e.g., hexadecane or xylene) and water [57,59]. The hydrophobicity of the strain is then determined by the degree of bacterial affinity for the hydrocarbon phase, expressed as absorbance at 600 nm [50]. Finally, direct adhesion assays with mammalian epithelial cell lines such as Caco-2, HT-29, or fetal I-407 cells are also widely used to ascertain the actual binding capacity of a strain [60,61].

Furthermore, the colonization of the gut by probiotics is usually preceded by adhesion to the intestinal mucosa leading to the formation of biofilms, which does not favor the binding of enteropathogens [43]. Probiotics isolated from traditional fermented foods have been shown to demonstrate adhesion properties which were linked to the synthesis of exopolysaccharides and enzymes that encourage interactions between probiotics and host-specific receptors. Furthermore, they function as a cell surface capsule during food preservation or during gastrointestinal tract passage, shielding the probiotics against harmful substances and stressful situations [21,31,38].

2.3. *Antipathogenic Activity*

For effective probiotic activity within the gut, following successful microbial adherence to the intestinal lining, probiotics can synthesize a range of antimicrobial compounds/bioactive metabolites which aids in host defense by engaging in direct antagonistic activities against pathogenic microorganisms [50]. This protective effect is a key factor when evaluating the performance of emerging microbial strains for safe and effective use in functional food. Probiotics synthesize these bioactive compounds by metabolizing available carbohydrates, proteins, and other dietary components into bioactive metabolites that inhibit or kill pathogens. These include the production of organic acids (which reduce local pH to create an unfavorable environment for most pathogenic microbes), digestive enzymes, hydrogen peroxide, bacteriocins, and other low-molecular-weight antimicrobial peptides [62]. Combined, these metabolites inhibit the growth and activity of pathogenic bacteria in the gut. In addition, probiotics also participate in other antagonistic activities such as direct competition with pathogens for nutrients, physical deterrence of occupation of attachment sites by pathogens on the gut epithelium, and even coaggregation with them, forming clumps that traps pathogens, thus, allowing for their expedited removal from the gut by normal digestion and excretion [22,29]. LAB strains such as *L. plantarum* 53, *L. fermentum* 60 and 296, and *L. paracasei* 106, isolated from fruit by-products, were found to demonstrate strong auto-aggregation abilities which enhanced their ability to create physical barriers in the gut, inhibiting the colonization of pathogenic microbes through the occupation of adhesion sites and the inhibition of their access to epithelial surfaces [47].

Interestingly, these beneficial activities are highly strain-specific. Species and strains vary considerably in the types and quantity of antimicrobial substances produced by them and in their effectiveness at competing-out or aggregating-out pathogenic bacteria [63]. To determine these traits in a candidate strain, basic laboratory assays including the agar inhibition zone assay is undertaken to measure the antimicrobial activity by observing clear zones where pathogen growth is restricted [64]. Furthermore, competition can be investigated by testing the capacity of a probiotic to interfere with pathogen adhesion to intestinal cell lines in vitro [65]. Similarly, capacity for coaggregation is assessed by co-cultivating the probiotic with common pathogens like *E. coli*, *S. aureus*, *Candida* spp., or *L.*

monocytogenes, and noting whether the probiotic will bind to and aggregate the pathogens, which will enable their expulsion from the gut [29,66].

The beneficial health-promoting potential of probiotic strains are directly related to these distinct biological properties, which allow them to adapt effectively to the body systems of the host and the gut microbiota [37]. The ability of probiotics to modulate the immune system is one of the primary mechanisms by which they achieve their positive impacts. Certain strains can boost the immune response of the host, making it easier for the body to defend itself against diseases and infection [37]. In addition, probiotics also play a significant role in neutralizing colonization of pathogenic microorganisms by competing for adhesion sites and nutrients in the gut. This defense mechanism is complemented by the production of various antimicrobial compounds, including organic acids (e.g., lactic acid), hydrogen peroxide, lysozymes, and bacteriocins [62]. Bacteriocins for instance, are peptide-based compounds with high potency and low toxicity, usually produced in situ by probiotics as an adaptation mechanism, and can exert a bactericidal or bacteriostatic effect against closely related strains of the producer organism and other bacteria genera [62]. These antimicrobial agents support the equilibrium of the gut environment by lowering the pH and preventing harmful microbes from colonizing the gut. Additionally, probiotics can synthesize basic digestive enzymes such as amylase, protease, lipase, pectinase, and endoglucanase, beneficial for the breakdown of complex food constituents, making nutrients more accessible and easier to digest. Most probiotics also synthesize vitamins, such as vitamin A, some B-group vitamins (e.g., B1, B2, B6, B12), C, E, K, PP (niacin), beneficial for the nutrition and health of the host [37].

2.4. Safety and Validation Assessment

In introducing live microbial strains into diets to be consumed by humans, especially from unconventional sources, safety assessment is a priority. Although probiotics are deemed safe, for new probiotic strains, they have to undergo stringent, evidence-based examination to confirm safety [2]. For the safety of consumers, some legislations worldwide such as the European Union Novel Food Regulation, Qualified Presumption of Safety (QPS) list, PROSAFE initiative, and U.S. FDA, WHO, and Canada's NHPR have established strict standards for probiotic safety assessment for use in humans. The standards typically require extensive documentation of the strain's origin and history of isolation; conclusive taxonomic classification; and proof that the strain is devoid of virulence determinants, infectivity, toxins, or antibiotic resistance gene transferability [50,67].

Traditionally, most probiotic strains that are used in commercial products were sourced from healthy human beings. This approach ensures the highest likelihood that the microbes are best adapted to survive and work within the human gut [2,10]. However, with ongoing efforts to identify new or emerging probiotic strains in unorthodox sources such as fermented foods, plants, marine environments, or soil, rigorous safety screening is now more essential. For any novel isolated strain, early and precise identification is a critical step in the selection process. This initial identification helps in eliminating any strains that possess pathogenic or undesirable traits [68]. Primary species-level distinction tends to combine classical biochemical tests such as metabolic profile and ability to grow on a variety of carbon sources, Gram stain, catalase, nitrate reductase, and urease tests with genetic methods like 16S rRNA gene sequencing to validate taxonomic allocation [50]. However, for the sake of ensuring probiotic safety and traceability at the strain level, such common techniques are not sufficient on their own because they might not be able to offer adequate genetic variability to differentiate between strains that are nearly related. Therefore, molecular fingerprinting techniques such as Repetitive Element Palindromic PCR (rep-PCR), Random Amplified Polymorphic DNA (RAPD), or Pulsed-Field Gel Electrophoresis (PFGE) are usually utilized in order to ensure precise strain-level discrimination [69,70]. While previous probiotic research has focused on routine safety testing like enterotoxin testing or hemolysis testing, the requirements now are more rigorous, with comprehensive safety screening of any emerging strain intended for human consumption. Beyond screening for basic characteristics, it has become essential to screen for other risk factors such as the production of undesirable metabolites like D(-)-lactate, deconjugation of bile salts, and virulence factors like gelatinase, DNAse, or hemolysins. Identification of such possible hazards relies on molecular techniques, such as PCR-based identification of virulence genes, cytotoxicity assays, and, where required, confirmatory in vivo testing [26,43,71].

One of the key issues with the safety of modern probiotics is the assessment of antibiotic resistance potential, as this has become a high priority [2]. This is because there is a risk for probiotics to pass on resistance genes to gut pathogenic bacteria through horizontal gene transfer, which would be an issue for public health [36]. Accordingly, QPS and GRAS evaluations now explicitly include checks for resistance traits [72]. Resistance in probiotic strains are screened by methods such as agar disk diffusion, E-tests, or minimum inhibitory concentration (MIC) assays that determine the minimum dose of an antibiotic to inhibit bacterial growth [73]. For LAB, specific media like LAB susceptibility test medium (LSM) are used to avoid deceptively false results by media incompatibilities during antibiotic susceptibility tests [74]. However, these are phenotypic tests and are not standardized between laboratories. Therefore, PCR has emerged as the method of choice for reproducibly detecting antibiotic resistance genes because of its sensitivity and reliability. Using PCR, DNA for probiotic strains is amplified with gene-specific primers, and products may be confirmed by sequencing to be assured of gene identity [50].

Interestingly, not all antibiotic resistance poses equal risk. If a probiotic's resistance is intrinsic, i.e., it is chromosomally mediated and not on transmissible elements, then there is little risk of passing it onto other gut bacteria. In such a case, resistant probiotics are also effective in restoring the microbiota after treatment with antibiotics. However, horizontally transferable acquired resistance genes on plasmids or transposons involve a significant risk of transfer and must be rigorously tested [36,72]. Filter mating experiments, in which a probiotic with a detected antibiotic resistance gene is cultured with recipient cells that do not have the gene, can be used to analyze gene transfer. Phenotypic and molecular methods can then be used to analyze the transfer ratio to the recipient cell [75]. In recent times, WGS which includes all extrachromosomal components, has been employed as a powerful and cost-effective "gold standard" approach for strain identification and screening of antibiotic resistance genes in probiotic candidates. Through full sequencing and mapping against available reference databases, such as the National Center for Biotechnology Information (NCBI), WGS enables identification of microbes down to the species and strain level and helps determine the likelihood of resistance gene transmission to other gut microbes [2,76].

Finally, beyond genomic and toxicological evaluations, the ultimate validation of probiotic safety and efficacy requires human clinical evidence. For a candidate microbial strain to qualify and be granted probiotic status, it must demonstrate a measurable health benefit in at least one well-designed human clinical trial, ideally followed by confirmation studies [2]. Randomized, placebo-controlled, and double-blind trial designs, compliant with recognized clinical research standards such as The Consolidated Standards of Reporting Trials (CONSORT), are strongly recommended to ensure reliability and reproducibility of outcomes. These trials help confirm not only the strain's tolerability and absence of adverse effects but also its ability to confer the intended physiological benefit in humans [77,78]. In addition to clinical safety, probiotics must remain alive and at an efficacious dose throughout product shelf life [46]. While no fixed dose is included in the formal definition of probiotics, probiotics must be taken in a quantity sufficient to provide a health effect. Although some microbes can exert effects at low dosages due to their capacity to replicate within the host, probiotics generally require sufficient viable counts to provide beneficial effects [10]. Unlike pharmaceutical agents, probiotics are living organisms, and dose-ranging or maximum tolerable dose (MTD) studies are rarely conducted. This is majorly due to the notion that food ingredients are presumed safe [2]. Given the general presumed safety profile of probiotics and the fact that they are mostly not subjected to MTD or dose ranging studies, most studies and products use doses ranging from 10^8 to 10^{11} colony-forming units (CFU), based on evidence from previous trials demonstrating efficacy and safety [2]. Finally, to ensure reliability and consistency, probiotic viability must be measured using standardized enumeration methods, such as CFU plating on selective media or flow cytometry [79]. These methods help confirm that the stated dose is maintained during clinical studies and that the probiotics remain viable and effective throughout a product's shelf-life. Ensuring this stability and viability is essential for both consumer safety and regulatory compliance.

3. Emerging Probiotics

Traditionally, probiotics have been derived from well-established genera like *Lactobacillus* and *Bifidobacterium*. However, the growing demand for evidence-based functional foods and alternatives suitable for lactose-intolerant individuals has driven interest in novel probiotic strains from non-dairy sources such as traditional fermented drinks, raw vegetables, fruits, and flowers [21,80]. Hence, in the last decade, these unconventional sources have been actively explored for microbial strains with both techno-functional properties and probiotic potential, offering promising opportunities for biotechnological applications and/or functionalization of foods.

3.1. Emerging Probiotics from Non-Dairy Fermented Foods

In recent times, emerging probiotics have frequently been isolated from non-dairy and unconventional sources such as traditional fermented foods, made with a broad range of raw materials of both plant and animal origins. These foods reflect rich culinary and cultural traditions, which are often passed on across generations and modified to suit local raw materials availability in different regions of the world. The microbial composition of such foods is determined by the type of raw material used and, consequently, the type of probiotic species and strains that may be isolated from them [80]. Non-dairy fermented foods, (especially those of vegetable, meat, and seafood origin), have proven to be rich sources of LAB and other beneficial microorganisms. For instance, LAB have been easily recovered from fermented fish products, cured or dried meats, fermented meats, salted crabs, and other seafood products [81,82]. Additionally, plant fermentations centered on substrates such as soybeans and vegetables (e.g., cabbage, carrots, and mustard greens) also harbor dense microbial populations, which are often dominated by LAB such as *Lactobacillus*, *Leuconostoc*, *Pediococcus*, and *Weissella* species.

LAB in traditional fermented foods vary depending on the substrate and fermentation environment. For instance, in fermented fish and crustacean products, *Enterococcus* is the dominant LAB genus present and is usually characterized by its high tolerance to salt [80]. On the other hand, *Lactobacillus* species are more commonly isolated from fermented meats and vegetable products. These microbial tendencies are due to adaptation to the salinity of the environment where fermentation occurs. The halotolerance of LAB strains from species such as *Enterococcus faecium* with the capacity to endure in media containing sodium chloride (NaCl) levels more than 22%, makes them well suited for performing fermentations of salted seafood and other highly saline products [83]. By comparison, plant fermentations tend to be less saline, and the prevailing LAB species that occur in these habitats are *L. fermentum* and *L. plantarum*, both of which can grow in media containing less than 6% NaCl [83].

In their study, Senthong et al. [82] isolated 306 LAB strains from samples of *Poo-Khem*, a traditional salted crab sold in Thailand. The isolates were screened for probiotic characteristics. They found that four strains including one strain of *Enterococcus thailandensis*, one strain of *L. plantarum*, and two strains of *L. fermentum* displayed notable probiotic traits such as tolerance to acid and bile salts, antimicrobial properties against common foodborne pathogens, and high cell surface hydrophobicity (an indicator of good adhesion ability to the human gastrointestinal tract). These probiotic traits validate their use as potential candidate LAB in starter cultures for controlled fermentation of seafood products like *Poo-Khem*. A similar study by Siripornadulsil et al. [84] found that strains of *Pediococcus pentosaceous* were the most isolated LAB from various traditional Thai fermented foods containing fish and pork. The strains exhibited probiotic properties such as tolerance to acidic conditions at pH 2, 0.3–0.5% bile salt and 1–14% NaCl. They also inhibited the growth of some pathogenic bacteria, including *Vibrio cholera*, *E. coli*, *Pseudomonas aeruginosa*, *Salmonella typhimurium*, *Bacillus cereus* and *Staphylococcus epidermidis*. Bacha et al. [85] evaluated the probiotic potential of LAB strains isolated from *Wakalim*, a traditional Ethiopian fermented beef sausage. Out of the 99 LAB strains that were evaluated, 44 tolerated pH 3.0 and bile salt concentration $\geq$ 0.3% for 3 h. The highest tolerance to pH 3.0 was observed among the pediococci (81.6%, 31/38) followed by lactobacilli (14.3%, 8/56), demonstrating their potential to enhance health when consumed through fermented sausages.

In India, Agaliya and Jeevaratnam [86] screened eight *L. plantarum* strains isolated from fermented idli batter for probiotic potential using in vitro assays such as bile tolerance, acid tolerance, transit tolerance in the upper human gastrointestinal tract, auto-aggregation, co-aggregation, hydrophobicity, susceptibility to various antibiotics, bile salt hydrolase assay, cholesterol assimilation and hemolysis. They found that the isolates could withstand pH values of between 2.5 and 8.5 as well as up to 0.3% bile for 4–6 h. Additionally, the isolates were able to withstand intestinal and stomach fluid expansion. Across all isolates, the auto-aggregation of the various strains of *L. plantarum* varied between 65 and 80% with co-aggregation ranging from 51 to 64% in pathogens like *Listeria monocytogenes* (MTCC 657) and *Escherichia coli* (MTCC 728). The isolates possessed â-galactosidase activity exhibiting 322 to 1000 milliunits of enzyme activity, and showed no hemolysis activity, indicating their probiotic potential which would attribute beneficial health effect to mankind [86]. Furthermore, Oluwajoba et al. [87] isolated probiotic LAB from Kunu-zaki, a traditional fermented beverage from Nigeria prepared from non-germinated sorghum and millet cereal grains. These LAB species demonstrated antibacterial activity against the reference strains of *Staphylococcus aureus*, *Escherichia coli*, *Pseudomonas aeruginosa*, and *Enterococcus faecalis*, as well as resistance to pH 3 and 3% bile. These species were mostly *Lactobacillus* species, and were identified *Pediococcus*, *Lactobacillus*, and *Lactococcus* species.

3.2. Emerging Probiotics from Dairy Fermented Foods

Dairy fermented foods have long been considered as the main source of probiotic products available in the market. Traditional dairy fermented products, such as yogurt, kefir, buttermilk, cheese, and cultured milk drinks, harbor a rich diversity of microorganisms, some of which have been found to possess exceptional health-enhancing properties beyond overall nutrition [5]. With the rise in the demand for functional foods, scientific interest in emerging probiotic strains from dairy fermented foods with improved survival in the gastrointestinal tract, enhanced adhesion, and distinct health benefits such as lactose digestion and immune modulation have also significantly increased [88]. In their study, Sharifi et al. [89] isolated 47 LAB from Iranian traditional yogurts, out of which 12 were considered potential probiotic candidates based on their ability to tolerate acidic pH and resistance to bile salts. Of these, six probiotic isolates belonged to *Pediococcus acidilacticii* and other six isolates to *Lactobacillus plantarum*, *L. brevis*, *L. fermentum* and *L. kefiri*. These organisms have been shown to occur naturally in cow milk [90]. Similarly, Hoque et al. [91] isolated *Lactobacillus* spp. from two regional yogurts in Bangladesh which demonstrated probiotic properties such as resistance to inhibitory substances like phenol (0.4%), NaCl (1–9%) and bile acid (0.05–0.3%). Additionally, the isolates demonstrated good growths in the presence of 1% NaCl and 0.3% bile acid.

Furthermore, Kefir, a traditional fermented dairy beverage made from the fermentation of different types of milk and kefir grains, is considered a healthy product with high nutritional value and has been associated with various health benefits [92]. This has spurred research towards the identification of potential novel probiotic strains present in kefir. Different studies have isolated and characterized microorganisms from kefir grains all over the world, some of them exhibiting probiotic properties. These include various fungi species such as 34 yeast strains isolated from 4 milky and 3 sugary kefir grains by Diosma et al. [93] which were identified as *Saccharomyces cerevisiae* (15 strains), *Saccharomyces unisporus* (6 strains), *Issatchenkia occidentalis* (4 strains), and *Kluyveromyces marxianus* (9 strains). In this research, *K. marxianus* CIDCA 8154 and *S. cerevisiae* CIDCA 8112 strains were selected and further studied. Both strains demonstrated the capacity to adhere to epithelial intestine-derived cells in vitro and to survive passage through the gastrointestinal tract of BALB/c mice. In another study, Chen et al. [71] isolated *Lactobacillus kefiranofaciens* M1 from Taiwanese milk kefir grain and investigated its effects on intestinal epithelial cells in vitro and on dextran sodium sulfate (DSS)-induced colitis in vivo. In vitro results showed that *L. kefiranofaciens* M1 could strengthen the epithelial barrier function by boosting transepithelial electrical resistance (TEER) and significantly increasing the level of the chemokine CCL-20, which plays a major role in immune signaling. In vivo, *L. kefiranofaciens* M1 reduced the severity of chemically induced colitis as shown by a significant attenuation of the bleeding score and colon length shortening. Additionally, it helped in balancing the immune response by decreasing the production of proinflammatory cytokines and increasing the production of the anti-inflammatory cytokine IL-10 demonstrating its potential to be used in fermented dairy products as an alternative therapy for intestinal disorders [71].

3.3. *Emerging Probiotics from Other Unconventional Sources*

Besides dairy and non-dairy fermented foods, other sources such as marine environments (e.g., seaweeds, fish gut microbiota, and marine sediments), fruit, flowers, and raw vegetables have also been explored as niches for the isolation of potential probiotic microbes for biotechnological application or functionalization of foods [21,32]. These include some flowers which are colonized by fructophilic bacteria (that is, a preference for D-fructose over D-glucose) belonging to the genus *Fructobacillus* [94]. These bacteria produce mannitol and acetate rather than ethanol which is useful for biotechnological applications considering that mannitol is a low-calorie polyol frequently used in the food sector [95]. The potential of these as probiotics have been further evaluated by recent studies like that of Patil et al. [29] that isolated strains with fructophilic characteristics such as *F. fructosus* MCC 3996 from the nectar of *Butea monosperma* flower. The strain was evaluated in vitro for probiotic characteristics and it was found to demonstrate resistance to acidic environments and gastric juice, improved auto-aggregation even when hydrolytic enzymes were present, hydrophobicity, co-aggregation with pathogenic bacteria, and the absence of hemolytic activity [29]. Similarly, another study conducted in China by Sakandar et al. [96] isolated fructophilic LAB (FLAB) from various flowers and fruits and evaluated them for probiotic characteristics. The isolated strains include *F. pseudoficulneus* JNGBKS1 and JNGBKS3 from peach and banana, respectively; *F. fructosus* JNGBKS2 and JNGBKS4 from narcissus and sunflower, respectively; *F. durionis* JNGBKS5 from kiwi fruit; *L. kunkeei* JNGBKS6, JNGBKS7, and JNGBKS8 from narcissus, yellow rose, and pink rose, respectively. In comparison to the other FLAB strains that were isolated, *L. kunkeei* strains demonstrated the maximum capacity to tolerate various low pH levels and bile acid concentrations together with maximal effects on antipathogenic activity, cholesterol assimilation, and hydrophobicity [96].

Similarly, *Weissella paramesenteroides* which has gained considerable attention in recent years as bacteriocin and EPS producers, was isolated in the study by Pabari et al. [43] from different fruits (cherry, sapota, banana, orange, and plum). In this study, five *W. paramesenteroides* strains, namely, FX1, FX2, FX5, FX9, and FX12, were isolated and screened for probiotic and prebiotic traits. Two strains FX5 and FX9 were chosen for prebiotic utilization investigations using thin layer chromatography (TLC) and high-performance liquid chromatography for the synthesis of short-chain fatty acids (SCFAs) based on their functionality. Since only the top low molecular weight fractions vanished from cell-free supernatants (CFS), the TLC profile demonstrated that these two strains were able to use low molecular weight fructooligosaccharides (FOS) and galactooligosaccharides (GOS). GOS utilization ability was demonstrated by increased β-galactosidase activity which is related to galactose buildup in the residual CFS of GOS. Both strains demonstrated antimicrobial activity against *S. aureus* and *E. coli*, as well as significant SCFA synthesis when prebiotic was present, indicating their potential as synbiotics [43].

From marine ecosystems, the exploration of probiotic LAB has also been reported. In their study, Das et al. [32] evaluated the probiotic potentials of three marine LAB isolates characterized as *L. casei* SB71, SB73 and SB93. They found that these isolates demonstrated high level of gastrointestinal survival, marked cholesterol assimilation, adherence to Caco-2 cells, and inability to form biogenic amines. The isolates were also tolerant to NaCl, bile and low pH. Notably, all three isolates produced bacteriocins that exhibited sustained antimicrobial activity against *V. cholerae*, whereas the EPS from the SB93 strain notably disrupted the adherence of bacteria and promoted remediation of cadmium and lead [32].

4. Application of Probiotic Strains in Functional Food Development

Functional application of probiotic strains depends not only on their benefits to health but also on their ease of application in real food products. As previously discussed in Section 2, the biological and safety characteristics of each strain, including stress tolerance, adhesion capacity, antipathogenic activity, and confirmed safety are critical determinants of its practical performance in food systems. In functional food design, these properties must align with formulation requirements, processing stability, and sensory quality to ensure that probiotics remain viable and efficacious throughout the shelf life of the product [2,10]. Hence, the development of probiotic-enriched foods therefore focuses on major aspects such as the compatibility of the strain with the food matrix, technological strategies such as encapsulation to enhance survival and stability, and innovative product formulation to maintain probiotic delivery and consumer acceptability. Together, these factors determine whether a probiotic product will deliver the assured health benefits to consumers.

4.1. Food Matrix Compatibility

Determining the compatibility of the probiotic strain with the selected food matrix is very crucial when developing functional foods. This is an important consideration which has encouraged the search for newer probiotic strains, amongst other reasons. The addition of traditional probiotic strains of dairy origin into plant-based foods typically involves significant hurdles. Most commercial probiotic strains fail to maintain adequate survival rates in non-dairy substrates because these environments can be harsher due to factors like reduced pH, different levels of nutrients, or the presence of antimicrobial phytochemicals [71]. The ability of a strain to maintain metabolic activity and cell viability during food processing and storage depends on these parameters, which can vary significantly across food types. Fermented fruit juices or cereal-based beverages for example, often present acidic conditions and oxygen exposure that can reduce the stability of probiotics. However, matrices rich in fibers, sugars, or polysaccharides (such as β-glucans or inulin) can act as protective agents, enhancing the survival of the microorganisms during storage and gastrointestinal transit [97]. By comparison, LAB found naturally in plant matrices tend to be more likely to survive and maintain life in such environments, increasing their probability to deliver probiotic benefit when applied to vegetable, fruit, or cereal-food products [98]. The use of plant-derived LAB probiotics can help to broaden the range of functional foods which can be fortified with probiotics to meet the demands of both traditional and emerging consumer markets.

Another key determinant for matrix compatibility is the influence of probiotic addition on sensory and textural properties. The interaction between the matrix and probiotic cells can impact flavor, aroma, viscosity, and appearance, which ultimately determines consumer acceptability. Studies have shown that supplementation of *L. plantarum*, *L. fermentum*, and *L. brevis* in non-dairy matrices such as oat, soy, and fruit-based beverages maintained viable counts above 10^7 CFU/mL under storage conditions preserved acceptable sensory characteristics [99]. Microencapsulation techniques, prebiotic enrichment, and co-fermentation using compatible starter cultures have also been explored further to maintain probiotic stability and incorporation into various matrices [100]. These approaches balance microbial performance with technological viability and customer acceptability, and these are critical for effective product development and marketing.

4.2. *Encapsulation of Probiotics to Improve Shelf-Life and Viability*

As indicated by the World Gastroenterology Organization, the effective dose of probiotics necessary to deliver health benefits differs widely with the microbial strain and type of product [101]. Although many commercial probiotic foods or supplements are typically in the 1 to 10 billion CFU (colony-forming units) per dose range, there are strains that are inoculated at lesser doses, and others that require higher concentrated doses to have an appreciable effect [100]. It is generally recommended that a probiotic food product has a minimum of 10^6 to 10^9 CFU per gram and daily intake equals at least 100 g or milliliters to have an effective therapeutic effect [102]. However, ensuring that live probiotic cells do, in fact, reach the consumer in efficacious numbers is challenging. During food storage and processing, probiotic organisms encounter several stressors such as heat, moisture, light, oxygen, osmotic stress, and exposure to high acidity and digestive enzymes when consumed [103]. In the dairy industry, for instance, procedures such as pasteurization, homogenization, agitation, freezing, drying, and even the addition of salt or other antimicrobial compounds can drastically reduce the viability of probiotic bacteria [104]. Studies have shown that to retain a shelf-life of up to 12 months at room temperature, the water activity of a probiotic product must be extremely low, i.e., preferably less than 0.25, to prevent microbial loss [105].

In view of these challenges, the initial cell count, strain selection, formulation pH, processing steps, and the food matrix itself are all significant parameters that dictate if probiotics survive and remain effective [106]. To overcome these limitations, advanced encapsulation technologies have been an effective approach in probiotic food design. Techniques like microencapsulation, which involves using methods like spray drying, freeze-drying, or nano-coating, gives a protective layer to sensitive probiotic cells [105]. The protective layer cushions the microbes against the stressful processing conditions and the acidic environment of the stomach and small intestine so that more viable cells can pass through to the gut to exert their effects [104]. Additionally, encapsulation may be employed to conceal any unfavorable taste or odor of live cultures and deliver a long shelf-life without compromising sensory qualities [106]. Importantly, selecting the right encapsulation material is essential to the success of these technologies. In addition to being safe for human consumption and cost-effective, the material should ensure that the probiotic cells are protected throughout food processing and storage, and during delivery in the gastrointestinal tract [107]. Materials with gel-like structures and greater dry matter content tend to be more effective since they create a stable matrix for the living cells. Various natural polymers and hydrocolloids including alginate, xanthan gum, chitosan, carrageenan, cellulose, and certain fats and proteins are often utilized in formulating microcapsules, each bearing specific characteristics that can increase encapsulation effectiveness and probiotic viability [108]. Hence, pairing encapsulation technologies with the careful screening of robust emerging strains would allow food companies to increase the bioavailability and stability of probiotics in dairy and non-dairy functional foods, widening product possibilities and consumer trust in probiotic health claims.

Studies have reported on the encapsulation of bacteriocins from LAB for effective activity against spoilage and pathogenic organisms, including spores [109,110]. However, recent studies have reported the use of encapsulated probiotic cells themselves in the production of functional foods to mitigate the challenges faced by free probiotic cells in high acidic environmental conditions. For instance, Afzaal et al. [111] evaluated the effect of microencapsulation on the viability and stability of probiotic bacteria in yogurt and simulated gastrointestinal conditions. They incorporated free probiotic (*L. acidophilus*) into the yogurt during its preparation and found that the survival of the bacteria decreased from 9.97 log CFU/mL on first day to 6.12 log CFU/mL on day 28. The free probiotic cells also showed very poor survival during in vitro gastrointestinal assay. To improve the viability of probiotics, they used the extrusion technique to encapsulate the bacteria using sodium alginate and carrageenan. They found that the encapsulation improved the viability of the probiotics in the prepared yogurt and gastrointestinal tract. At zero-day, the probiotics encapsulated with sodium alginate and carrageenan had cell counts of 9.91 and 9.89 logs CFU/mL, respectively, which dropped to 8.74 and 8.39 log CFU/mL by the end of the day. The study also demonstrated that sodium alginate was a better encapsulating material for probiotics in comparison to carrageenan. [111]. A similar study by Chen et al. [112] evaluated the effect of xanthan-chitosan-xanthan (XCX) encapsulation system on the viability of *Bifidobacterium bifidum* BB01 in yogurt during 21 days storage at 4 °C and 25 °C, respectively. They used chitosan as an inner layer to coat the xanthan-probiotic (XC) and xanthan was again used as an outer layer for coating chitosan-xanthan-probiotic particles. The results showed that the *B. bifidum* BB01 encapsulated with these multilayer hydrogels remained viable for up to 21 days in yogurt stored at both refrigeration and room temperatures, with significantly ($p < 0.05$) better survival than free, unencapsulated cells. In comparison to the free cells, all the microcapsules showed higher cell survival of probiotic in the simulated gastric fluid and bile salt solution. Notably, the XC microcapsules exhibited better release profile than XCX microcapsules in the simulated intestinal fluid. These findings demonstrate how carefully designed biopolymer encapsulation can maintain probiotic viability during storage and release them in the gut, facilitating the development of robust functional foods.

In non-dairy functional foods like fruit juices, the use of encapsulated probiotics has also been reported. Dias et al. [113] produced a powdered probiotic passion fruit juice through microencapsulation of *Bifidobacterium animalis* subsp. *lactis* BB-12 with maltodextrin and/or inulin as encapsulating agents. Powders were tested for physical, thermal, and microbial stability for 30 days of storage at 25 °C and 4 °C. Their results showed that inulin, either alone or with maltodextrin, was a better protectant for maintaining the probiotic viability, especially at 25 °C, compared with maltodextrin alone. All the powders maintained flow characteristics and solubility, though differences in carrier materials influenced moisture content, particle shape, and color over time stability. Overall, the study demonstrated that probiotic encapsulants combined with controlled storage conditions may be utilized to promote the shelf life of probiotics in non-dairy food systems like fruit juices and enable the development of stable, functional probiotic drinks. In another study, internal gelation was used to encapsulate probiotic bacteria, including *L. acidophilus* and *B. bifidum*, in alginate beads with a mean diameter of 54.25 ± 0.18 μm. The pasteurized grape juice was then mixed with both encapsulated cells and free cells as control samples, and the mixture was kept for 60 days. The bacteria in the encapsulated samples had significantly ($p < 0.05$) higher survivability at the end of the storage period than the sample with free bacteria cells (8.67 ± 0.12 and 7.57 ± 0.08 log CFU/mL for *L. acidophilus* and 8.27 ± 0.05 and 7.53 ± 0.07 log CFU/mL for *B. bifidum* for the encapsulated and free forms, respectively) [114].

4.3. Role of Artificial Intelligence in the Discovery, Characterization, and Application of New Probiotics

Artificial intelligence (AI) and machine learning (ML) methods are becoming increasingly applied across the probiotic pipeline, from rapid genome and metagenome screening to prediction of colonization outcome, risk assessment of safety, formulation and delivery optimization. These technologies support traditional wet-lab screening by (i) discovering candidate strains with desirable genetic and functional properties, (ii) automating detection of risk factors (e.g., antimicrobial resistance genes, mobile elements), and (iii) modeling host–microbe and community-level responses for improved strain selection and formulation [115,116]. For example, the study by Sadeghi et al. [117] demonstrated the use of ML models for the selection and prioritization of 15 out of the 144 LAB strains screened in their study, based on their combined potential probiotic properties and antimicrobial activity for application in plant tissue cultures. This study highlights the ability of AI-driven models to enhance the precision and efficiency of probiotic screening and selection processes. Another study by Zhang et al. [118] analyzed 31,977 LAB genomes and identified over 130,000 secondary metabolite biosynthetic gene clusters (BGCs) using machine learning models to predict their antimicrobial potential. Through the integration of metagenomic and metatranscriptomic analyses, the researchers linked LAB-derived metabolites to protective and regulatory functions in the human microbiome, demonstrating how AI and omics integration can accelerate the discovery and functional characterization of novel probiotic metabolites with therapeutic potential.

Bacteriocin/RiPP mining software such as BAGEL (and BAGEL4) are still needed to identify putative antimicrobial peptides encoded in genomes and are used routinely in combination with ML pipelines for functional annotation [119]. AI classifiers improve the detection of antibiotic resistance genes (ARGs) and mobile genetic elements in high-throughput sequencing data. For instance, DeepARG, a deep-learning program trained on hand-curated ARG databases, has been shown to provide sensitive and accurate ARG annotation from both isolate and metagenome-derived genomes, a crucial step in ruling out probiotic candidates that may be responsible for horizontal transfer of resistance [120]. Similarly, ML tools like PlasFlow, PlasClass and Deeplasmid help identify plasmid-derived contigs and differentiate them from chromosomal sequences, which is important in assessing the mobility risk of identified resistance or virulence genes [121]. Furthermore, newer specialized tools such as metaProbiotics has incorporated language-model to directly mine metagenomes rapidly for probiotic-like genome bins, facilitating recovery and priority of candidate strains from large data sets and performing better on test benchmarks [122]. ML can also be used in the prediction of functional outputs such as short-chain fatty acid production or other metabolites from strain mixtures and environmental conditions, allowing in silico optimization of strain consortia prior to laboratory validation. For example, Westfall et al. [116] developed an artificial model of the human gastrointestinal tract called ABIOME (A Bioreactor Imitation of the Microbiota Environment), for the prediction of metabolic activity of probiotic formulations and hence therapeutic potential with machine learning tools. The combination of ABIOME and the multivariate adaptive regression splines (MARS) model led to the identification of several probiotic combinations that stimulated synergistic production of bioavailable metabolites, each with a different therapeutic capacity. Collectively, these examples demonstrate how AI and ML approaches significantly enhance the precision, scalability, and predictive power of probiotic discovery and formulation, bridging computational insights with practical applications in food and health. However, this is not without limitations. While the integration of AI and ML has led to the acceleration of probiotic candidate selection and hypothesis generation, outputs should be interpreted cautiously. ML predictions are strongly dependent on training data quality and representativeness; models trained from sparse or biased training data may generate false positives/negatives [123]. Experimental validation (in vitro, in vivo, and human trials) is still required to confirm predicted safety, functionality, and efficacy [2]. Furthermore, regulatory acceptance of AI evidence is more in its infancy; open publication of model structures, training data, and performance metrics will be crucial to industry and regulator uptake. Finally, the integration of clinical and individual microbiome data creates data-privacy and ethics problems that must be addressed before the wide implementation of personalized probiotic products [124].

5. Health Benefits and Mechanisms of Action of Probiotics

The beneficial effects of probiotics on the human body are usually because of one or a combination of several processes, including altering the intestinal microbiota, blocking pathogen attachment sites, production of metabolites, modifying the host immune response, and nutrient competition [21].

5.1. Immunomodulatory and Anti-Inflammatory Effects

Probiotics play a crucial role in enhancing the host's immune defenses by influencing how the body recognizes and responds to invading pathogens in the host. Studies have shown that probiotics can navigate through the intestinal mucus layer, in which they survive, interact with the gut epithelial cells and promote the modulation of innate and adaptive immune processes. In their study, Song et al. [125] found that *Lactobacillus brevis* B13-2 strain isolated from kimchi exhibited immunomodulatory activity by inducing the expression of some cytokines (IL-1β, TNF α, and IL-6) and iNOS. Interestingly, the observed immunomodulatory effect was in addition to its probiotic effects such as tolerance against artificial gastric acid and bile salts, adherence to HT-29 cells, and absence of β-glucuronidase production. In an in vivo experiment, male Swiss albino mice that consumed 10 CFU/mL of fruit-derived *S. cerevisiae* IFST 062013 showed increased humoral and cell-mediated immunity, and a stimulation of T-lymphocyte-specific proliferative response. Additionally, the probiotic improved host immunity by inducing pro- and anti-inflammatory mediators and preserving the equilibrium of Th1 and Th2 cytokines [126]. In another study by Ornellas et al. [127], it was demonstrated that mice that were given a single dosage of *L. plantarum* 81 and *L. plantarum* 90 isolated from cupuaçu (*Theobroma grandiflorum*) fermentation (10^8 CFU/mL) produced more of the regulatory cytokine IL-10 and less of the proinflammatory cytokines IFN-γ and IL-6, leading to an exhibition of anti inflammatory effect.

5.2. Effect on Cardiovascular Health

There is growing evidence of the role of probiotics in cardiovascular health maintenance, particularly their cholesterol-lowering property. Certain probiotic strains have been found to help decrease serum cholesterol levels, thus lowering heart disease risk [128]. One of the major ways probiotics achieve this is by decreasing the solubility of cholesterol in the intestine, which decreases the quantity absorbed into the blood [50]. In addition, many probiotics produce bile salt hydrolase (BSH) enzymes that breaks/deconjugate the bond between a conjugated bile acid and its attached amino acid (glycine or taurine), releasing the bile acid in its free (deconjugated) form. Such deconjugated bile salts are readily removed from the body through feces. Additionally, since cholesterol is a major precursor of bile acid synthesis, this increased removal of bile acids provokes the body to convert more circulating cholesterol into new bile salts, resulting in lower total serum cholesterol levels [129]. The in vivo study by Cavalcante et al. [130] demonstrated that male Wistar rats fed with high-fat diets treated with fruit-derived *L. fermentum* 296 had improved biochemical and cardiovascular parameters that are altered in cardiometabolic disorders. Similarly, another in vivo study by Costabile et al. [131] evaluated the cholesterol reducing capacity of *Lactobacillus plantarum* ECGC 13110402 in 49 adults with normal to mild hypercholesterolemia. The study was run in a parallel, double blind, placebo controlled, randomized design in which the active group of volunteers ingested 2×10^9 CFU encapsulated *Lactobacillus plantarum* ECGC 13110402 twice daily for a period of 6–12 weeks. Results showed that this probiotic strain reduced LDL cholesterol by 13.9% among the subjects who had relatively higher initial cholesterol levels and reduced total cholesterol by as high as 37.6% among subjects with higher initial levels. Also, participants over 60 years experienced a significant decrease in triglycerides (53.9%) and a rise in healthy HDL cholesterol (14.7%). Importantly, lowering of systolic blood pressure (6.6%) was also observed in the trial, and the probiotic proved to be tolerant without any gastrointestinal side effects. This demonstrates that probiotic strains like *L. plantarum* ECGC 13110402 hold promise in having a facilitative role to play in cholesterol management and reduction in cardiovascular risk, either as add-on or standalone in place of conventional therapy.

5.3. Anti-Anxiety and Anti-Depression

There is increasing evidence that symptoms of depression and anxiety can be eased by certain probiotic strains. This hypothesis was first evidenced in the 1970s by Tannock and Savage [132], who reported that stress in mice markedly reduced their levels of gut *Lactobacilli*, implying a connection between mental status and gut microbiome. Since then, the gut–brain axis has been a key focus of psychobiotic research. Studies have shown that probiotics may exert a positive effect on mood and mental health via numerous mechanisms, such as the enhancement of immune response and extension of production or availability of mood-regulating neurotransmitters like serotonin [133]. Animal models, for instance, rats in forced swim tests, have been used for the preliminary screening of probiotic strains for antidepressant-like effects. These experiments also measure changes in behavior alongside biochemical markers like neurotransmitter levels of serotonin, dopamine, and noradrenaline, or their precursors like tryptophan [50].

In humans, clinical trials have demonstrated that probiotic supplementation reduces symptoms of depression, anxiety, and stress as assessed using conventional scales like the Beck Depression Inventory (BDI), the Beck Anxiety Inventory (BAI), or the Hamilton

Anxiety Rating Scale (HAM-A). Reductions in the levels of stress hormones like cortisol and adrenocorticotropic hormone have also been reported in probiotic-treated groups compared to controls, indicating a modulating effect on the body's stress response [50,134,135]. In their study, Akkasheh et al. [136] evaluated the impact of probiotic supplementation on metabolic profiles, serum high-sensitivity C-reactive protein (hs-CRP), indicators of oxidative stress, and depression symptoms in patients with major depressive disorder (MDD). The study was run in a randomized, double-blind, placebo-controlled design in which patients were randomly allocated into two groups to receive either probiotic supplements ($n = 20$) or placebo ($n = 20$) for 8 weeks. The probiotic capsule consisted of three viable and freeze-dried strains: *Lactobacillus acidophilus* (2×10^9 CFU/g), *Lactobacillus casei* (2×10^9 CFU/g), and *Bifidobacterium bifidum* (2×10^9 CFU/g). Results showed that the BDI scores of patients who received the probiotic supplements were considerably lower than those of the placebo group (-5.7 ± 6.4 vs. -1.5 ± 4.8, $p = 0.001$). Furthermore, when compared to the placebo, there were notable drops in serum hs-CRP concentrations (-1138.7 ± 2274.9 vs. 188.4 ± 1455.5 ng/mL, $p = 0.03$), insulin resistance as determined by the homeostasis model (-0.6 ± 1.2 vs. 0.6 ± 2.1, $p = 0.03$), and serum insulin levels (-2.3 ± 4.1 vs. 2.6 ± 9.3 µIU/mL, $p = 0.03$). This highlights the potential of probiotics to serve as an accessible, complementary approach for supporting mental health, especially when used in combination with conventional therapies. However, more robust, large-scale human studies are still needed to confirm which specific strains and dosages provide the most consistent benefits for mental well-being.

5.4. Antimicrobial Activity and Modulation of Gut Microbiota

One of the most important functional properties of probiotics is their ability to produce antimicrobial compounds, including organic acids, short-chain fatty acids (SCFAs), and bacteriocins that help to create an unfavorable environment for pathogenic microorganisms [62]. Nisin, for instance, which is produced by *L. lactis*, has been reported to inactivate growth of vegetative spore-forming Gram-positive bacteria by binding to lipid II, disrupting cell wall biosynthesis and facilitating the formation of pores. This antimicrobial action has also been reported in emerging probiotics. For example, Mabeku et al. [137] demonstrated in their study that LAB strains isolated from fermented *Theobroma cacao* fruit juice successfully inhibited the bacterium *Helicobacter pylori*, which is implicated in gastritis, gastric ulcer, and even certain types of stomach cancer. The authors attributed the antimicrobial activity of the LAB strains to their ability to produce bacteriocins that led to the inhibition of the pathogen's growth. This antimicrobial activity is not only a welcome alternative to traditional antibiotics, but a necessity in the extension of the shelf life and safety of foods.

In addition to their direct inhibition of pathogenic microbes, probiotics are also significantly crucial in balancing gut microbiota. A healthy colon is dominated by beneficial anaerobic bacteria. However, when dysbiosis (microbial imbalance) occurs, undesirable microbes such as hemolytic *Staphylococcus* species have the potential to overgrow and induce inflammation or impair gut barrier function [50]. Studies have shown that probiotics in food supplements play crucial roles in reestablishing gut microbiota balance by stimulating growth of beneficial anaerobes while suppressing the proliferation of unwanted bacteria. For instance, yeasts like *Wickerhamomyces subpelliculosus* and *Pichia kudriavzevii* used as a starter culture in a cornelian cherry beverage demonstrated in vitro to be effective at improving the numbers of beneficial anaerobic bacteria and inhibiting *Staphylococcus* spp. growth, even better than noted prebiotics like fructooligosaccharides [138]. However, more stringent, heterogeneous human trials are needed to establish comprehensively the scope and uniformity of the antimicrobial and microbiota-modifying activity of emerging probiotics.

6. Translational Insights: Market Trends and Consumer Acceptance of Probiotics in Functional Foods

The technological and biological attributes of probiotics strains that enable them to withstand food processing, storage, and gastrointestinal transit also enhance their commercial potential/viability. This becomes more relevant as scientific understanding in this field grows and is utilized to elucidate mechanisms of stress resistance, adhesion, and safety. These insights increasingly inform the strategic development and positioning of novel functional food products, fast-tracking modern probiotics from laboratory characterization to commercial product development. This is further shaped by the evolving consumer preferences, regulatory frameworks, and innovations in food formulation.

As highlighted, functional foods fortified with probiotics have moved from a niche category centered on fermented dairy to a diversified global marketplace spanning beverages, snacks, infant and medical nutrition, and shelf-stable categories. This has increased the market reach of these products. Furthermore, with the growing scientific interest in gut microbiome research, combined with advances in formulation and quality control, product formats which can be fortified with live microorganisms have broadened, together with delivery methods and a rapid emergence of personalized, microbiome-based products. However, there are still some concerns on how scientific validation can be standardized and streamlining of regulatory guardrails to shape consumer acceptance.

6.1. Market Growth of Functional Foods and Probiotics

Over the years, the global probiotic market has grown significantly. Industry analyses estimate the global probiotics market to be on the order of tens of billions USD today, with reported valuations ranging from ~USD 70–90 billion in the mid-2020s and projected growth to >USD 100 billion within the next five to ten years depending on scope, with one estimate placing the value at USD 87.70 billion as of 2023, with a projected annual growth rate up to 14% in 2030 (USD 220 billion) [139]. This growth is driven by the converging consumer demands for diverse classes of probiotic-fortified functional foods, a shifting interest in digestive comfort, immune support, and holistic wellness through better-for-you snacking and beverages; and the proliferation of plant-based matrices that invite fermentation and fortification [140]. Conventional strains of probiotics including *Lactobacillus* and *Bifidobacterium* species are still utilized in the bulk of food applications because their safety is well-documented, manufacturing is scalable, and regulatory pathways are comparatively clear [106]. These strains dominate categories where cold-chain distribution is practical and consumer familiarity is high, especially yogurt, kefir, fermented milks, and cultured plant-based alternatives. In parallel, with encapsulation technologies and the adoption of heat- and acid-tolerant spore formers such as *Bacillus coagulans*, more conventional foods can be fortified with these emerging probiotics, when properly characterized at the strain level and manufactured to avoid toxigenicity [24]. This allows for the use of probiotics in ambient beverages, nutrition bars, baked goods, confectionery, and cereals, which are food forms that were historically resistant to live-culture inclusion. The growth in the probiotic market varies with regions, with the Asia-Pacific region being the fastest-growing and often largest regional market for probiotics and probiotic-fortified foods, driven by large population bases, rising disposable income, and cultural acceptance of fermented products [141].

6.2. Consumer Awareness and Preferences

Consumer awareness has shifted from generic "good bacteria" messaging to more nuanced expectations about strain identity, dose, and evidence. Many consumers now look for labeled genus-species-strain designations, guaranteed CFUs through end of shelf life, and clear benefits linked to everyday outcomes such as regularity, tolerance of lactose and Fermentable Oligo-, Di-, Mono-saccharides and Polyols (FODMAPs), reduced bloating, or immune support during seasonal challenges [142,143]. Trust is built through strategies that allow consumers to make informed choices. This is promoted through third-party certifications of good manufacturing practices, and QR-code transparency that links to certificates of analysis and study summaries [144]. Sensory performance remains critical: despite health motivations, products must deliver desirable taste, texture, and aroma. Fermented dairy continues to set the benchmark for probiotic palatability, but growth in plant-based matrices has diversified preferences toward oat, coconut, and soy substrates, where fermentation can mitigate off-notes, reduce sugars, or improve protein digestibility, provided starter cultures and probiotics are harmonized to avoid sensory defects. Furthermore, cultural factors also shape consumer perception and acceptance of fortified foods: in East Asian markets, traditional fermented foods and beverages create a receptive baseline, whereas in parts of Europe regulatory conservatism raises the bar for claims but fosters trust in products that pass it [145]. Across geographies, clean-label expectations (short ingredient lists, minimal processing, non-GMO where relevant) and sustainability credentials (low-impact packaging, responsible sourcing of prebiotics) increasingly influence purchase decisions [146].

One of the key determinants of the success of functional foods with probiotics, particularly with emerging strains recently, is consumer education. It has been identified that early adopters of probiotic and other health-related food products are younger, health-conscious consumers who are extremely trend-sensitive as a result of what they observe on social media and who place an extremely high value on the use of naturalness in foods [147–149]. Alongside this, consumer attitudes towards food, health, and nutrition have become more complex and fragmented [1]. Part of the reason is internet connectivity that has empowered people to instantly search for information on food products, ingredients, and health claims on their phone. This easy access to (potentially contradictory) information has emboldened consumers to be more skeptical of expert advice, especially younger consumers who can be skeptical of traditional health professionals [150]. Importantly, many consumers confuse probiotics, prebiotics, postbiotics, and fermented foods; others assume any fermentation confers probiotic effects, which is not the case unless specific strains and doses are delivered alive where benefits occur. There is need to provide plain-language explanations, distinguishing live probiotic strains from general starter cultures, clarifying that benefits are strain specific, and explaining why CFUs must be guaranteed through the end of shelf life. Equally, responsible communication includes reminders that probiotics complement, not replace, medical care for diagnosed conditions. For example, variations in manufacturing processes may influence the ratio of live to dead bacteria in a probiotic product. It is certain that during manufacture, storage, or freeze-drying, some probiotic cells will die. To compensate and meet the given live cell count on the label, companies will often add extra live cells, a process called "overfilling", in such a manner that the product will continue to deliver the minimum indicated CFU at the time of consumption [151]. However, current labeling conventions usually report the number of live, viable bacteria per dose as CFU but not the total number of bacterial cells consumed, including dead cells. Consumers are therefore sometimes unknowingly consuming more total microbial load than reported on the label. While dead cells might infrequently exert beneficial health effects [125], this lack of clear information can affect consumer confidence and raise concerns about transparency and informed choice. Hence, it has become important for clear, simple labeling and consumer education to be considered when introducing new probiotic strains or new microbial ingredients. This is because trust that product labeling is clear and accurate is another key determinant of the way consumers perceive emerging probiotic products, since consumers would like to know what strains were included, what specific health benefit they may provide, and if the statements are supported by trustworthy science. Stricter guidelines and improved communication are therefore needed to help them understand exactly what CFU means, how production methods influence viability, and the role of both live and dead cells. While companies and regulators face the challenge of balancing scientific accuracy with simple, understandable communication that builds trust, fulfilling this need for clarification will be key to winning more acceptance of new probiotic products in an increasingly skeptical but health-conscious market.

6.3. *Personalized Nutrition and Microbiome-Based Products*

The rapidly expanding catalogue of next-generation and engineered probiotic organisms amplifies personalization potential of functional foods. Candidate commensals such as *Akkermansia* and *Faecalibacterium* spp. are being explored for the amelioration of metabolic and inflammatory conditions, although most remain investigational for general food use [42]. Engineered LAB that secrete enzymes (for example, lactase for lactose maldigestion) or other bioactive peptides foreshadow a future of function-specific foods. Even within conventional species, personalization can occur by selecting strains with distinct bile-salt hydrolase activity, exopolysaccharide profiles that influence immune signaling, or carbohydrate utilization pathways matched to an individual's habitual diet [152]. Synbiotic co-design takes this further by pairing a strain with the prebiotic it best metabolizes, improving engraftment and persistence. Personalization is reshaping the probiotic food landscape through three intersecting pathways: diagnostics, targeted strain selection, and adaptive delivery [153]. Consumer microbiome testing, typically 16S rRNA sequencing with increasing adoption of shotgun metagenomics has become a gateway to individualized recommendations. While methodological limitations remain, these tests provide relative abundance profiles, diversity metrics, and functional potential proxies that can be mapped to diet and probiotic suggestions. Of recent, it is now possible to translate such readouts into tailored food plans and curated assortments of probiotic-fortified products, sometimes combined with wearable-derived lifestyle data and short-term biomarkers (e.g., breath hydrogen or glycemic responses) [154]. A growing number of programs use iterative algorithms where consumers submit periodic samples and symptom diaries. These are utilized together with advanced technologies including blockchain and algorithms to update the strain mix, prebiotic pairing, and intake timing to nudge the gut microbiome toward desired functional states [155,156].

Despite promise, personalized probiotic foods face scientific and ethical hurdles. Inter-individual variability in microbiome ecology and host genetics complicates attribution: the same strain may have divergent effects depending on baseline community structure, diet, and transit time [157]. Also, many endpoints of interest such as mood, stress resilience, skin appearance are multidimensional and require careful study design to avoid over-interpreting placebo-responsive signals. Also, data privacy and consent frameworks must evolve to protect genomic and health-related information collected through testing programs. These are issues which need to be addressed as the market for personalized probiotic foods grows.

7. Challenges and Future Perspectives

While the global interest in probiotics has continued to grow over the years, driven by consumer demand for health-promoting foods, the use of novel and emerging microbial strains presents new challenges and raises questions about safety, harmonized regulatory frameworks, and scientific rigor.

7.1. Regulatory and Safety Considerations

As research continues the exploration of emerging microbial strains from different unconventional sources, it is important to ensure the safety of these products for human use. Regulatory frameworks like the Generally Recognized as Safe (GRAS) status in the United States and the Qualified Presumption of Safety (QPS) list in the European Union provide structured approaches for assessing the safety of microbial strains used in food and supplements. The QPS concept for instance, was developed by the EFSA as a generic risk assessment method for biological agents used in the production of food and feed to streamline and standardize the evaluation of notified biological agents across the EFSA's numerous Scientific Panels and Units [36]. This technique makes it easier to evaluate and approve low-risk microorganisms with a pre-established history of safe use, while allowing for a more targeted use of the available resources for agents with higher risk potential [158]. If under the QPS system a strain can be unequivocally placed in a taxonomic group already present on the QPS list, then a very limited number of additional safety checks could be required. Such a system would rely on four broad pillars namely: correct taxonomic identification; sufficient knowledge and literature related to safe use; evidence ruling out pathogenic or virulent nature; and assurance of the end use and effectiveness of the microorganism in the final product [36,72,159].

In addition to QPS, the EU-funded PROSAFE project (Biosafety Assessment of Probiotic Lactic Acid Bacteria for Human Consumption) was launched to increase evidence-based recommendations to assess the safety of probiotics for use in human beings [72]. In 2006, PROSAFE specialists outlined operational measures to guide future probiotic development. These included calling for the application of proper molecular identification methods, such as ribosomal DNA sequencing by qualified laboratories, to ensure taxonomic LAB classification [160]. Strains should also be deposited in established public collections for reference, with access being controlled. One of the key suggestions was that strains with higher or unusual antibiotic resistance should not be put into the food chain for human or animal consumption without extensive risk evaluation [74]. Similarly, probiotic strains with documented possession of established virulence genes for example, some *Enterococcus* species, are to be avoided at all costs. It was also ascertained that fringe properties such as bile acid deconjugation or broad adhesion capability, were insufficient metrics to base probiotic safety decisions on independently [72]. Furthermore, PROSAFE outlines that the use of animals for safety was deemed inferior to human trials as in vivo safety tests of probiotic ability to adhere to human gut should preferably be evaluated by well-designed human clinical trials which ideally should be randomized, placebo-controlled and double-blind. However, if animal models are to be used, rat endocarditis model was suggested as being most informative because it allows for the assessment of potential pathogenicity and translocation of probiotic strains to sterile sites, such as cardiac tissue, thereby providing a sensitive and reproducible system to detect virulence traits under conditions closely simulating the pathophysiology of infective endocarditis in humans [161].

Globally, the 2002 FAO/WHO guidelines remain authoritative in laying down overall principles for safe food use of probiotics [46]. These expert guidelines coincide mostly with those of PROSAFE. They highlight the importance of having a documented history of safe use, absence of transferable antibiotic resistance and pathogenic virulence factors, and additional testing for some metabolic functions such as D-lactate production or toxin production in species known to produce toxins. They also recommend checking hemolytic activity if appropriate and observing potential side effects throughout human trials. Lastly, the guideline notes the necessity of continuing post-market surveillance to monitor any side effects when products reach consumers. However, while these guidelines emphasize key elements of safety, they lack full details on how to conduct these tests or interpret their results in practice.

In the United States, the regulation on probiotics remains fragmented as there is no official legal definition of "probiotic" at the federal level [72]. The FDA rather governs probiotic products within established paradigms based on their intended use, that is, as foods, food ingredients, dietary supplements, or drugs. Each of these has its own standards regarding safety, claims, efficacy, and manufacturing practices [50]. Under the current system, probiotics marketed as dietary supplements are treated as "foods" and must comply with the Dietary Supplement Health and Education Act (DSHEA) as well as FDA guidelines. If a probiotic is intended for a medical therapeutic purpose, it is being treated as a "drug" and must undergo the formal FDA drug approval process. Probiotics used as biological products require authorization through a Biologic License Application (BLA) under FDA regulation [162]. In addition to product classification, the FDA also monitors the types of claims permitted for traditional probiotic foods and beverages. Although this is rare, health claims need to be pre-approved by the FDA through a formal petitioning process. However, structure, function, or nutrient content claims can be made without premarket approval, though companies are required by the FDA to provide notice to the agency 30 days before marketing. To make such claims viable, the Federal Trade Commission (FTC) requires evidence based on proper research and testing, which the FDA adopts to ensure that the claims are true and supported by good scientific research [162].

To augment the evidence base, the National Institutes of Health (NIH) and the FDA approved thorough systematic reviews in the auspices of the Agency for Healthcare Research and Quality (AHRQ). This review revolves around evaluating the safety profiles of the most used probiotic genera such as *Lactobacillus*, *Bifidobacterium*, *Saccharomyces*, *Streptococcus*, *Enterococcus*, and *Bacillus*, with special emphasis on when used for prevention, management, or reduction in disease risk [72,163]. The objective is to chart what is currently known about probiotic safety; determine the gaps; and provide guidance to practitioners, researchers, and regulators on best practices for assessing probiotic risk and benefit. Remarkably, early findings suggest that while many studies focus on probiotic efficacy, relatively few comprehensively address safety endpoints in sufficient detail [72]. This suggests the need for more systematic safety testing in future clinical trials to build strong regulatory agreement about what safety measures need to be standardized for probiotics in the American marketplace.

Collectively, these initiatives illustrate both the progress and persistent challenges in ensuring the safe use of probiotics, especially for novel and emerging strains. While there is agreement on core safety principles such as robust molecular identification, accurate functional screening, transparent strain documentation, and human trial data, the lack of harmonized global standards and clear, practical testing protocols remains a barrier. As the field of probiotics research continues to advance, it is crucial to align these efforts and close any gaps to balance innovation with consumer protection.

7.2. Research Gaps and Future Directions

By definition, a probiotic must deliver proven health benefit to the host. To meet this standard, there must be evidence of at least one rigorously performed human trial with an acceptable benefit on a relevant health outcome, ideally confirmed by follow-up studies of similar quality [10]. While there have been some promising findings, most of the probiotic research to date relies on short-term trials involving small sample sizes and insufficient follow-up, making it difficult to draw definitive conclusions about long-term safety, reliability, and real-world benefits. Moreover, demonstrating a true health benefit also depends not only on whether a trial is designed and conducted well, but on clear reporting and, ultimately, the ability of the scientific community to critically review results [2]. Therefore, the future generation of probiotic research must prioritize large-scale, randomized controlled trials that are methodologically rigorous, well reported, and for long durations. Such evidence is required to validate strain-specific benefit, establish dose–response, and rule out potential adverse effects with long-term application; all of which are critical to convincingly position probiotics as effective functional food ingredients backed by sound science.

As previously described, it is well recognized that probiotic benefits are highly strain-specific, with wide genetic differences between two probiotic strains of the same species [45]. However, this variability has broader implications for the functional classification of probiotics. For example, Douillard et al. [164] compared 100 isolates of *L. rhamnosus* from human and dairy origin and found that the strains displayed significant variations in essential functional traits such as resistance to bile acids, metabolism of carbohydrates, and possession of mucus-binding pili. This demonstrates that taxonomic classification alone is not an efficient predictor of the way a probiotic will perform; rather, the distinct functional attributes of each strain must be rigorously tested [2,45]. Hence, probiotic research cannot be limited to physiological classification, but each strain must be identified correctly, thoroughly characterized, and chosen based on both its genetic attributes, good evidence of its functional characteristics and its health impacts. It is this more nuanced understanding that constitutes the foundation of "precision probiotics" (strains that are selectively designed or bioengineered to produce predictable, specific effects in specific states of health).

Future studies must enhance the effort to define these strain-specific mechanisms of action and test the interaction of different strains with each other and the host in vivo. A deeper understanding is also needed on how probiotics and the human body interact, which include the immune system, gut barrier, and resident microbiota. While in vitro and in vivo studies have provided valuable information, more human-centric research will be required to map these interactions at the cellular and molecular levels and delineate with certainty how they deliver real health outcomes. Insight into these dynamic microbiome–host interactions could allow for the development of next-generation probiotics that specifically target the gut microbiome with more precision and deliver consistent, quantifiable health benefits. This also provides the foundation to explore multi-strain or synbiotic products designed to have synergistic or complementary effects. With the advancement in next-generation genomics, transcriptomics, and functional screening tools, precision probiotics will play a leading role in next-generation functional foods designed to meet individualized health requirements with scientific accuracy.

8. Conclusions

The growing global interest in functional foods continues to drive the discovery and application of emerging probiotic strains from unconventional sources such as fruits, vegetables, marine environments, etc. Different studies have demonstrated that these sources are valuable reservoirs of viable LAB and yeast species that contain unique techno-functional attributes. These emerging strains not only exhibit significant probiotic characteristics such as resistance to acid and bile, pathogen inhibition, and survival in the gut but also hold greater promise for additional health benefits such as anti-hypertensive, cholesterol-lowering, immunomodulatory, and anti-diabetic effects. However, to maximize the potential of these novel probiotic sources, it is necessary that every probable candidate strain is thoroughly screened and properly characterized, both phenotypically and genotypically. Care must be taken to verify safety attributes, including absence of hemolysis, no breakdown of mucin, and absence of transmissible antibiotic resistance genes, amongst other attributes that must be demonstrated and not assumed. Furthermore, while most of these novel strains have shown promising functional outcomes both in vitro and in animal models, robust human clinical trials must be performed to validate these positive effects across diverse groups such as the elderly, those with metabolic disorders, or individuals with specific dietary restrictions such as lactose intolerance. The integration of new technologies such as advanced molecular methods, culture-independent techniques, and omics-based technologies will augment the cataloguing of strain-specific effects, enabling specific probiotic development towards health outcomes. Importantly, to foster consumer trust, increased transparent regulatory guidance, labeling, and consumer education will be required as these emerging microbial strains are introduced into commercial markets. The future of probiotics in functional foods lies in exploring the wealth of microbial diversity in nature beyond the conventional dairy-based sources, and collaborative efforts are needed to drive this innovation for the benefit of global health and well-being.

Author Contributions: Conceptualization, C.K.A.; methodology, C.K.A. and C.C.U.; writing—original draft, C.C.U. and C.K.A.; writing—review and editing, C.C.U. and C.K.A. validation, C.C.U.; Supervision, C.K.A. All authors have read and agreed to the published version of the manuscript.

Funding: This research received no external funding.

Institutional Review Board Statement: Not applicable.

Informed Consent Statement: Not applicable.

Data Availability Statement: No new data were created or analyzed in this study. Data sharing is not applicable to this article.

Conflicts of Interest: The authors declare no conflicts of interest.

References

1. Birch, C.S.; Bonwick, G.A. Ensuring the future of functional foods. *Int. J. Food Sci. Technol.* **2019**, *54*, 1467–1485. [CrossRef]
2. Binda, S.; Hill, C.; Johansen, E.; Obis, D.; Pot, B.; Sanders, M.E.; Tremblay, A.; Ouwehand, A.C. Criteria to qualify microorganisms as "probiotic" in foods and dietary supplements. *Front. Microbiol.* **2020**, *11*, 1662. [CrossRef]
3. Shimizu, M. History and current status of functional food regulations in Japan. In *Nutraceutical and Functional Food Regulations in the United States and Around the World*; Elsevier: Amsterdam, The Netherlands, 2019; pp. 337–344.
4. Manach, C.; Milenkovic, D.; Van de Wiele, T.; Rodriguez-Mateos, A.; de Roos, B.; Garcia-Conesa, M.T.; Landberg, R.; Gibney, E.R.; Heinonen, M.; Tomás-Barberán, F. Addressing the inter-individual variation in response to consumption of plant food bioactives: Towards a better understanding of their role in healthy aging and cardiometabolic risk reduction. *Mol. Nutr. Food Res.* **2017**, *61*, 1600557. [CrossRef] [PubMed]
5. Dong, Y.; Li, M.; Yue, X. Current research on probiotics and fermented products. *Foods* **2024**, *13*, 1406. [CrossRef]
6. Oyeniran, A.; Gyawali, R.; Aljaloud, S.O.; Krastanov, A.; Ibrahim, S.A. Probiotic characteristics and health benefits of the yogurt bacterium Lactobacillus delbrueckii sp. bulgaricus. In *Current Issues and Challenges in the Dairy Industry*; IntechOpen: Rijeka, Croatia, 2020.
7. Khan, R.S.; Grigor, J.; Winger, R.; Win, A. Functional food product development–Opportunities and challenges for food manufacturers. *Trends Food Sci. Technol.* **2013**, *30*, 27–37. [CrossRef]
8. Vlaicu, P.A.; Untea, A.E.; Varzaru, I.; Saracila, M.; Oancea, A.G. Designing nutrition for health—Incorporating dietary by-products into poultry feeds to create functional foods with insights into health benefits, risks, bioactive compounds, food component functionality and safety regulations. *Foods* **2023**, *12*, 4001. [CrossRef]
9. Latif, A.; Shehzad, A.; Niazi, S.; Zahid, A.; Ashraf, W.; Iqbal, M.W.; Rehman, A.; Riaz, T.; Aadil, R.M.; Khan, I.M. Probiotics: Mechanism of action, health benefits and their application in food industries. *Front. Microbiol.* **2023**, *14*, 1216674; Erratum in *Front. Microbiol.* **2024**, *14*, 1216674. [CrossRef] [PubMed]
10. Hill, C.; Guarner, F.; Reid, G.; Gibson, G.R.; Merenstein, D.J.; Pot, B.; Morelli, L.; Canani, R.B.; Flint, H.J.; Salminen, S. Expert consensus document: The International Scientific Association for Probiotics and Prebiotics consensus statement on the scope and appropriate use of the term probiotic. *Nat. Rev. Gastroenterol. Hepatol.* **2014**, *11*, 506–514. [CrossRef]
11. ISAPP. *Minimum Criteria for Probiotics [Online]*; International Scientific Association for Probiotics and Prebiotics: Sacramento, CA, USA, 2018.
12. Champagne, C.P.; da Cruz, A.G.; Daga, M. Strategies to improve the functionality of probiotics in supplements and foods. *Curr. Opin. Food Sci.* **2018**, *22*, 160–166. [CrossRef]

13. Granato, D.; Branco, G.F.; Nazzaro, F.; Cruz, A.G.; Faria, J.A. Functional foods and nondairy probiotic food development: Trends, concepts, and products. *Compr. Rev. Food Sci. Food Saf.* **2010**, *9*, 292–302. [CrossRef]
14. Corbo, M.R.; Campaniello, D.; Speranza, B.; Altieri, C.; Sinigaglia, M.; Bevilacqua, A. Neutralisation of toxins by probiotics during the transit into the gut: Challenges and perspectives. *Int. J. Food Sci. Technol.* **2018**, *53*, 1339–1351. [CrossRef]
15. Mokhtari, S.; Khomeiri, M.; Jafari, S.M.; Maghsoudlou, Y.; Ghorbani, M. Descriptive analysis of bacterial profile, physicochemical and sensory characteristics of grape juice containing *Saccharomyces cerevisiae* cell wall-coated probiotic microcapsules during storage. *Int. J. Food Sci. Technol.* **2017**, *52*, 1042–1048. [CrossRef]
16. Campaniello, D.; Speranza, B.; Petruzzi, L.; Bevilacqua, A.; Corbo, M.R. How to routinely assess transition, adhesion and survival of probiotics into the gut: A case study on propionibacteria. *Int. J. Food Sci. Technol.* **2018**, *53*, 484–490. [CrossRef]
17. Moumita, S.; Das, B.; Sundaray, A.; Satpathi, S.; Thangaraj, P.; Marimuthu, S.; Jayabalan, R. Study of soy-fortified green tea curd formulated using potential hypocholesterolemic and hypotensive probiotics isolated from locally made curd. *Food Chem.* **2018**, *268*, 558–566. [CrossRef]
18. Mohammadi-Sartang, M.; Bellissimo, N.; de Zepetnek, J.T.; Brett, N.; Mazloomi, S.; Fararouie, M.; Bedeltavana, A.; Famouri, M.; Mazloom, Z. The effect of daily fortified yogurt consumption on weight loss in adults with metabolic syndrome: A 10-week randomized controlled trial. *Nutr. Metab. Cardiovasc. Dis.* **2018**, *28*, 565–574. [CrossRef] [PubMed]
19. Ephrem, E.; Najjar, A.; Charcosset, C.; Greige-Gerges, H. Encapsulation of natural active compounds, enzymes, and probiotics for fruit juice fortification, preservation, and processing: An overview. *J. Funct. Foods* **2018**, *48*, 65–84. [CrossRef]
20. Wardill, H.R.; Van Sebille, Y.Z.; Ciorba, M.A.; Bowen, J.M. Prophylactic probiotics for cancer therapy-induced diarrhoea: A meta-analysis. *Curr. Opin. Support. Palliat. Care* **2018**, *12*, 187–197. [CrossRef]
21. Pimentel, T.C.; de Oliveira, L.I.G.; Macedo, E.d.L.C.; Costa, G.N.; Dias, D.R.; Schwan, R.F.; Magnani, M. Understanding the potential of fruits, flowers, and ethnic beverages as valuable sources of techno-functional and probiotics strains: Current scenario and main challenges. *Trends Food Sci. Technol.* **2021**, *114*, 25–59. [CrossRef]
22. Lakra, A.K.; Domdi, L.; Hanjon, G.; Tilwani, Y.M.; Arul, V. Some probiotic potential of *Weissella confusa* MD1 and *Weissella cibaria* MD2 isolated from fermented batter. *LWT* **2020**, *125*, 109261. [CrossRef]
23. Vasiee, A.; Alizadeh Behbahani, B.; Tabatabaei Yazdi, F.; Mortazavi, S.A.; Noorbakhsh, H. Diversity and probiotic potential of lactic acid bacteria isolated from horreh, a traditional Iranian fermented food. *Probiotics Antimicrob. Proteins* **2018**, *10*, 258–268. [CrossRef]
24. Cao, J.; Yu, Z.; Liu, W.; Zhao, J.; Zhang, H.; Zhai, Q.; Chen, W. Probiotic characteristics of *Bacillus coagulans* and associated implications for human health and diseases. *J. Funct. Foods* **2020**, *64*, 103643. [CrossRef]
25. Lee, N.-K.; Kim, S.-Y.; Han, K.J.; Eom, S.J.; Paik, H.-D. Probiotic potential of *Lactobacillus* strains with anti-allergic effects from kimchi for yogurt starters. *LWT-Food Sci. Technol.* **2014**, *58*, 130–134. [CrossRef]
26. Aarti, C.; Khusro, A. Functional and technological properties of exopolysaccharide producing autochthonous *Lactobacillus plantarum* strain AAS3 from dry fish based fermented food. *LWT* **2019**, *114*, 108387. [CrossRef]
27. Son, S.-H.; Yang, S.-J.; Jeon, H.-L.; Yu, H.-S.; Lee, N.-K.; Park, Y.-S.; Paik, H.-D. Antioxidant and immunostimulatory effect of potential probiotic Lactobacillus paraplantarum SC61 isolated from Korean traditional fermented food, jangajji. *Microb. Pathog.* **2018**, *125*, 486–492. [CrossRef] [PubMed]
28. Woraprayote, W.; Pumpuang, L.; Tosukhowong, A.; Roytrakul, S.; Perez, R.H.; Zendo, T.; Sonomoto, K.; Benjakul, S.; Visessanguan, W. Two putatively novel bacteriocins active against Gram-negative food borne pathogens produced by *Weissella hellenica* BCC 7293. *Food Control* **2015**, *55*, 176–184. [CrossRef]
29. Patil, M.; Jadhav, A.; Patil, U. Functional characterization and in vitro screening of *Fructobacillus fructosus* MCC 3996 isolated from *Butea monosperma* flower for probiotic potential. *Lett. Appl. Microbiol.* **2020**, *70*, 331–339. [CrossRef]
30. Lee, N.-K.; Hong, J.-Y.; Yi, S.-H.; Hong, S.-P.; Lee, J.-E.; Paik, H.-D. Bioactive compounds of probiotic *Saccharomyces cerevisiae* strains isolated from cucumber jangajji. *J. Funct. Foods* **2019**, *58*, 324–329. [CrossRef]
31. Bonatsou, S.; Karamouza, M.; Zoumpopoulou, G.; Mavrogonatou, E.; Kletsas, D.; Papadimitriou, K.; Tsakalidou, E.; Nychas, G.-J.E.; Panagou, E.Z. Evaluating the probiotic potential and technological characteristics of yeasts implicated in cv. Kalamata natural black olive fermentation. *Int. J. Food Microbiol.* **2018**, *271*, 48–59. [CrossRef]
32. Das, P.; Khowala, S.; Biswas, S. In vitro probiotic characterization of *Lactobacillus casei* isolated from marine samples. *LWT* **2016**, *73*, 383–390. [CrossRef]
33. Kim, Y.; Choi, S.-I.; Jeong, Y.; Kang, C.-H. Evaluation of safety and probiotic potential of *Enterococcus faecalis* MG5206 and *Enterococcus faecium* MG5232 isolated from kimchi, a Korean fermented cabbage. *Microorganisms* **2022**, *10*, 2070. [CrossRef] [PubMed]
34. Zhang, T.; Li, Q.; Cheng, L.; Buch, H.; Zhang, F. Akkermansia muciniphila is a promising probiotic. *Microb. Biotechnol.* **2019**, *12*, 1109–1125. [CrossRef] [PubMed]

5. Martín, R.; Miquel, S.; Benevides, L.; Bridonneau, C.; Robert, V.; Hudault, S.; Chain, F.; Berteau, O.; Azevedo, V.; Chatel, J.M. Functional characterization of novel *Faecalibacterium prausnitzii* strains isolated from healthy volunteers: A step forward in the use of *F. prausnitzii* as a next-generation probiotic. *Front. Microbiol.* **2017**, *8*, 1226. [CrossRef]
6. Brodmann, T.; Endo, A.; Gueimonde, M.; Vinderola, G.; Kneifel, W.; de Vos, W.M.; Salminen, S.; Gómez-Gallego, C. Safety of novel microbes for human consumption: Practical examples of assessment in the European Union. *Front. Microbiol.* **2017**, *8*, 1725. [CrossRef]
7. Bubnov, R.V.; Babenko, L.P.; Lazarenko, L.M.; Mokrozub, V.V.; Spivak, M.Y. Specific properties of probiotic strains: Relevance and benefits for the host. *EPMA J.* **2018**, *9*, 205–223. [CrossRef] [PubMed]
8. García-Ruiz, A.; de Llano, D.G.; Esteban-Fernández, A.; Requena, T.; Bartolomé, B.; Moreno-Arribas, M.V. Assessment of probiotic properties in lactic acid bacteria isolated from wine. *Food Microbiol.* **2014**, *44*, 220–225. [CrossRef]
9. Salmerón, I. Fermented cereal beverages: From probiotic, prebiotic and synbiotic towards Nanoscience designed healthy drinks. *Lett. Appl. Microbiol.* **2017**, *65*, 114–124. [CrossRef] [PubMed]
0. Gupta, S.; Abu-Ghannam, N. Probiotic fermentation of plant based products: Possibilities and opportunities. *Crit. Rev. Food Sci. Nutr.* **2012**, *52*, 183–199. [CrossRef]
1. Betoret, N.; Puente, L.; Dıaz, M.; Pagán, M.; Garcıa, M.; Gras, M.; Martínez-Monzó, J.; Fito, P. Development of probiotic-enriched dried fruits by vacuum impregnation. *J. Food Eng.* **2003**, *56*, 273–277. [CrossRef]
2. Verhoog, S.; Taneri, P.E.; Roa Díaz, Z.M.; Marques-Vidal, P.; Troup, J.P.; Bally, L.; Franco, O.H.; Glisic, M.; Muka, T. Dietary factors and modulation of bacteria strains of *Akkermansia muciniphila* and *Faecalibacterium prausnitzii*: A systematic review. *Nutrients* **2019**, *11*, 1565. [CrossRef]
3. Pabari, K.; Pithva, S.; Kothari, C.; Purama, R.K.; Kondepudi, K.K.; Vyas, B.R.M.; Kothari, R.; Ambalam, P. Evaluation of probiotic properties and prebiotic utilization potential of Weissella paramesenteroides isolated from fruits. *Probiotics Antimicrob. Proteins* **2020**, *12*, 1126–1138. [CrossRef]
4. Mei, L.; Wang, J.; Hao, Y.; Zeng, X.; Yang, Y.; Wu, Z.; Ji, Y. A comprehensive update on the immunoregulatory mechanisms of *Akkermansia muciniphila*: Insights into active ingredients, metabolites, and nutrient-driven modulation. *Crit. Rev. Food Sci. Nutr.* **2024**, *65*, 5487–5504. [CrossRef]
5. Day, R.L.; Harper, A.J.; Woods, R.M.; Davies, O.G.; Heaney, L.M. Probiotics: Current landscape and future horizons. *Future Sci. OA* **2019**, *5*, FSO391. [CrossRef]
6. FAO/WHO. Guidelines for the evaluation of probiotics in food. In *Food and Agriculture Organization of the United Nations and World Health Organization Working Group Report*; WHO: Geneva, Switzerland, 2002.
7. de Albuquerque, T.M.R.; Garcia, E.F.; de Oliveira Araújo, A.; Magnani, M.; Saarela, M.; de Souza, E.L. In vitro characterization of Lactobacillus strains isolated from fruit processing by-products as potential probiotics. *Probiotics Antimicrob. Proteins* **2018**, *10*, 704–716. [CrossRef]
8. Wattam, A.R.; Davis, J.J.; Assaf, R.; Boisvert, S.; Brettin, T.; Bun, C.; Conrad, N.; Dietrich, E.M.; Disz, T.; Gabbard, J.L. Improvements to PATRIC, the all-bacterial bioinformatics database and analysis resource center. *Nucleic Acids Res.* **2017**, *45*, D535–D542. [CrossRef] [PubMed]
9. Kõll-Klais, P.; Mändar, R.; Leibur, E.; Marcotte, H.; Hammarström, L.; Mikelsaar, M. Oral lactobacilli in chronic periodontitis and periodontal health: Species composition and antimicrobial activity. *Oral Microbiol. Immunol.* **2005**, *20*, 354–361. [CrossRef]
0. de Melo Pereira, G.V.; de Oliveira Coelho, B.; Júnior, A.I.M.; Thomaz-Soccol, V.; Soccol, C.R. How to select a probiotic? A review and update of methods and criteria. *Biotechnol. Adv.* **2018**, *36*, 2060–2076. [CrossRef] [PubMed]
1. Xu, Y.; Zhou, T.; Tang, H.; Li, X.; Chen, Y.; Zhang, L.; Zhang, J. Probiotic potential and amylolytic properties of lactic acid bacteria isolated from Chinese fermented cereal foods. *Food Control* **2020**, *111*, 107057. [CrossRef]
2. Garcia, E.F.; Luciano, W.A.; Xavier, D.E.; da Costa, W.C.; de Sousa Oliveira, K.; Franco, O.L.; de Morais Junior, M.A.; Lucena, B.T.; Picao, R.C.; Magnani, M. Identification of lactic acid bacteria in fruit pulp processing byproducts and potential probiotic properties of selected *Lactobacillus* strains. *Front. Microbiol.* **2016**, *7*, 1371. [CrossRef] [PubMed]
3. Fernández-Pacheco, P.; García-Béjar, B.; Jiménez-del Castillo, M.; Carreño-Domínguez, J.; Briones Pérez, A.; Arévalo-Villena, M. Potential probiotic and food protection role of wild yeasts isolated from pistachio fruits (*Pistacia vera*). *J. Sci. Food Agric.* **2021**, *101*, 2201–2209. [CrossRef]
4. Maragkoudakis, P.A.; Zoumpopoulou, G.; Miaris, C.; Kalantzopoulos, G.; Pot, B.; Tsakalidou, E. Probiotic potential of *Lactobacillus* strains isolated from dairy products. *Int. Dairy J.* **2006**, *16*, 189–199. [CrossRef]
5. Ogunremi, O.; Sanni, A.; Agrawal, R. Probiotic potentials of yeasts isolated from some cereal-based Nigerian traditional fermented food products. *J. Appl. Microbiol.* **2015**, *119*, 797–808. [CrossRef] [PubMed]
6. Boonaert, C.J.; Rouxhet, P.G. Surface of lactic acid bacteria: Relationships between chemical composition and physicochemical properties. *Appl. Environ. Microbiol.* **2000**, *66*, 2548–2554. [CrossRef]
7. Duary, R.K.; Rajput, Y.S.; Batish, V.K.; Grover, S. Assessing the adhesion of putative indigenous probiotic lactobacilli to human colonic epithelial cells. *Indian J. Med. Res.* **2011**, *134*, 664–671. [CrossRef]

58. Tuo, Y.; Yu, H.; Ai, L.; Wu, Z.; Guo, B.; Chen, W. Aggregation and adhesion properties of 22 *Lactobacillus* strains. *J. Dairy Sci.* **2013**, *96*, 4252–4257. [CrossRef] [PubMed]
59. Chelliah, R.; Ramakrishnan, S.R.; Prabhu, P.R.; Antony, U. Evaluation of antimicrobial activity and probiotic properties of wild-strain *Pichia kudriavzevii* isolated from frozen idli batter. *Yeast* **2016**, *33*, 385–401. [CrossRef]
60. Leite, A.M.; Miguel, M.; Peixoto, R.; Ruas-Madiedo, P.; Paschoalin, V.; Mayo, B.; Delgado, S. Probiotic potential of selected lactic acid bacteria strains isolated from Brazilian kefir grains. *J. Dairy Sci.* **2015**, *98*, 3622–3632. [CrossRef]
61. Ramos, C.L.; Thorsen, L.; Schwan, R.F.; Jespersen, L. Strain-specific probiotics properties of *Lactobacillus fermentum*, *Lactobacillus plantarum* and *Lactobacillus brevis* isolates from Brazilian food products. *Food Microbiol.* **2013**, *36*, 22–29. [CrossRef]
62. Mihaylova-Garnizova, R.; Davidova, S.; Hodzhev, Y.; Satchanska, G. Antimicrobial peptides derived from bacteria: Classification, sources, and mechanism of action against multidrug-resistant bacteria. *Int. J. Mol. Sci.* **2024**, *25*, 10788. [CrossRef]
63. Vera-Pingitore, E.; Jimenez, M.E.; Dallagnol, A.; Belfiore, C.; Fontana, C.; Fontana, P.; von Wright, A.; Vignolo, G.; Plumed-Ferrer, C. Screening and characterization of potential probiotic and starter bacteria for plant fermentations. *LWT-Food Sci. Technol.* **2016**, *71*, 288–294. [CrossRef]
64. Divya, J.B.; Varsha, K.K.; Nampoothiri, K.M. Newly isolated lactic acid bacteria with probiotic features for potential application in food industry. *Appl. Biochem. Biotechnol.* **2012**, *167*, 1314–1324. [CrossRef] [PubMed]
65. Inturri, R.; Stivala, A.; Furneri, P.; Blandino, G. Growth and adhesion to HT-29 cells inhibition of Gram-negatives by *Bifidobacterium longum* BB536 e *Lactobacillus rhamnosus* HN001 alone and in combination. *Eur. Rev. Med. Pharmacol. Sci.* **2016**, *20*, 4943–4949. [PubMed]
66. Benítez-Cabello, A.; Calero-Delgado, B.; Rodríguez-Gómez, F.; Garrido-Fernández, A.; Jiménez-Díaz, R.; Arroyo-López, F.N. Biodiversity and multifunctional features of lactic acid bacteria isolated from table olive biofilms. *Front. Microbiol.* **2019**, *10*, 836. [CrossRef]
67. Sanders, M.E.; Shane, A.L.; Merenstein, D.J. Advancing probiotic research in humans in the United States: Challenges and strategies. *Gut Microbes* **2016**, *7*, 97–100. [CrossRef] [PubMed]
68. Yadav, R.; Shukla, P. An overview of advanced technologies for selection of probiotics and their expediency: A review. *Crit. Rev. Food Sci. Nutr.* **2017**, *57*, 3233–3242. [CrossRef]
69. Bagheripoor-Fallah, N.; Mortazavian, A.; Hosseini, H.; Khoshgozaran-Abras, S.; Rad, A.H. Comparison of molecular techniques with other methods for identification and enumeration of probiotics in fermented milk products. *Crit. Rev. Food Sci. Nutr.* **2015**, *55*, 396–413. [CrossRef]
70. Temmerman, R.; Huys, G.; Swings, J. Identification of lactic acid bacteria: Culture-dependent and culture-independent methods. *Trends Food Sci. Technol.* **2004**, *15*, 348–359. [CrossRef]
71. Chen, Y.; Hsiao, P.; Hong, W.; Dai, T.; Chen, M. Lactobacillus kefiranofaciens M1 isolated from milk kefir grains ameliorates experimental colitis in vitro and in vivo. *J. Dairy Sci.* **2012**, *95*, 63–74. [CrossRef]
72. Sanders, M.E.; Akkermans, L.M.; Haller, D.; Hammerman, C.; Heimbach, J.T.; Hörmannsperger, G.; Huys, G. Safety assessment of probiotics for human use. *Gut Microbes* **2010**, *1*, 164–185. [CrossRef]
73. Ashraf, R.; Shah, N.P. Antibiotic resistance of probiotic organisms and safety of probiotic dairy products. *Int. Food Res. J.* **2011**, *18*, 837–853.
74. Klare, I.; Konstabel, C.; Werner, G.; Huys, G.; Vankerckhoven, V.; Kahlmeter, G.; Hildebrandt, B.; Müller-Bertling, S.; Witte, W.; Goossens, H. Antimicrobial susceptibilities of *Lactobacillus*, *Pediococcus* and *Lactococcus* human isolates and cultures intended for probiotic or nutritional use. *J. Antimicrob. Chemother.* **2007**, *59*, 900–912. [CrossRef]
75. Nawaz, M.; Wang, J.; Zhou, A.; Ma, C.; Wu, X.; Moore, J.E.; Cherie Millar, B.; Xu, J. Characterization and transfer of antibiotic resistance in lactic acid bacteria from fermented food products. *Curr. Microbiol.* **2011**, *62*, 1081–1089. [CrossRef]
76. Senan, S.; Prajapati, J.; Joshi, C. Feasibility of genome-wide screening for biosafety assessment of probiotics: A case study of Lactobacillus helveticus MTCC 5463. *Probiotics Antimicrob. Proteins* **2015**, *7*, 249–258. [CrossRef]
77. Schulz, K.F.; Altman, D.G.; Moher, D.; The CONSORT Group. CONSORT 2010 statement: Updated guidelines for reporting parallel group randomized trials. *Ann. Intern. Med.* **2010**, *152*, 726–732. [CrossRef]
78. Hutchinson, A.N.; Bergh, C.; Kruger, K.; Sűsserová, M.; Allen, J.; Améen, S.; Tingö, L. The effect of probiotics on health outcomes in the elderly: A systematic review of randomized, placebo-controlled studies. *Microorganisms* **2021**, *9*, 1344. [CrossRef] [PubMed]
79. Zielińska, D.; Ołdak, A.; Rzepkowska, A.; Zieliński, K. Enumeration and identification of probiotic bacteria in food matrices. In *Advances in Biotechnology for Food Industry*; Elsevier: Amsterdam, The Netherlands, 2018; pp. 167–196.
80. Sornplang, P.; Piyadeatsoontorn, S. Probiotic isolates from unconventional sources: A review. *J. Anim. Sci. Technol.* **2016**, *58*, 26. [CrossRef] [PubMed]
81. Nanasombat, S.; Phunpruch, S.; Jaichalad, T. Screening and identification of lactic acid bacteria from raw seafoods and Thai fermented seafood products for their potential use as starter cultures. *Songklanakarin J. Sci. Technol.* **2012**, *34*, 255–262.

82. Senthong, R.; Chanthachum, S.; Sumpavapol, P. Screening and identification of probiotic lactic acid bacteria isolated from Poo-Khem, A traditional salted crab. In Proceedings of the International Conference on Nutrition and Food Sciences, Singapore, 23–24 July 2012; pp. 111–115.
83. Miyashita, M.; Yukphan, P.; Chaipitakchonlatarn, W.; Malimas, T.; Sugimoto, M.; Yoshino, M.; Potacharoen, W.; Tanasupawat, S.; Nakagawa, Y.; Kirtikara, K. 16S rRNA gene sequence analysis of lactic acid bacteria isolated from fermented foods in Thailand. *Microbiol. Cult. Coll.* **2012**, *28*, 1–9.
84. Siripornadulsil, W.; Tasaku, S.; Buahorm, J.; Siripornadulsil, S. Probiotic properties of lactic acid bacteria isolated from fermented food. *Intl. J. Biol. Food Vet. Agri. Eng.* **2014**, *8*, 364–366.
85. Bacha, K.; Mehari, T.; Ashenafi, M. In-vitro probiotic potential of lactic acid bacteria isolated from 'Wakalim', a traditional Ethiopian fermented beef sausage. *Ethiop. J. Health Sci.* **2009**, *19*, 21–27.
86. Agaliya, P.J.; Jeevaratnam, K. Screening of Lactobacillus plantarum isolated from fermented idli batter for probiotic properties. *Afr. J. Biotechnol.* **2012**, *11*, 12856–12864. [CrossRef]
87. Oluwajoba, S.O.; Akinyosoye, F.A.; Oyetayo, V.O. In vitro screening and selection of probiotic lactic acid bacteria isolated from spontaneously fermenting Kunu-Zaki. *Adv. Microbiol.* **2013**, *3*, 309. [CrossRef]
88. Sarıtaş, S.; Mondragon Portocarrero, A.d.C.; Miranda, J.M.; Witkowska, A.M.; Karav, S. Functional Yogurt: Types and Health Benefits. *Appl. Sci.* **2024**, *14*, 11798. [CrossRef]
89. Sharifi Yazdi, M.K.; Davoodabadi, A.; Khesht Zarin, H.R.; Tajabadi Ebrahimi, M.; Soltan Dallal, M.M. Characterisation and probiotic potential of lactic acid bacteria isolated from Iranian traditional yogurts. *Ital. J. Anim. Sci.* **2017**, *16*, 185–188. [CrossRef]
90. Anumudu, C.K.; Ikimi, C.G.; Zige, D.V.; Omeje, F.I.; Gbodo, E.E. Production of Bacteriocins by *Lactobacillus plantarum* and Pediococcus acidilactici Isolated from Cow Milk. *Niger. J. Microbiol.* **2019**, *33*, 4373–4379.
91. Hoque, M.; Akter, F.; Hossain, K.; Rahman, M.; Billah, M.; Islam, K. Isolation, identification and analysis of probiotic properties of *Lactobacillus* spp. from selective regional yoghurts. *World J. Dairy Food Sci.* **2010**, *5*, 39–46.
92. Plessas, S.; Nouska, C.; Mantzourani, I.; Kourkoutas, Y.; Alexopoulos, A.; Bezirtzoglou, E. Microbiological exploration of different types of kefir grains. *Fermentation* **2017**, *3*, 1. [CrossRef]
93. Diosma, G.; Romanin, D.E.; Rey-Burusco, M.F.; Londero, A.; Garrote, G.L. Yeasts from kefir grains: Isolation, identification, and probiotic characterization. *World J. Microbiol. Biotechnol.* **2014**, *30*, 43–53. [CrossRef]
94. Maeno, S.; Kajikawa, A.; Dicks, L.; Endo, A. Introduction of bifunctional alcohol/acetaldehyde dehydrogenase gene (adhE) in *Fructobacillus fructosus* settled its fructophilic characteristics. *Res. Microbiol.* **2019**, *170*, 35–42. [CrossRef]
95. Behare, P.V.; Mazhar, S.; Pennone, V.; McAuliffe, O. Evaluation of lactic acid bacteria strains isolated from fructose-rich environments for their mannitol-production and milk-gelation abilities. *J. Dairy Sci.* **2020**, *103*, 11138–11151. [CrossRef]
96. Sakandar, H.A.; Kubow, S.; Sadiq, F.A. Isolation and in-vitro probiotic characterization of fructophilic lactic acid bacteria from Chinese fruits and flowers. *Lwt* **2019**, *104*, 70–75. [CrossRef]
97. Adetunji, C.O.; Akram, M.; Michael, O.S.; Shahzad, K.; Ayeni, A.E.; Hasan, S.; Adetunji, J.B.; Hasan, S.M.; Inamuddin; Olaniyan, M. Polysaccharides derived from natural sources: A panacea to health and nutritional challenges. In *Polysaccharides: Properties and Applications*; Wiley: Hoboken, NJ, USA, 2021; pp. 701–738. [CrossRef]
98. Rodrigues, N.P.A.; Garcia, E.F.; de Souza, E.L. Selection of lactic acid bacteria with promising probiotic aptitudes from fruit and ability to survive in different food matrices. *Braz. J. Microbiol.* **2021**, *52*, 2257–2269. [CrossRef] [PubMed]
99. Tenea, G.N.; Perugachi, E. Innovative functional juice enriched with native probiotics for enhanced nutrition and antimicrobial properties. *Front. Nutr.* **2025**, *12*, 1552745. [CrossRef] [PubMed]
100. Sarao, L.K.; Arora, M. Probiotics, prebiotics, and microencapsulation: A review. *Crit. Rev. Food Sci. Nutr.* **2017**, *57*, 344–371. [CrossRef]
101. WGO. *Global Guidelines: Probiotics and Prebiotics*; WGO: Hong Kong, China, 2017.
102. Flach, J.; van der Waal, M.B.; van den Nieuwboer, M.; Claassen, E.; Larsen, O.F. The underexposed role of food matrices in probiotic products: Reviewing the relationship between carrier matrices and product parameters. *Crit. Rev. Food Sci. Nutr.* **2018**, *58*, 2570–2584. [CrossRef]
103. Yao, M.; Xie, J.; Du, H.; McClements, D.J.; Xiao, H.; Li, L. Progress in microencapsulation of probiotics: A review. *Compr. Rev. Food Sci. Food Saf.* **2020**, *19*, 857–874. [CrossRef] [PubMed]
104. Reque, P.M.; Brandelli, A. Encapsulation of probiotics and nutraceuticals: Applications in functional food industry. *Trends Food Sci. Technol.* **2021**, *114*, 1–10. [CrossRef]
105. Fenster, K.; Freeburg, B.; Hollard, C.; Wong, C.; Rønhave Laursen, R.; Ouwehand, A.C. The production and delivery of probiotics: A review of a practical approach. *Microorganisms* **2019**, *7*, 83. [CrossRef]
106. Terpou, A.; Papadaki, A.; Lappa, I.K.; Kachrimanidou, V.; Bosnea, L.A.; Kopsahelis, N. Probiotics in food systems: Significance and emerging strategies towards improved viability and delivery of enhanced beneficial value. *Nutrients* **2019**, *11*, 1591. [CrossRef]

107. Singh, S.; Gupta, R.; Chawla, S.; Gauba, P.; Singh, M.; Tiwari, R.K.; Upadhyay, S.; Sharma, S.; Chanda, S.; Gaur, S. Natural sources and encapsulating materials for probiotics delivery systems: Recent applications and challenges in functional food development. *Front. Nutr.* **2022**, *9*, 971784. [CrossRef]
108. Ta, L.P.; Bujna, E.; Antal, O.; Ladányi, M.; Juhász, R.; Szécsi, A.; Kun, S.; Sudheer, S.; Gupta, V.K.; Nguyen, Q.D. Effects of various polysaccharides (alginate, carrageenan, gums, chitosan) and their combination with prebiotic saccharides (resistant starch, lactosucrose, lactulose) on the encapsulation of probiotic bacteria Lactobacillus casei 01 strain. *Int. J. Biol. Macromol.* **2021**, *183*, 1136–1144. [CrossRef]
109. Anumudu, C.; Miri, T.; Onyeaka, H. Sporicidal Activity of Micro-Encapsulated Nisin-like Bacteriocins Obtained from *Lactococcus lactis* against *Bacillus cereus* spores. In Proceedings of the 2024 European Symposium on Food Safety, Geneva, Switzerland, 30 April–2 May 2024; IAFP: Urbandale, IA, USA, 2024.
110. Heunis, T.; Botes, M.; Dicks, L. Encapsulation of *Lactobacillus plantarum* 423 and its bacteriocin in nanofibers. *Probiotics Antimicrob. Proteins* **2010**, *2*, 46–51. [CrossRef]
111. Afzaal, M.; Khan, A.U.; Saeed, F.; Ahmed, A.; Ahmad, M.H.; Maan, A.A.; Tufail, T.; Anjum, F.M.; Hussain, S. Functional exploration of free and encapsulated probiotic bacteria in yogurt and simulated gastrointestinal conditions. *Food Sci. Nutr.* **2019**, *7*, 3931–3940. [CrossRef] [PubMed]
112. Chen, L.; Yang, T.; Song, Y.; Shu, G.; Chen, H. Effect of xanthan-chitosan-xanthan double layer encapsulation on survival of Bifidobacterium BB01 in simulated gastrointestinal conditions, bile salt solution and yogurt. *LWT-Food Sci. Technol.* **2017**, *81*, 274–280. [CrossRef]
113. Dias, C.O.; de Almeida, J.d.S.O.; Pinto, S.S.; de Oliveira Santana, F.C.; Verruck, S.; Müller, C.M.O.; Prudêncio, E.S.; Amboni, R.D.d.M.C. Development and physico-chemical characterization of microencapsulated bifidobacteria in passion fruit juice: A functional non-dairy product for probiotic delivery. *Food Biosci.* **2018**, *24*, 26–36. [CrossRef]
114. Mokhtari, S.; Jafari, S.M.; Khomeiri, M. Survival of encapsulated probiotics in pasteurized grape juice and evaluation of their properties during storage. *Food Sci. Technol. Int.* **2019**, *25*, 120–129. [CrossRef]
115. Asar, R.; Erenler, S.; Devecioglu, D.; Ispirli, H.; Karbancioglu-Guler, F.; Ozturk, H.I.; Dertli, E. Understanding the Functionality of Probiotics on the Edge of Artificial Intelligence (AI) Era. *Fermentation* **2025**, *11*, 259. [CrossRef]
116. Westfall, S.; Carracci, F.; Estill, M.; Zhao, D.; Wu, Q.-l.; Shen, L.; Simon, J.; Pasinetti, G.M. Optimization of probiotic therapeutics using machine learning in an artificial human gastrointestinal tract. *Sci. Rep.* **2021**, *11*, 1067. [CrossRef]
117. Sadeghi, M.; Panahi, B.; Mazlumi, A.; Hejazi, M.A.; Komi, D.E.A.; Nami, Y. Screening of potential probiotic lactic acid bacteria with antimicrobial properties and selection of superior bacteria for application as biocontrol using machine learning models. *LWT* **2022**, *162*, 113471. [CrossRef]
118. Zhang, D.; Zhang, J.; Kalimuthu, S.; Liu, J.; Song, Z.-M.; He, B.-b.; Cai, P.; Zhong, Z.; Feng, C.; Neelakantan, P. A systematically biosynthetic investigation of lactic acid bacteria reveals diverse antagonistic bacteriocins that potentially shape the human microbiome. *Microbiome* **2023**, *11*, 91. [CrossRef] [PubMed]
119. van Heel, A.J.; de Jong, A.; Song, C.; Viel, J.H.; Kok, J.; Kuipers, O.P. BAGEL4: A user-friendly web server to thoroughly mine RiPPs and bacteriocins. *Nucleic Acids Res.* **2018**, *46*, W278–W281. [CrossRef]
120. Arango-Argoty, G.; Garner, E.; Pruden, A.; Heath, L.S.; Vikesland, P.; Zhang, L. DeepARG: A deep learning approach for predicting antibiotic resistance genes from metagenomic data. *Microbiome* **2018**, *6*, 23. [CrossRef]
121. Krawczyk, P.S.; Lipinski, L.; Dziembowski, A. PlasFlow: Predicting plasmid sequences in metagenomic data using genome signatures. *Nucleic Acids Res.* **2018**, *46*, e35. [CrossRef]
122. Wu, S.; Feng, T.; Tang, W.; Qi, C.; Gao, J.; He, X.; Wang, J.; Zhou, H.; Fang, Z. metaProbiotics: A tool for mining probiotic from metagenomic binning data based on a language model. *Brief. Bioinform.* **2024**, *25*, bbae085. [CrossRef]
123. González-Sendino, R.; Serrano, E.; Bajo, J. Mitigating bias in artificial intelligence: Fair data generation via causal models for transparent and explainable decision-making. *Future Gener. Comput. Syst.* **2024**, *155*, 384–401. [CrossRef]
124. D'Urso, F.; Broccolo, F. Applications of artificial intelligence in microbiome analysis and probiotic interventions—An overview and perspective based on the current state of the art. *Appl. Sci.* **2024**, *14*, 8627. [CrossRef]
125. Song, M.W.; Chung, Y.; Kim, K.-T.; Hong, W.S.; Chang, H.J.; Paik, H.-D. Probiotic characteristics of *Lactobacillus brevis* B13-2 isolated from kimchi and investigation of antioxidant and immune-modulating abilities of its heat-killed cells. *LWT* **2020**, *128*, 109452. [CrossRef]
126. Fakruddin, M.; Hossain, M.N.; Ahmed, M.M. Antimicrobial and antioxidant activities of Saccharomyces cerevisiae IFST062013, a potential probiotic. *BMC Complement. Altern. Med.* **2017**, *17*, 64. [CrossRef]
127. Ornellas, R.M.S.; Santos, T.T.; Arcucio, L.B.; Sandes, S.H.C.; Oliveira, M.M.; Dias, C.V.; de Carvalho Silva, S.; Uetanabaro, A.P.T.; Vinderola, G.; Nicoli, J.R. Selection of lactic acid bacteria with probiotic potential isolated from the fermentation process of "Cupuaçu"(*Theobroma grandiflorum*). *Adv. Microbiol. Infect. Dis. Public Health* **2017**, *7*, 1–16.
128. Ooi, L.-G.; Liong, M.-T. Cholesterol-lowering effects of probiotics and prebiotics: A review of in vivo and in vitro findings. *Int. J. Mol. Sci.* **2010**, *11*, 2499–2522. [CrossRef] [PubMed]

29. Kumar, M.; Nagpal, R.; Kumar, R.; Hemalatha, R.; Verma, V.; Kumar, A.; Chakraborty, C.; Singh, B.; Marotta, F.; Jain, S. Cholesterol-lowering probiotics as potential biotherapeutics for metabolic diseases. *J. Diabetes Res.* **2012**, *2012*, 902917; Erratum in *J. Diabetes Res.* **2022**, *2022*, 3952529. [CrossRef]
30. Cavalcante, R.G.; de Albuquerque, T.M.; de Luna Freire, M.O.; Ferreira, G.A.; Dos Santos, L.A.C.; Magnani, M.; Cruz, J.C.; Braga, V.A.; de Souza, E.L.; de Brito Alves, J.L. The probiotic Lactobacillus fermentum 296 attenuates cardiometabolic disorders in high fat diet-treated rats. *Nutr. Metab. Cardiovasc. Dis.* **2019**, *29*, 1408–1417. [CrossRef]
31. Costabile, A.; Buttarazzi, I.; Kolida, S.; Quercia, S.; Baldini, J.; Swann, J.R.; Brigidi, P.; Gibson, G.R. An in vivo assessment of the cholesterol-lowering efficacy of *Lactobacillus plantarum* ECGC 13110402 in normal to mildly hypercholesterolaemic adults. *PLoS ONE* **2017**, *12*, e0187964.
32. Tannock, G.W.; Savage, D.C. Influences of dietary and environmental stress on microbial populations in the murine gastrointestinal tract. *Infect. Immun.* **1974**, *9*, 591–598. [CrossRef]
33. Luna, R.A.; Foster, J.A. Gut brain axis: Diet microbiota interactions and implications for modulation of anxiety and depression. *Curr. Opin. Biotechnol.* **2015**, *32*, 35–41. [CrossRef]
34. Collins, S.M.; Surette, M.; Bercik, P. The interplay between the intestinal microbiota and the brain. *Nat. Rev. Microbiol.* **2012**, *10*, 735–742. [CrossRef] [PubMed]
35. Dinan, T.G.; Cryan, J.F. Melancholic microbes: A link between gut microbiota and depression? *Neurogastroenterol. Motil.* **2013**, *25*, 713–719. [CrossRef]
36. Akkasheh, G.; Kashani-Poor, Z.; Tajabadi-Ebrahimi, M.; Jafari, P.; Akbari, H.; Taghizadeh, M.; Memarzadeh, M.R.; Asemi, Z.; Esmaillzadeh, A. Clinical and metabolic response to probiotic administration in patients with major depressive disorder: A randomized, double-blind, placebo-controlled trial. *Nutrition* **2016**, *32*, 315–320. [CrossRef] [PubMed]
37. Kouitcheu Mabeku, L.B.; Ngue, S.; Bonsou Nguemo, I.; Leundji, H. Potential of selected lactic acid bacteria from *Theobroma cacao* fermented fruit juice and cell-free supernatants from cultures as inhibitors of Helicobacter pylori and as good probiotic. *BMC Res. Notes* **2020**, *13*, 64. [CrossRef]
38. Di Cagno, R.; Filannino, P.; Cantatore, V.; Polo, A.; Celano, G.; Martinovic, A.; Cavoski, I.; Gobbetti, M. Design of potential probiotic yeast starters tailored for making a cornelian cherry (*Cornus mas* L.) functional beverage. *Int. J. Food Microbiol.* **2020**, *323*, 108591. [CrossRef] [PubMed]
39. Sardar, D.; Morol, I.; Bari, J.; Sarkar, A.; Habib, A. Optimization of cryoprotectants and storage temperatures for preserving viability and probiotic properties of lyophilized bacterial strains from chicken gut. *PLoS ONE* **2025**, *20*, e0328216. [CrossRef]
40. Abdul Manan, M. Progress in Probiotic Science: Prospects of Functional Probiotic-Based Foods and Beverages. *Int. J. Food Sci.* **2025**, *2025*, 5567567. [CrossRef] [PubMed]
41. Gore, A. Probiotics Market Report 2025 (Global Edition). Available online: https://www.cognitivemarketresearch.com/probiotics-market-report (accessed on 3 September 2025).
42. Hill, P.; Muir, J.G.; Gibson, P.R. Controversies and Recent Developments of the Low-FODMAP Diet. *Gastroenterol. Hepatol.* **2017**, *13*, 36–45.
43. Murillo, A.Z.; Arévalo, F.E.; Jáuregui, E.P. Diet low in fermentable oligosaccharides, disaccharides, monosaccharides and polyols (FODMAPs) in the treatment of irritable bowel syndrome: Indications and design. *Endocrinol. Nutr. (Engl. Ed.)* **2016**, *63*, 132–138. [CrossRef]
44. Fredua-Agyeman, M.; Larbi, E.A. Inaccurate labelling practices in probiotic products: A regulatory shortfall in Accra, Ghana. *PLoS ONE* **2025**, *20*, e0322194. [CrossRef]
45. Siegrist, M.; Shi, J.; Giusto, A.; Hartmann, C. Worlds apart. Consumer acceptance of functional foods and beverages in Germany and China. *Appetite* **2015**, *92*, 87–93. [CrossRef]
46. Chang, M.Y.; Chen, H.S. Understanding Consumers' Intentions to Purchase Clean Label Products: Evidence from Taiwan. *Nutrients* **2022**, *14*, 3684. [CrossRef]
47. Bimbo, F.; Bonanno, A.; Nocella, G.; Viscecchia, R.; Nardone, G.; De Devitiis, B.; Carlucci, D. Consumers' acceptance and preferences for nutrition-modified and functional dairy products: A systematic review. *Appetite* **2017**, *113*, 141–154. [CrossRef]
48. Kraus, A.; Annunziata, A.; Vecchio, R. Sociodemographic factors differentiating the consumer and the motivations for functional food consumption. *J. Am. Coll. Nutr.* **2017**, *36*, 116–126. [CrossRef]
49. Roman, S.; Sánchez-Siles, L.M.; Siegrist, M. The importance of food naturalness for consumers: Results of a systematic review. *Trends Food Sci. Technol.* **2017**, *67*, 44–57. [CrossRef]
50. Rendahl, J.; Korp, P.; Ekström, M.P.; Berg, C. Adolescents' trust in food messages and their sources. *Br. Food J.* **2017**, *119*, 2712–2723. [CrossRef]
51. De Simone, C. The unregulated probiotic market. *Clin. Gastroenterol. Hepatol.* **2019**, *17*, 809–817. [CrossRef]
52. Liang, L.; Yi, Y.; Lv, Y.; Qian, J.; Lei, X.; Zhang, G. A comprehensive genome survey provides novel insights into bile salt hydrolase (BSH) in *Lactobacillaceae*. *Molecules* **2018**, *23*, 1157. [CrossRef]

153. Abdul Manan, M. The role of probiotics in personalized therapeutics: Advances in gut microbe-driven interventions. *Microb* **2025**, *8*, 100497. [CrossRef]
154. Abouelela, M.E.; Helmy, Y.A. Next-Generation Probiotics as Novel Therapeutics for Improving Human Health: Current Trends and Future Perspectives. *Microorganisms* **2024**, *12*, 430. [CrossRef] [PubMed]
155. McCoubrey, L.E.; Elbadawi, M.; Orlu, M.; Gaisford, S.; Basit, A.W. Harnessing machine learning for development of microbiome therapeutics. *Gut Microbes* **2021**, *13*, 1872323. [CrossRef]
156. Li, P.; Luo, H.; Ji, B.; Nielsen, J. Machine learning for data integration in human gut microbiome. *Microb. Cell Fact.* **2022**, *21*, 241. [CrossRef]
157. Pedroza Matute, S.; Iyavoo, S. Exploring the gut microbiota: Lifestyle choices, disease associations, and personal genomics. *Front Nutr* **2023**, *10*, 1225120. [CrossRef]
158. Leuschner, R.G.; Robinson, T.P.; Hugas, M.; Cocconcelli, P.S.; Richard-Forget, F.; Klein, G.; Licht, T.R.; Nguyen-The, C.; Querol, A. Richardson, M. Qualified presumption of safety (QPS): A generic risk assessment approach for biological agents notified to the European Food Safety Authority (EFSA). *Trends Food Sci. Technol.* **2010**, *21*, 425–435. Erratum in *Trends Food Sci. Technol.* **2011**, *22* 51–52. [CrossRef]
159. EFSA-BIOHAZ; Koutsoumanis, K.; Allende, A.; Alvarez-Ordóñez, A.; Bolton, D.; Bover-Cid, S.; Chemaly, M.; Davies, R.; De Cesare, A.; Hilbert, F. Update of the list of QPS-recommended biological agents intentionally added to food or feed as notified to EFSA 10: Suitability of taxonomic units notified to EFSA until March 2019. *EFSA J.* **2019**, *17*, e05753. [PubMed]
160. Huys, G.; Vancanneyt, M.; d'Haene, K.; Vankerckhoven, V.; Goossens, H.; Swings, J. Accuracy of species identity of commercial bacterial cultures intended for probiotic or nutritional use. *Res. Microbiol.* **2006**, *157*, 803–810. [CrossRef]
161. Vankerckhoven, V.; Moreillon, P.; Piu, S.; Giddey, M.; Huys, G.; Vancanneyt, M.; Goossens, H.; Entenza, J.M. Infectivity of *Lactobacillus rhamnosus* and *Lactobacillus paracasei* isolates in a rat model of experimental endocarditis. *J. Med. Microbiol.* **2007**, *56* 1017–1024. [CrossRef] [PubMed]
162. Koirala, S.; Anal, A.K. Probiotics-based foods and beverages as future foods and their overall safety and regulatory claims. *Future Foods* **2021**, *3*, 100013. [CrossRef]
163. Sanders, M.E. How do we know when something called "probiotic" is really a probiotic? A guideline for consumers and health care professionals. *Funct. Food Rev.* **2009**, *1*, 3–12.
164. Douillard, F.P.; Ribbera, A.; Kant, R.; Pietilä, T.E.; Järvinen, H.M.; Messing, M.; Randazzo, C.L.; Paulin, L.; Laine, P.; Ritari, J. Comparative genomic and functional analysis of 100 Lactobacillus rhamnosus strains and their comparison with strain GG. *PLoS Genet.* **2013**, *9*, e1003683. [CrossRef] [PubMed]

microorganisms

Review

Microbial Food Safety and Antimicrobial Resistance in Foods: A Dual Threat to Public Health

Ayman Elbehiry [1,*], Eman Marzouk [1], Adil Abalkhail [1], Husam M. Edrees [2], Abousree T. Ellethy [3], Abdulaziz M. Almuzaini [4], Mai Ibrahem [5], Abdulrahman Almujaidel [1], Feras Alzaben [6], Abdullah Alqrni [7] and Akram Abu-Okail [8]

1 Department of Public Health, College of Applied Medical Sciences, Qassim University, P.O. Box 6666, Buraydah 51452, Saudi Arabia; e.marzouk@qu.edu.sa (E.M.)
2 Department of Physiology, Faculty of Medicine, University of Tabuk, Tabuk 71491, Saudi Arabia
3 Department of Basic Oral Sciences and Dental Education, College of Dentistry, Qassim University, Buraydah 51452, Saudi Arabia
4 Department of Veterinary Preventive Medicine, College of Veterinary Medicine, Qassim University, Buraydah 51452, Saudi Arabia
5 Department of Public Health, College of Applied Medical Science, King Khalid University, Abha 61421, Saudi Arabia
6 Department of Food Service, King Fahad Armed Hospital, Jeddah 23311, Saudi Arabia
7 Department of Family Medicine, King Fahad Armed Hospital, Jeddah 23311, Saudi Arabia
8 Department of Pathology and Laboratory Diagnosis, College of Veterinary Medicine, Qassim University, Buraydah 51452, Saudi Arabia
* Correspondence: ar.elbehiry@qu.edu.sa

Abstract: The intersection of microbial food safety and antimicrobial resistance (AMR) represents a mounting global threat with profound implications for public health, food safety, and sustainable development. This review explores the complex pathways through which foodborne pathogens—such as *Salmonella* spp., *Escherichia coli* (*E. coli*), *Listeria monocytogenes* (*L. monocytogenes*), and *Campylobacter* spp.—acquire and disseminate resistance within human, animal, and environmental ecosystems. Emphasizing a One Health framework, we examine the drivers of AMR across sectors, including the misuse of antibiotics in agriculture, aquaculture, and clinical settings, and assess the role of environmental reservoirs in sustaining and amplifying resistance genes. We further discuss the evolution of surveillance systems, regulatory policies, and antimicrobial stewardship programs (ASPs) designed to mitigate resistance across the food chain. Innovations in next-generation sequencing, metagenomics, and targeted therapeutics such as bacteriophage therapy, antimicrobial peptides (AMPs), and CRISPR-based interventions offer promising alternatives to conventional antibiotics. However, the translation of these advances into practice remains uneven, particularly in low- and middle-income countries (LMICs) facing significant barriers to diagnostic access, laboratory capacity, and equitable treatment availability. Our analysis underscores the urgent need for integrated, cross-sectoral action—anchored in science, policy, and education—to curb the global spread of AMR. Strengthening surveillance, investing in research, promoting responsible antimicrobial use, and fostering global collaboration are essential to preserving the efficacy of existing treatments and ensuring the microbiological safety of food systems worldwide.

Keywords: antimicrobial resistance (AMR), foodborne pathogens; One Health; surveillance systems; microbial food safety; Public Health

 https://doi.org/10.3390/microorganisms13071592

1. Introduction

Ensuring the microbial safety of food remains a paramount concern in public health, as foodborne illnesses continue to pose significant challenges globally. Pathogens such as *Salmonella* spp., *Escherichia coli* (*E. coli*), *Listeria monocytogenes* (*L. monocytogenes*), and *Campylobacter* spp. are frequently implicated in outbreaks, which are often linked to contaminated meat, dairy, and ready-to-eat (RTE) products [1,2]. The emergence and proliferation of antimicrobial-resistant (AMR) strains among these pathogens exacerbate the issue, rendering conventional treatments less effective and leading to persistent infections [2,3]. As illustrated in Figure 1, key drivers such as antimicrobial misuse and environmental contamination contribute to the development of AMR in foodborne pathogens. This leads to significant global public health concerns and necessitates mitigation strategies within a One Health framework. The figure provides a conceptual summary of these pathways.

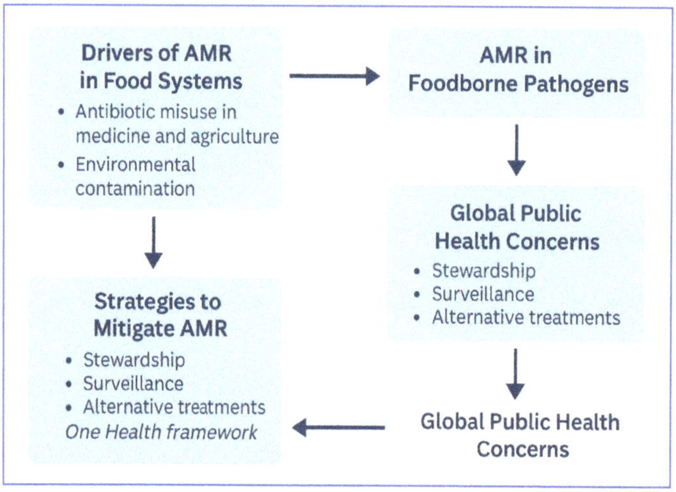

Figure 1. Conceptual diagram illustrating the progression from drivers of AMR in food systems—including antibiotic misuse and environmental contamination—to the emergence of AMR in foodborne pathogens, associated global public health concerns, and key mitigation strategies within a One Health framework.

The misuse and overuse of antibiotics in both human medicine and agriculture have accelerated the development of AMR. In the agricultural sector, antibiotics are frequently employed not only for therapeutic purposes but also for growth promotion and disease prevention in livestock [4,5]. This practice contributes to the emergence of multidrug-resistant (MDR) bacteria, which can be transmitted to humans through the food chain [6–8]. Notably, MDR strains of *Staphylococcus aureus* (*S. aureus*), *E. coli*, and *Salmonella* have been isolated from meat and dairy products, posing significant health risks [9,10].

RTE foods are particularly concerning, as they often bypass cooking processes that would eliminate harmful bacteria. Studies have detected MDR pathogens in various RTE products, highlighting the need for stringent hygiene practices and monitoring [11,12]. The World Health Organization (WHO) recognizes AMR as one of the top global public health threats, estimating that by 2050, AMR could cause up to 10 million deaths annually if no action is taken [13,14]. This underscores the urgency of addressing AMR through a One Health approach that considers the interconnectedness of human, animal, and environmental health [15,16]. Lammie and Hughes [17] identified AMR as a complex global health challenge closely linked to food systems and emphasized that addressing it requires a multifaceted strategy and a One Health approach involving improved communication, cooperation, and collaboration across sectors.

In recent years, the intensification of global food production and international trade has heightened the complexity of microbial risk management across the food supply chain. The increasing movement of agricultural goods across borders facilitates the transboundary spread of antimicrobial-resistant bacteria and resistance genes [9,10]. Environmental factors, including the contamination of water sources used in irrigation or livestock production, contribute to the persistence and amplification of AMR in food ecosystems [11]. Moreover, metagenomic studies have revealed that resistance determinants are not only prevalent in clinically relevant bacteria but also in commensal organisms, which may act as reservoirs and vectors for horizontal gene transfer [15,18]. This underscores the need for comprehensive monitoring strategies that go beyond routine pathogen surveillance to include resistome and mobilome analyses across the entire farm-to-fork continuum.

Technological advances offer promising avenues for addressing the intertwined challenges of foodborne pathogens and AMR. Novel molecular approaches, including real-time PCR, next-generation sequencing (NGS), and metagenomic shotgun sequencing, have significantly improved the speed and sensitivity of pathogen detection and resistance profiling in complex food matrices [2,19]. These tools not only facilitate the early detection of contamination but also enable the identification of emerging resistance traits before they become widespread. Furthermore, innovative control measures such as bacteriophage therapy, antimicrobial peptides (AMPs), and microbiome modulation are being explored as alternatives to conventional antibiotics [8,20]. However, the implementation of these approaches requires coordinated efforts among policymakers, food producers, and public health authorities to ensure feasibility, safety, and consumer acceptance.

Efforts to combat AMR in the food sector include implementing antimicrobial stewardship programs, enhancing the surveillance of resistant strains, and promoting research into alternative treatments [21,22]. Additionally, public education on proper food handling and the risks associated with antibiotic misuse is crucial [20,23]. Recent studies have also emphasized the role of consumers in mitigating AMR by making informed choices and advocating for responsible antibiotic use in food production [24–26]. This review aims to explore the current state of microbial safety and AMR in foods, examining the prevalence of MDR pathogens in meat, dairy, and RTE products, and discussing strategies to mitigate these risks.

Moreover, this review addresses a critical gap in the current literature by integrating the implications of AMR for food safety. While previous reviews have focused primarily on AMR's impact on human and animal health, this work uniquely highlights how AMR in foodborne pathogens threatens food availability, access, and stability. It synthesizes recent data on resistant strains in foods, contamination sources, prevention strategies, detection methodologies, and legislation, all within the context of food safety. The manuscript presents a cohesive, evidence-based narrative linking data, analysis, and practical strategies under a unified One Health framework, concluding with key implications for global food safety.

2. Methodology

This narrative review was conducted to synthesize current knowledge on foodborne AMR under the One Health framework. Literature searches were performed in PubMed, Scopus, and Web of Science databases up to 2025. The following keywords were used in various combinations: AMR, food safety, foodborne pathogens, One Health, surveillance, AMR policy, and control strategies. Inclusion criteria comprised peer-reviewed articles, systematic reviews, meta-analyses, surveillance reports, and major international agency documents published in English. Priority was given to studies from the last 10 years, but older seminal works were also considered where relevant. Data extraction focused on AMR prevalence, drivers, surveillance, mitigation strategies, and policy frameworks. The selection process was performed by the authors independently, with consensus achieved through discussion.

3. Foodborne Pathogens and the Burden of AMR

Foodborne pathogens such as *Salmonella* spp., *E. coli*, *L. monocytogenes*, and *Campylobacter* spp. are major contributors to foodborne illnesses worldwide. According to the Antimicrobial Resistance Collaborators [27], bacterial AMR was directly responsible for 1.27 million deaths and contributed to nearly 5 million global deaths in 2019. These pathogens, particularly when resistant to multiple antibiotics, pose serious health risks due to prolonged illness, increased hospitalization, and higher mortality [28,29]. The misuse and overuse of antibiotics in agriculture play a substantial role in accelerating resistance. Farmers often use antimicrobials not only therapeutically but also for growth promotion and prophylaxis, especially in poultry and livestock [5,30]. These practices create reservoirs of MDR bacteria in the food chain, including *E. coli* and *Salmonella*, which have been isolated from meat, dairy, and RTE products [6,31]. Figure 2 presents a schematic of the One Health framework, illustrating the interconnected roles of human health, animal health (including aquaculture and agriculture), food systems, and the environment in addressing AMR. The figure highlights how these sectors contribute to and are impacted by AMR challenges in food production and safety.

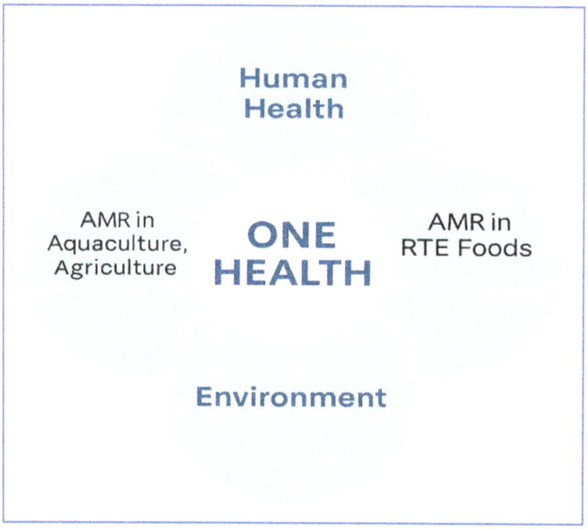

Figure 2. One Health schematic illustrating the pathways of AMR transmission between animals, food, environment, and humans. Key resistance mechanisms, such as plasmid-mediated transfer and environmental reservoirs, are adapted from Hendriksen et al. [32] and Van Boeckel et al. [30]. The figure highlights the interconnectedness of foodborne AMR across sectors.

Aquaculture is a rapidly growing sector in global food production and is increasingly being recognized as a hotspot for AMR. The intensive use of antibiotics in fish and shrimp farming to prevent disease and enhance productivity fosters the selection of resistant bacterial strains. Deng et al. [33] analyzed 23,165 antimicrobial resistance gene (ARG) contigs from aquaculture isolates in 75 countries. They reported that *blaCMY-2* occurred in approximately 10–15% of samples, cphA in 30–35%, and mcr-3 in 5–10%, highlighting the widespread presence of mobile ARGs in aquaculture environments, which poses substantial risks for transfer to human pathogens. Yuan et al. [34] corroborated this by documenting widespread contamination with antibiotics, resistant bacteria, and ARGs across water, sediment, and aquatic species. Similarly, Hossain et al. [35] identified elevated antibiotic concentrations in Bangladeshi aquaculture environments, posing ecological and public health risks. Together, these findings underscore the urgent need for surveillance, judicious antimicrobial use, and effective regulatory frameworks to contain AMR emergence from aquaculture systems.

L. monocytogenes, a significant concern in cold-stored RTE foods, has exhibited resistance to clinically important antimicrobials including ciprofloxacin and erythromycin, though resistance to first-line treatments like penicillin G and tetracycline remains relatively rare [36]. Similarly, *Campylobacter jejuni* has demonstrated widespread fluoroquinolone resistance, particularly in poultry systems, driven by chromosomal mutations and horizontal gene transfer [37,38]. Surveillance reports by the European Food Safety Authority (EFSA) [39] and the U.S. Centers for Disease Control and Prevention [40] confirm a significant increase in AMR, as demonstrated by rising multidrug-resistant *Salmonella* and extended-spectrum β-lactamase (ESBL)-producing *E. coli* rates in food products globally [41]. Notably, MDR *Salmonella enterica* (*S. enterica*) and ESBL-producing *E. coli* have been found in poultry, pork, and beef, with frequent resistance to ciprofloxacin and third-generation cephalosporins [42,43]. In support of these findings, Shafiq et al. [44] documented the presence of ESBL-encoding plasmids along China's pork production chain, illustrating the global scale of this threat.

Environmental pathways also contribute significantly to AMR transmission. Zhang et al. [45] demonstrated that ARGs from manure-amended soils can transfer to the microbial communities of vegetables such as lettuce, revealing a direct route for resistance to enter the food chain. In addition, studies by Zhu et al. [46] and Berendonk et al. [11] emphasized the role of agricultural runoff and wastewater as reservoirs and vectors of ARGs in both environmental and agricultural ecosystems. Resistance genes such as *blaCTX-M*, *mcr-1*, and *tetA* are not only found in pathogens but are also common among commensal bacteria, often carried on mobile genetic elements that promote interspecies dissemination [16]. Recently, Keet and Rip [47] reported MDR *L. monocytogenes* strains isolated from raw meats and ready-to-eat foods in the Western Cape, South Africa, underscoring the regional challenges in managing antimicrobial resistance in foodborne pathogens.

An increasing burden of AMR in foodborne pathogens: EFSA and ECDC [41] estimate approximately 7% of confirmed human salmonellosis and 17% of campylobacteriosis cases in the EU are caused by resistant strains. In 2019, there were over 300 confirmed outbreaks linked to multidrug-resistant (MDR) Salmonella in EU member states, resulting in approximately 2500 reported hospitalizations. To curb the growing AMR burden in food systems, a comprehensive One Health approach is essential—one that integrates food safety protocols, veterinary antimicrobial stewardship, and environmental controls. Non-traditional interventions such as bacteriophage therapy [48] and bacteriocins [49] show promise as future alternatives to antibiotics. Public education, policy enforcement, and consumer-driven advocacy for responsible antibiotic use remain pivotal elements of long-term containment strategies [49].

4. One Health Approach to Mitigating AMR

AMR is a multifaceted global health challenge that transcends the boundaries of human medicine, encompassing animal health, agriculture, and environmental ecosystems. The One Health approach, which recognizes the interconnectedness of these domains, has emerged as a pivotal strategy in addressing the complex dynamics of AMR. As illustrated in Figure 3, the One Health approach provides a comprehensive framework that integrates human, animal, and environmental health sectors to address the multifaceted nature of AMR. This model guides the subsequent sections by mapping key drivers, surveillance mechanisms, interventions, innovations, educational strategies, and global policy responses across all domains.

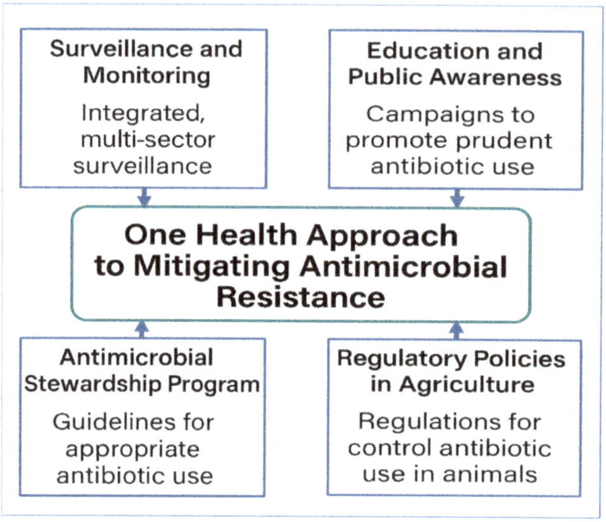

Figure 3. One Health approach to mitigating AMR. The diagram illustrates the integration of surveillance, education, stewardship, and regulatory policies required to control AMR across human, animal, and environmental sectors.

4.1. Conceptual Framework of One Health in AMR

The One Health paradigm emphasizes the integration of human, animal, and environmental health sectors to develop comprehensive strategies against AMR. This approach acknowledges that the misuse and overuse of antimicrobials in any of these sectors can contribute to the emergence and dissemination of resistant pathogens, necessitating coordinated interventions across all domains.

4.2. Drivers of AMR Across Sectors

4.2.1. Human Health Sector

In the human health domain, inappropriate antibiotic prescribing remains a major contributor to the emergence of AMR. Numerous studies have revealed that a substantial proportion of antibiotic prescriptions are unnecessary, often issued for self-limiting viral infections or under patient pressure. In a systematic review, Sijbom et al. [50] identified multiple determinants influencing inappropriate antibiotic prescriptions in primary care, including diagnostic uncertainty, perceived patient expectations, and lack of rapid diagnostic tools. Browne et al. [51] further demonstrated that global antibiotic consumption rose sharply between 2000 and 2018, particularly in LMICs where access and stewardship programs are often limited. In a pivotal U.S. study, Fleming-Dutra et al. [52] reported that approximately 30% of outpatient antibiotic prescriptions were unnecessary, underscoring the systemic overuse in ambulatory care.

To address the persistent problem of antibiotic misuse, ASPs have been widely implemented in clinical settings. In a national survey of U.S. hospitals, Pollack et al. [53] reported that only 39% had fully adopted all seven Centers for Disease Control and Prevention (CDC) Core Elements of Hospital ASPs as of 2014, underscoring both progress and the need for ongoing institutional commitment. These core elements—ranging from leadership commitment and accountability to education and tracking—have since become the standard framework for promoting judicious antibiotic use in healthcare facilities. Complementary to this, the CDC continues to update and disseminate stewardship guidelines through its Core Elements of Hospital Antibiotic Stewardship Programs [54].

Beyond the frequency of antibiotic use, inappropriate selection of antibiotics for common infections is also a growing concern. Ardillon et al. [55] conducted a multicentric cohort study in three LMICs and found high rates of inappropriate antibiotic prescribing in outpatient pediatric care. The study highlighted that such practices not only contribute to resistance but also increase the risk of treatment failure and adverse outcomes, particularly in settings with limited diagnostic capacity. Their findings reinforce the need for tailored stewardship interventions, especially in community-level primary healthcare systems.

4.2.2. Animal Health and Agriculture

The agricultural sector, particularly intensive livestock farming, has been identified as a major contributor to AMR due to the prophylactic and growth-promoting use of antibiotics. Van Boeckel et al. [56] reported that approximately 73% of all antimicrobials sold globally are used in animals raised for food. This extensive use facilitates the selection of resistant bacteria, which can be transmitted to humans through direct contact or the consumption of animal products. Recent projections indicate that, under a business-as-usual scenario, global antibiotic use in livestock could reach approximately 143,481 tons by 2040, representing a 29.5% increase from the 2019 baseline of around 110,777 tons. This underscores the urgency for coordinated efforts to reduce antibiotic use intensity and manage livestock biomass effectively [57].

In low- and middle-income countries, the situation is particularly concerning. A study analyzing 901 point-prevalence surveys revealed that China and India are significant hotspots for AMR in animals, with emerging hotspots in Brazil and Kenya. From 2000 to 2018, the proportion of antimicrobials showing resistance above 50% increased from 15% to 41% in chickens and from 13% to 34% in pigs [56]. The misuse of antibiotics in animal agriculture not only affects animal health but also poses a significant threat to human health. The CDC highlights that antimicrobial-resistant germs can spread between animals and humans through various pathways, including direct contact, consumption of contaminated food, and environmental routes such as water runoff from farms [54].

Furthermore, the environmental impact of antibiotic use in agriculture cannot be overlooked. Antibiotics and antifungals used in animals can contaminate the environment through animal waste, which may carry drug residues and resistant germs. This contamination can affect soil and water sources, contributing to the development and spread of resistant germs in the environment [54]. To mitigate these risks, it is essential to implement stringent regulations on antibiotic use in agriculture, promote good animal husbandry practices, and enhance surveillance of antibiotic use and resistance patterns in the agricultural sector. Adopting a One Health approach that integrates human, animal, and environmental health strategies is crucial in combating the spread of AMR.

4.2.3. Environmental Sector

The environment acts as both a reservoir and a conduit for AMR, facilitating the dissemination of resistant bacteria and genes through soil, water, and air. Pharmaceutical effluents, agricultural runoff, and improper waste disposal introduce antibiotics and resistant organisms into environmental compartments, accelerating the spread of resistance. Berendonk et al. [11] highlighted the persistence of ARGs in aquatic and terrestrial ecosystems, advocating for environmental surveillance as a critical component of AMR mitigation strategies. Recent global assessments have reinforced the environmental dimension of AMR. The United Nations Environment Programme has stressed that pollution from human activities—including pharmaceutical manufacturing, agriculture, and waste mismanagement—is a key driver of AMR emergence and transmission. This underscores the necessity for pollution control and ecosystem health in AMR policy [58].

Aquatic systems, in particular, have emerged as hotspots for ARG accumulation. Cheung et al. [59] conducted a scoping review and found that fecal pollution in surface waters leads to significant concentrations of resistant bacteria, posing ecological and public health risks. The review emphasized the importance of adequate sanitation infrastructure and improved wastewater treatment technologies in controlling the spread of resistance. Furthermore, climate change has been increasingly associated with AMR dynamics. Fernández Salgueiro et al. [60] reviewed the influence of environmental factors such as rising temperatures, flooding, and droughts on microbial evolution and horizontal gene transfer. These climate-related stressors are believed to exacerbate AMR propagation, especially in vulnerable ecosystems.

Furthermore, the presence of antibiotic residues in the environment has been recognized as a driver for AMR. Larsson and Flach [61] discussed the impact of antibiotic pollution on the emergence of resistant bacteria, emphasizing the need for stringent regulations on antibiotic disposal and wastewater treatment. Implementing such measures is crucial to prevent the proliferation of resistance genes in environmental settings. Together, these findings highlight the need to include the environmental sector in AMR surveillance and response strategies. A One Health approach that integrates environmental protection, sanitation infrastructure, and antibiotic waste regulation is essential to prevent the unchecked spread of resistance beyond clinical and agricultural settings.

4.3. *Surveillance and Monitoring*

Surveillance and monitoring are critical pillars in addressing AMR, enabling early detection of resistance trends, identifying emerging threats, and evaluating the effectiveness of interventions. A robust surveillance system must adopt a One Health approach, integrating data from human health, animal health, agriculture, and the environment to comprehensively map resistance dynamics and inform policy decisions. The Global Antimicrobial Resistance Surveillance System (GLASS), launched by the WHO, exemplifies this integrated framework. GLASS promotes standardized collection and reporting of AMR and antimicrobial consumption data across member states, thereby enhancing the comparability and utility of global surveillance outputs [62]. As of the latest GLASS report, over 120 countries are enrolled, contributing data on priority pathogens such as *E. coli*, *Klebsiella pneumoniae*, *Salmonella* spp., and *S. aureus*, alongside their resistance patterns and treatment challenges.

In Europe, the European AMR Surveillance Network, managed by the European CDC, focuses on AMR surveillance in invasive bacterial infections. The network utilizes data from over 30 countries to generate harmonized resistance trends, offering policymakers timely insights into the effectiveness of national and EU-wide antimicrobial stewardship programs [63]. Complementing this, the European Surveillance of Veterinary Antimicrobial Consumption collects data on veterinary antimicrobial sales, enabling assessment of trends in antimicrobial usage in food-producing animals [64]. In the United States, the National AMR Monitoring System provides integrated surveillance across human, animal, and food sectors. Jointly coordinated by the CDC, FDA, and United States Department of Agriculture (USDA), National Antimicrobial Resistance Monitoring System (NARMS) tracks resistance in pathogens such as *Salmonella*, *Campylobacter*, and *E. coli* from clinical, retail meat, and animal sources. These findings guide regulatory action, including the restriction of critical antimicrobials in livestock [65].

For LMICs, surveillance implementation often faces logistical and technical barriers. To bridge these gaps, the Fleming Fund, funded by the UK government, supports laboratory strengthening, data harmonization, and human–animal surveillance integration in over 20 LMICs [66]. Similarly, the Food and Agriculture Organization of the United Nations (FAO)'s ATLASS tool (Assessment Tool for Laboratories and AMR Surveillance Systems) provides a structured approach to evaluating and enhancing national AMR surveillance in food and agriculture systems [67]. Environmental surveillance is an emerging frontier in AMR monitoring. Wastewater-based epidemiology has shown promise as a cost-effective method to capture community-wide AMR burdens. A landmark study by Hendriksen et al. [32] demonstrated that metagenomic analysis of urban sewage could accurately reflect local resistance gene prevalence and trends, offering a scalable model for global surveillance.

Advances in whole-genome sequencing (WGS) and bioinformatics have further revolutionized AMR tracking. By enabling high-resolution pathogen typing and identification of resistance genes, WGS supports outbreak detection, resistance mechanism elucidation, and real-time tracking of resistant clones. Several national AMR programs, including those in the UK, Australia, and the Philippines, have incorporated WGS into their routine surveillance, enhancing data precision and intervention targeting [68]. Nonetheless, critical challenges persist: data fragmentation, limited real-time sharing, disparities in capacity, and a lack of standardized methodologies across sectors and regions. To address these gaps, international collaboration, open-access data platforms, harmonized protocols, and investment in surveillance infrastructure remain essential.

4.4. Interventions and Strategies

4.4.1. Antimicrobial Stewardship Programs (ASPs)

ASPs are structured, evidence-based initiatives aimed at promoting the responsible use of antimicrobials to achieve optimal clinical outcomes while minimizing the selection and spread of antimicrobial-resistant organisms. These programs are essential pillars of national and global strategies to combat AMR, with direct implications for both public health and food system resilience. The primary goals of ASPs include ensuring the appropriate selection, dosage, route, and duration of antimicrobial therapy, thereby reducing unnecessary or inappropriate use. These programs often operate through multidisciplinary teams—comprising infectious disease specialists, clinical microbiologists, pharmacists, and infection control experts—who develop local antibiotic guidelines, monitor prescribing behavior, and provide prescriber feedback.

Dyar et al. [69] showed that ASP implementation across European hospitals led to a significant reduction in broad-spectrum antibiotic use and improved compliance with prescribing standards. A systematic review by Baur et al. [70] covering 32 hospital-based ASPs revealed a 35% reduction in antibiotic consumption and a 10% decline in AMR rates, with no adverse effects on patient mortality or clinical outcomes. In the United States, ASPs are now mandated in all acute care hospitals under federal regulations. The CDC's Core Elements of Hospital Antibiotic Stewardship Programs guide healthcare institutions in building sustainable ASP infrastructure, emphasizing leadership support, tracking, reporting, and staff education [54]. Importantly, the relevance of ASPs extends far beyond hospital walls. In veterinary medicine and food animal production, ASPs aim to curtail the overuse and misuse of antimicrobials that contribute to the emergence of resistant foodborne pathogens. The use of critically important antimicrobials in livestock—often for growth promotion or prophylaxis—has been strongly linked to AMR-positive isolates of

Salmonella, *E. coli*, *L. monocytogenes*, and *Campylobacter* detected in meat and dairy products. Accordingly, stewardship practices in the agricultural sector are increasingly viewed as central to the prevention of foodborne AMR transmission.

In LMICs, ASP implementation faces logistical and economic barriers. Nevertheless, promising examples have emerged. In India, several studies have documented the progress and impact of ASPs. Singh et al. [71] demonstrated that ASP implementation in a tertiary care hospital significantly reduced the use of high-end antibiotics and improved patient outcomes. Debnath et al. [72] reported a successful introduction of ASPs in primary and secondary care hospitals, emphasizing that even resource-limited settings can benefit from tailored interventions. Sahni et al. [73] conducted a systematic review highlighting the gaps in ASP implementation across Indian healthcare facilities and calling for strengthened institutional policies, local antibiogram development, and routine training of healthcare personnel. When coupled with infection prevention and control measures, ASPs have been shown to amplify reductions in resistant infections. Furthermore, digital tools—such as real-time decision support systems, electronic prescribing, and automated alerts—are increasingly enhancing the efficacy and scalability of stewardship initiatives. In conclusion, ASPs represent a foundational strategy in the One Health response to AMR. Expanding their reach beyond clinical settings to include animal health and food production is essential to controlling the spread of resistant bacteria, protecting food safety, and preserving antibiotic efficacy for future generations.

4.4.2. Regulatory Policies in Agriculture

The regulation of antimicrobial use in agriculture is a critical intervention in the global effort to combat AMR, especially given the established link between antibiotic use in food-producing animals and the emergence of resistant bacteria in the human population. These regulations aim to restrict non-therapeutic uses of antibiotics, improve veterinary oversight, and promote responsible antimicrobial practices across the food production chain. A landmark intervention was the European Union (EU)'s 2006 ban on the use of antibiotics as growth promoters in livestock. This policy was followed by substantial reductions in resistance rates among zoonotic bacteria such as *Salmonella* and *Campylobacter*, with no negative impact on animal productivity when good farming practices were maintained [20]. Countries like Denmark and the Netherlands have further advanced these efforts by enforcing strict antimicrobial usage monitoring systems, mandating veterinary prescriptions, and incentivizing low-antibiotic farming [30,74].

In the United States, the FDA issued Guidance for Industry #213, which led to the withdrawal of medically important antimicrobials for growth promotion and introduced veterinary oversight for therapeutic use. The FDA's 2022 report shows a continued decline in overall sales of such antibiotics in food animals, reflecting the positive impact of regulatory reforms [75]. In emerging economies, the enforcement of antimicrobial policies remains inconsistent. However, countries like China and India have begun to implement restrictions on the use of critically important antimicrobials. For example, China banned colistin as a feed additive following the discovery of the plasmid-mediated colistin resistance gene *mcr-1*, which raised global concern due to its potential for rapid spread [76]. India has also issued national action plans and advisories to regulate antibiotic use in poultry and aquaculture, although implementation and surveillance remain in early stages [77].

International agencies, including the WHO, the FAO, and the World Organization for Animal Health (WOAH), advocate for harmonized global actions. These include phasing out antibiotics for growth promotion, banning over-the-counter antimicrobial sales, and establishing surveillance systems for antimicrobial use and resistance in the animal sector. Ultimately, effective regulatory policies—when combined with enhanced farm biosecurity, vaccination programs, and improved animal husbandry—can substantially reduce antibiotic dependence in food production, lower AMR transmission risk to humans, and contribute to both public health protection and food system sustainability.

4.4.3. Environmental Management

The environment is a critical but often overlooked component in the transmission and persistence of AMR. Environmental compartments—such as soil, surface water, groundwater, and sediments—act as reservoirs and conduits for antimicrobial residues, resistant bacteria, and resistance genes originating from human, animal, and industrial sources. To mitigate environmental contamination, a multipronged approach is essential. One key strategy is enhancing wastewater treatment infrastructure, particularly in hospitals, pharmaceutical manufacturing sites, and agricultural operations. Conventional wastewater treatment plants are not specifically designed to remove antibiotics or resistance genes, allowing these elements to enter natural water bodies and spread across ecosystems [78]. Upgrading treatment processes with advanced oxidation technologies, membrane filtration, and constructed wetlands can significantly reduce the environmental AMR burden.

Another important intervention involves regulating pharmaceutical discharges. Industrial effluents from drug manufacturing facilities—especially in countries with weak environmental oversight—are often heavily contaminated with antibiotics, creating hotspots for resistance gene selection and horizontal gene transfer [79]. Implementing strict discharge limits, enforcing zero-liquid-discharge technologies, and ensuring routine environmental monitoring can curtail these emissions. Sustainable agricultural practices also play a crucial role in minimizing the release of antibiotics and resistant microbes into the environment. This includes the safe management of animal manure, composting, limiting antibiotic use in aquaculture, and promoting integrated pest and disease management. Improved handling of farm waste reduces the leaching of contaminants into soils and water bodies, thereby lowering the environmental load of AMR determinants.

Larsson et al. [79] emphasized the necessity of global environmental regulations as part of a One Health response to AMR. They argued for the inclusion of AMR-relevant targets in environmental protection frameworks and greater collaboration between the health, environment, and agriculture sectors. Recent initiatives, such as the AMR Industry Alliance's manufacturing framework and the Stockholm Statement on Pharmaceuticals in the Environment, advocate for sustainable antibiotic production and improved environmental accountability. Ultimately, reducing environmental contributions to AMR requires coordinated global action—combining policy enforcement, technological innovation, industry responsibility, and international funding to strengthen environmental stewardship as a central pillar in the AMR response.

4.5. Research and Innovation

Continued investment in research and innovation is essential for tackling the growing challenge of AMR and safeguarding microbial food safety. Traditional antibiotics are losing their efficacy at an alarming rate, and the development pipeline for new antimicrobials remains limited. In response, research efforts have increasingly focused on novel therapeutics, next-generation diagnostics, and genomic tools that support precision intervention. Bacteriophage (phage) therapy has re-emerged as a promising alternative to conventional antibiotics, particularly against MDR pathogens. Phages are viruses that specifically target and lyse bacterial cells without affecting the host microbiota or contributing to resistance selection. Recent studies have demonstrated the successful use of phage cocktails to treat severe MDR infections in humans, and their potential application in veterinary medicine and food safety is gaining interest [80]. Phages can be applied to decontaminate food surfaces, control Listeria in dairy products, and reduce *Salmonella* in poultry processing.

AMPs, derived from microbial or host immune systems, represent another avenue of therapeutic innovation. These molecules exhibit broad-spectrum activity against bacteria, fungi, and viruses, and some are being explored for use in food preservation and livestock treatment with reduced resistance potential. Probiotics, prebiotics, and postbiotics are also under investigation for their role in enhancing gut health, competitively excluding pathogens, and reducing reliance on antibiotics in animal husbandry. For instance, certain strains of *Lactobacillus* and *Bacillus* have shown efficacy in improving animal productivity while minimizing pathogen shedding in feces. From a diagnostic standpoint, advances in genomics, metagenomics, and bioinformatics have significantly enhanced our ability to detect, characterize, and monitor resistance determinants. High-throughput sequencing platforms and resistome analyses allow researchers to track the evolution of resistance genes across human, animal, and environmental reservoirs. This is particularly important for understanding the transmission of AMR along the food chain and designing targeted interventions [81].

CRISPR-Cas systems offer highly specific antimicrobial activity by targeting resistance genes, providing a promising alternative to broad-spectrum antibiotics [82,83]. In diagnostics, biosensors, microfluidic devices, and AI-powered platforms enable real-time pathogen detection in food products, improving surveillance and response [84,85]. Advances in genomics and resistome analysis support tracking AMR transmission across the food chain and environment [81]. Global initiatives such as the Global Antibiotic Research and Development Partnership (GARDP) and Combating Antibiotic-Resistant Bacteria Biopharmaceutical Accelerator (CARB-X) support the development and equitable distribution of new antimicrobials and diagnostics, particularly in resource-limited settings. Sustained investment, cross-sector collaboration, and equitable access to innovation are essential to combat AMR and protect food systems.

While these innovative approaches offer promising solutions for AMR mitigation in food systems, their industrial implementation faces notable challenges. For bacteriophages, regulatory approval is evolving: Omar et al. [86] detail the current U.S. landscape, where only a few phage-based products (e.g., ListShield™, SalmoFresh) have received FDA/USDA approval. Challenges include establishing consistent manufacturing under Good Manufacturing Practices, ensuring formulation stability, and addressing industry concerns regarding integration into standard processing workflows. For AMPs, a recent review conducted by Pérez de la Lastra et al. [87] highlight their potential in food preservation through encapsulation and nanotechnology. However, large-scale production remains expensive and technically complex, with additional challenges related to maintaining activity in complex food matrices and securing regulatory acceptance. CRISPR-based tools likewise remain largely confined to research and pilot projects, with no current food industry applications. Acceptance by food producers and consumers hinges on a clear demonstration of efficacy, safety, and cost-competitiveness. To overcome these hurdles, harmonized regulatory frameworks, industrial-scale production technologies, and targeted stakeholder communication are essential [86,87].

4.6. Education and Public Awareness

Raising awareness about AMR across healthcare, agriculture, and the general public is a cornerstone of effective AMR mitigation. Misconceptions about antibiotic use and poor hygiene practices continue to fuel resistance, particularly in resource-limited settings. Targeted educational interventions among healthcare professionals improve prescribing behaviors and stewardship compliance. McCullough et al. [88] reported that structured awareness campaigns reduced misconceptions about resistance and promoted appropriate antibiotic use. Similarly, public health campaigns in high-income countries, such as those reviewed by Huttner et al. [89], led to measurable declines in outpatient antibiotic prescriptions.

In the livestock and poultry sectors, educational outreach and farmer training have shown strong impacts. A study by Chapot et al. [90] in Bangladesh demonstrated that farmer awareness programs reduced the misuse of antibiotics and improved animal husbandry practices. Global initiatives such as the World Antimicrobial Awareness Week, coordinated by WHO, have amplified public engagement through school-based programs, media outreach, and cross-sector collaboration [91]. Integrating AMR education into professional training, national curricula, and food safety regulations remains essential. Public awareness, driven by accurate communication and community participation, complements policy and research to ensure sustainable behavior change across sectors.

4.7. Global Collaboration and Policy Frameworks

AMR is a transboundary threat that requires coordinated action at local, national, and international levels. The interconnected nature of food systems, trade, human and animal health, and environmental exposures underscores the need for a unified, global response guided by strong policy frameworks and strategic collaboration. The WHO, in partnership with the FAO and the WOAH, has established the Tripartite Alliance, which serves as the cornerstone of the One Health Global Action Plan on AMR. This plan encourages nations to develop and implement National Action Plans tailored to local contexts while adhering to global objectives related to antimicrobial stewardship, surveillance, and capacity building [92]. To enhance food safety and reduce the spread of resistant pathogens through the food chain, these agencies promote regulations on antimicrobial use in agriculture, improvement of biosecurity and hygiene standards, and integrated AMR monitoring. The Codex Alimentarius Commission (Codex), co-managed by WHO and FAO, has issued Guidelines on Integrated Monitoring and Surveillance of Foodborne AMR [93], enabling member states to standardize practices in AMR data collection, risk assessment, and mitigation in food production.

Globally, funding initiatives such as the Fleming Fund, CARB-X, and the Global AMR R&D Hub are critical in strengthening AMR surveillance, diagnostics, and innovation, particularly in low- and middle-income countries. These programs support equitable access to laboratory infrastructure, early-stage research, and cross-sectoral data sharing to address gaps in implementation and capacity [66,94]. The GLASS, launched by WHO in 2015, now includes over 120 participating countries, standardizing AMR data in human health with plans to expand into food and environmental sectors [95]. Multilateral declarations by the United Nations, G7, and G20 have further positioned AMR as a global development and security threat, emphasizing the need for coordinated policy action. Effective frameworks must prioritize integration across health, agriculture, and the environment, while supporting countries to implement context-specific, enforceable, and sustainable AMR strategies.

Despite improvements in AMR surveillance globally, substantial gaps persist in foodborne AMR monitoring in vulnerable regions, including Latin America and sub-Saharan Africa. Severino et al. [96] conducted a systematic review of bacterial foodborne diseases in Central America and the Caribbean and identified critical deficiencies, including inadequate laboratory capacity, fragmented reporting systems, and limited integration of AMR data. In sub-Saharan Africa, Hendriksen et al. [32] highlighted that although genomic surveillance initiatives like SeqAfrica have enhanced AMR monitoring in human and animal sectors, One Health-aligned food AMR surveillance remains limited and fragmented. Addressing these gaps is essential for global AMR mitigation.

The integration of bioinformatics tools and predictive models into AMR surveillance is advancing rapidly. Metagenomic sequencing, combined with machine learning algorithms, allows for the detection and characterization of resistance genes directly from complex food and environmental samples without the need for culturing [32]. These approaches enable the identification of novel resistance determinants and provide early warnings for emerging threats. Machine learning models trained on genomic, epidemiological, and environmental data have been used to predict AMR outbreak risks and identify resistance hotspots in food systems [97]. Moreover, platforms such as the Global Sewage Surveillance Project and SeqAfrica have demonstrated the feasibility of using large-scale metagenomics for AMR monitoring across regions. Continued investment in bioinformatics capacity building, particularly in low- and middle-income countries, is critical to realize the full potential of these technologies for foodborne AMR surveillance. Table 1 provides a consolidated overview of key agencies and surveillance programs cited in this review, summarizing their primary roles in AMR mitigation and their interactions within One Health frameworks.

Table 1. Key national and international agencies and surveillance programs involved in foodborne AMR mitigation, summarizing their main roles and interactions within One Health frameworks.

Agency/Program	Key Role(s)	Coordination and Interaction
WHO	Oversees the GLASS, which standardizes AMR and antimicrobial consumption data—including for food and environment—across participating countries.	Works with FAO, WOAH, and United Nations Environment Programme (UNEP) under the Quadripartite/Tripartite One Health alliance.
FAO	Implements the Progressive Management Pathway for AMR (PMP-AMR) to support agriculture in managing AMR and food safety.	Engages in One Health collaboration with WHO and WOAH and supports national AMR Action Plans.
WOAH	Develops veterinary antimicrobial use standards, monitors AMR in animals, and promotes stewardship.	Partners with WHO, FAO, and member countries through One Health initiatives.
GLASS	Collects harmonized data on AMR and antimicrobial consumption from human, animal, and environmental sectors.	Integrates regional networks (e.g., European Antimicrobial Resistance Surveillance Network (EARS-Net), Central Asian and Eastern European Surveillance of Antimicrobial Resistance) and feeds data to WHO.
EFSA	Monitors AMR in food- and animal-origin samples within the EU.	Coordinates with European Centre for Disease Prevention and Control (ECDC) to support EU risk assessments and policy.
ECDC	Tracks AMR in human pathogens through EARS-Net and contributes to EU-level surveillance.	Collaborates with EFSA and national agencies under One Health.
NARMS	Monitors AMR in human, food, and animal bacterial samples via a One Health approach.	Led jointly by CDC, FDA, and USDA; supports data-driven interventions.
CDC Global AMR Lab and Response Network	Strengthens AMR detection capacity and rapid response through global lab networks.	Partners with international agencies; supports ~50 countries.
Global Leaders Group on AMR	Provides high-level advocacy and policy guidance to spur AMR action with a One Health lens.	Supported by a Quadripartite secretariat (WHO, FAO, WOAH, UNEP).

5. AMR and Food Safety: A Neglected Link in One Health

AMR in food systems represents an escalating challenge that threatens not only public health but also global food safety. The spread of resistant pathogens in food-producing environments can reduce access to safe, nutritious food, disrupt supply chains, and impose significant economic burdens on producers and consumers [30]. Food safety experts have emphasized that AMR in foodborne pathogens exacerbates food insecurity, particularly in LMICs, where food systems are most vulnerable [98].

5.1. Prevalence of Resistant Strains in Foods

A growing body of evidence demonstrates that foods of animal and aquatic origin serve as significant reservoirs for AMR bacteria, contributing to the global AMR burden and posing serious public health risks. In China, Tang et al. [99] examined retail pork and chicken meat samples from Zhejiang Province and found that 13% of *E. coli* isolates carried the colistin-resistance gene *mcr-1*; most of these isolates also produced extended-spectrum β-lactamases (ESBLs), indicating co-resistance to critical antibiotics. In Vietnam, Nhung et al. [100] reported that over 50% of *Salmonella* isolates from retail meat were MDR, with high levels of resistance to fluoroquinolones and extended-spectrum β-lactams, raising significant food safety concerns in the region. Phung et al. [101] also identified *L. monocytogenes* strains in RTE meat products, with several isolates displaying resistance to ampicillin and erythromycin—antibiotics commonly used in the clinical management of listeriosis.

In South Africa, Kayode and Okoh [102] analyzed *L. monocytogenes* isolates from RTE foods and reported high levels of AMR. Among 194 isolates, 53% were resistant to ceftriaxone, 61.9% to trimethoprim, and 62.9% to oxytetracycline. Notably, 83.5% of the isolates were multidrug-resistant, reflecting significant risks associated with RTE foods. In Egypt, Adel et al. [103] investigated *S. enterica* from retail meats and slaughterhouses, finding that 82.4% of isolates were multidrug-resistant. Among these, 41.2% produced ESBLs, and 67.6% carried plasmid-mediated quinolone resistance genes such as *qnrA*, *qnrB*, and *qnrS*, contributing to reduced efficacy of fluoroquinolone therapies. In Spain, Bort et al. [104] assessed *Campylobacter* spp. along the broiler production chain. Of 125 isolates, 97.6% exhibited resistance to at least one antibiotic, with particularly high levels of resistance to fluoroquinolones and macrolides—two key classes used in the treatment of campylobacteriosis. Collectively, these findings demonstrate that AMR hazards are widespread across the food chain and underscore the urgent need for integrated monitoring, prudent antimicrobial use in agriculture, and comprehensive food safety strategies at the global level.

5.2. Sources of Contamination

AMR contamination of foods can originate at multiple critical points across the agricultural, processing, and distribution continuum, posing significant challenges to food safety and public health. On farms, resistant bacteria can emerge and proliferate due to the widespread use of antibiotics not only for therapeutic purposes but also for prophylaxis and as growth promoters in food-producing animals [30]. Such practices exert selective pressure on microbial communities, facilitating the emergence and persistence of multidrug-resistant strains that can be transmitted through animal products. Crops may also serve as vehicles for AMR contamination. The application of animal manure as fertilizer, particularly when inadequately treated, introduces resistant bacteria and mobile genetic elements into agricultural soils [11]. Contaminated irrigation water—often carrying residues of antibiotics or resistant microorganisms from human and animal waste—represents an additional significant source of AMR dissemination in horticultural production [105]. These resistant bacteria can colonize plant surfaces or internal tissues, making their removal challenging during post-harvest handling.

At the slaughterhouse and food processing stages, poor hygiene practices can lead to direct cross-contamination of carcasses and meat products. Equipment, surfaces, and workers' hands may act as vectors for the spread of resistant pathogens between contaminated and uncontaminated products [106]. The risk is compounded by the formation of biofilms on processing equipment, drains, and other hard-to-clean surfaces. Biofilms can harbor AMR bacteria and serve as reservoirs of resistance genes that persist despite standard sanitation measures [107]. These biofilms not only protect bacteria from environmental stressors but also enhance horizontal gene transfer among microbial populations. Furthermore, the distribution and retail environment can introduce additional contamination points. For example, inadequate temperature control, poor handling, and cross-contact between raw and ready-to-eat foods can facilitate the survival and spread of AMR bacteria [108]. Packaging materials and food contact surfaces may also contribute if not properly managed. Taken together, these diverse sources illustrate the complex and interconnected pathways through which AMR can enter and propagate within the food chain, underscoring the need for integrated surveillance and control measures at every stage from farm to fork.

5.3. Biochemical and Molecular Mechanisms of AMR in Foodborne Pathogens

AMR in foodborne pathogens is driven by several key biochemical and molecular mechanisms that enable bacteria to survive antibiotic exposure. Efflux pumps play a major role in multidrug resistance. The AcrAB-TolC system in *E. coli* and *Salmonella* and the CmeABC system in Campylobacter actively export antibiotics such as tetracyclines, macrolides, and fluoroquinolones, reducing their intracellular concentration [109]. Additional pumps such as EmrAB and MdfA contribute to resistance in *Salmonella* against a broader range of antibiotics.

β-lactamase production, particularly ESBLs like CTX-M and TEM variants, hydrolyze β-lactam antibiotics, rendering them ineffective. ESBL-producing *E. coli* and *Salmonella* have been frequently detected in retail meats and food handlers [110]. These enzymes are often plasmid-mediated and co-occur with other resistance genes. Plasmid-mediated quinolone resistance (PMQR) genes such as qnrA, qnrB, qnrS, and oqxAB protect DNA gyrase or enhance efflux, contributing to fluoroquinolone resistance in Enterobacterales from food sources [111].

Target site modifications also contribute to AMR. Mutations in the quinolone resistance-determining regions (QRDRs) of gyrA and parC confer resistance to fluoroquinolones in *Salmonella* and *Campylobacter* [112]. Ribosomal target alterations, such as methylation of 23S rRNA, contribute to macrolide resistance. Colistin resistance mediated by mcr genes (e.g., *mcr-1*) has emerged globally. These genes encode a phosphoethanolamine transferase that modifies lipid A, reducing colistin binding and efficacy. They are plasmid-borne and widespread in Enterobacterales from food sources [113].

Integrons and transposons facilitate the capture and spread of multiple resistance genes. Class 1 integrons are commonly identified in *Salmonella* and *E. coli* from animal and food sources, often linked to multidrug resistance [114]. Environmental co-selection mechanisms are gaining attention. Disinfectant use and heavy metal exposure in food production environments may co-select for AMR, enhancing efflux activity and persistence of resistance genes. Understanding these mechanisms is crucial for guiding surveillance strategies, designing targeted diagnostics, and developing effective AMR mitigation measures within One Health frameworks.

5.4. Prevention of Contamination

To reduce AMR contamination in foods, integrated farm-to-fork interventions are essential and must involve coordinated actions at every stage of the food production chain. On the farm level, the adoption of good agricultural practices and good veterinary practices plays a fundamental role. These practices emphasize the prudent and responsible use of antimicrobials, limiting their use to genuine clinical needs, and promoting alternatives such as vaccination, enhanced biosecurity, improved housing conditions, and better nutrition to reduce disease burden [30]. Biosecurity measures are also vital for limiting the spread of AMR. These include controlling access to farms, enforcing strict sanitation protocols for equipment and personnel, providing clean feed and water, and ensuring that the introduction of new animals or materials does not bring resistant bacteria into the farm environment [115]. Regular monitoring of antimicrobial use and resistance patterns supports early detection of emerging threats and informs better management practices.

At the slaughterhouse and food processing levels, the implementation of Hazard Analysis and Critical Control Point systems and adherence to good hygiene practices are indispensable. These systems help identify critical points where contamination could occur and set out measures to prevent or control hazards, including those related to AMR [98]. Proper cleaning and disinfection of equipment, maintenance of processing lines, and staff training in hygienic handling all contribute to minimizing cross-contamination. Finally, education of food handlers, retailers, and consumers about safe food handling, appropriate cooking practices, and the risks associated with AMR is an important supporting measure to limit the spread of resistant bacteria from foods to humans.

5.5. Detection Methods for AMR in Foods

Advances in diagnostic technologies now enable the rapid, sensitive, and precise detection of AMR determinants in various food matrices, supporting surveillance, outbreak investigations, and risk assessments. Among molecular tools, quantitative real-time PCR (qPCR) remains one of the most widely used methods for detecting and quantifying specific resistance genes, such as *blaCTX-M*, *mecA*, and *mcr-1*, with high specificity and sensitivity [116]. qPCR's advantages include rapid turnaround time, cost-effectiveness, and applicability to high-throughput screening, though it is limited to detecting known target genes. For comprehensive profiling, WGS is increasingly employed to characterize the full resistome of foodborne pathogens, providing detailed insights into resistance genes, mobile genetic elements, and their potential for horizontal gene transfer. WGS is now a cornerstone in national and international AMR surveillance programs, enabling source attribution and monitoring of resistance trends [117].

Metagenomic approaches allow for culture-independent detection of AMR genes in complex microbial communities, including those present in ready-to-eat foods, raw produce, or environmental samples associated with food production. These methods provide a holistic view of both culturable and unculturable organisms and the resistome they carry, making them powerful tools in risk assessment frameworks [118]. Emerging biosensor technologies, such as electrochemical sensors and nanomaterial-based platforms, are under active development for on-site AMR detection in food products. These methods offer the promise of rapid, point-of-care screening with minimal sample preparation and could play a key role in future food safety monitoring systems [119]. Together, these complementary approaches greatly enhance our ability to detect, monitor, and respond to AMR contamination along the food chain.

5.6. Legislation and Policy

Policy frameworks at national and international levels are essential for addressing AMR in food systems, as they provide the regulatory backbone for surveillance, antimicrobial use, and risk mitigation. At the global level, integrated approaches based on the One Health concept are increasingly recognized as critical to tackling AMR across human, animal, and environmental interfaces [29]. This includes establishing clear guidelines and standards for antimicrobial use in food-producing animals and promoting harmonized monitoring and reporting systems. In the EU, efforts to address AMR include the promotion of integrated surveillance and risk management strategies that span human health, veterinary, and food production sectors. This approach aims to reduce antimicrobial use in agriculture while strengthening food safety and security [120]. National action plans in countries such as Thailand and the United Kingdom explicitly link AMR mitigation strategies to food safety objectives, combining regulatory controls with incentives for best practices, surveillance, and public awareness initiatives [15].

The development and enforcement of such policies require sustained commitment, inter-sectoral collaboration, and adequate resources to ensure meaningful reductions in AMR risks along the food chain. A synthesis of recent studies (2018–2024) on the prevalence of antimicrobial-resistant pathogens in various food products across different regions is presented in Table 2. These studies illustrate the global distribution of MDR strains in meat, poultry, ready-to-eat foods, aquaculture products, and animal sources, highlighting the widespread nature of AMR and its direct relevance to food safety and security. The table summarizes key findings regarding resistance patterns, food types, and geographical coverage, supporting the critical need for integrated surveillance and control measures.

Table 2. Recent studies (2018–2024) on the prevalence of antimicrobial-resistant pathogens in food products across different regions.

Study (Author, Year)	Country/Region	Food Type	Key Findings on AMR
Tang et al. [99]	China	Retail pork and chicken meat	13% *E. coli* carried *mcr-1*; most co-produced ESBLs, indicating co-resistance to critical antibiotics
Nhung et al. [100]	Vietnam	Retail meat	>50% *Salmonella* MDR; high resistance to fluoroquinolones and ESBLs
Adel et al. [103]	Egypt	Retail meats and slaughterhouse isolates	82.4% *S. enterica* MDR; 41.2% ESBL producers; 67.6% carried plasmid-mediated quinolone resistance genes (*qnrA*, *qnrB*, *qnrS*)
Kayode & Okoh [102]	South Africa	RTE foods	83.5% *L. monocytogenes* MDR; 53% resistant to ceftriaxone; 61.9% to trimethoprim; 62.9% to oxytetracycline
Bort et al. [104]	Spain	Broiler production chain	97.6% *Campylobacter* resistant to ≥1 antibiotic; high resistance to fluoroquinolones and macrolides
Shafiq et al. [44]	Pakistan	Food-producing animals	ESBL-producing, colistin-resistant *E. coli* isolated from healthy animals
Zhang et al. [42]	China	Retail aquatic products	High prevalence of MDR *E. coli*; first report of *mcr-1*-positive ESBL-producing *E. coli* ST2705 and ST10 in fish

5.7. *Implications for Food Security*

AMR in foodborne pathogens presents a serious and growing threat to global food security. Contamination of food products with MDR bacteria can lead to supply chain disruptions through product recalls, trade bans, and loss of consumer confidence, directly reducing the availability of safe food [99,103]. The economic consequences of AMR-related food safety incidents, including increased production costs and market losses, disproportionately affect vulnerable populations and limit equitable access to nutritious foods, particularly in LMICs [15,100].

Furthermore, AMR reduces the effectiveness of veterinary treatments essential for animal health and productivity, contributing to lower yields and threatening the stability of food systems. Food utilization is also impacted, as AMR-related foodborne infections increase illness burden, impair nutrient absorption, and elevate healthcare costs [102]. The persistence of AMR in the food chain exacerbates vulnerabilities during crises (e.g., pandemics, natural disasters), further undermining the stability pillar of food security. Addressing AMR is therefore essential not only for food safety but also as a cornerstone for ensuring resilient, equitable, and sustainable food systems. This calls for coordinated One Health actions that integrate AMR mitigation into food security, agriculture, and public health policies at local, national, and global levels [29].

Foodborne AMR threatens not only infectious disease control but also nutritional security, particularly in vulnerable populations. Outbreaks associated with AMR pathogens, precautionary culling of livestock, and trade restrictions can reduce the availability and affordability of animal-source foods. For example, Grace [98] highlighted how livestock-associated AMR reduces productivity and increases costs in low-income countries, indirectly impacting protein access. Van Boeckel et al. [30] demonstrated that rising AMR trends in food animals correlate with increased production costs and trade restrictions, affecting protein supplies. Additionally, Schar et al. [121] emphasized that AMR-related restrictions on aquaculture products in Asia have affected fish availability in local diets, contributing to nutritional insecurity. These nutritional shocks are compounded in settings where alternative protein sources are scarce or unaffordable. Foodborne AMR thus represents a dual threat to public health and food security, underscoring the need for integrated AMR mitigation strategies that also safeguard equitable nutrition.

6. Strategic Priorities and Challenges in One Health AMR Mitigation

6.1. *Operationalizing One Health: Bridging Policy and Practice*

The One Health approach is widely recognized as a crucial strategy to address AMR, yet translating it into operational frameworks remains a challenge [122]. In LMICs, fragmented governance, limited resources, and weak surveillance infrastructure hinder multisectoral integration. Cella et al. [122] highlight that national action plans often struggle with implementation due to a lack of political prioritization and institutional capacity. Even when action plans are in place, many countries lack the mechanisms to translate strategic goals into enforceable legislation, coordinated inter-agency action, or sustainable funding models. Moreover, institutional silos between ministries of health, agriculture, and environment often lead to parallel, uncoordinated efforts that undermine the spirit of One Health. Strengthening inter-ministerial coordination through dedicated One Health platforms and policy coherence mechanisms is essential. Capacity-building efforts must also prioritize the training of personnel in multisectoral thinking, data sharing, and integrated risk assessment. Successful examples, such as Denmark's integrated surveillance model [123], demonstrate that cross-sectoral collaboration, stakeholder engagement, and consistent investment are key enablers. Tailoring such models to fit local governance contexts can help improve implementation and ownership at the national level.

6.2. Innovation and Equity: Addressing the Access Gap

While advances in diagnostics, bioinformatics, and therapeutics have opened new avenues for AMR mitigation, these innovations are largely concentrated in high-income countries. In LMICs, for example, limited access to rapid molecular diagnostics significantly delays pathogen identification and susceptibility testing, leading to empirical broad-spectrum antibiotic use and poorer clinical outcomes. A case study from Kenya revealed that only 15% of tertiary hospitals had functioning microbiology laboratories capable of culture and sensitivity testing, forcing clinicians to prescribe antibiotics based on syndromic diagnosis rather than confirmed data [124]. Similarly, a WHO-supported assessment in Southeast Asia found that genomic surveillance capacity was either absent or underutilized in over 70% of surveyed laboratories, resulting in critical delays in resistance tracking and outbreak containment [125].

Organizations such as the GARDP and the WHO Global Access Initiative for New Antibiotics (SECURE) initiative advocate for equitable access to life-saving antibiotics and diagnostics. A recent study published in The Lancet Infectious Diseases revealed that fewer than 7% of people with severe drug-resistant infections in LMICs receive the antibiotics they need [81]. This lack of access not only causes significant suffering and high mortality but may also contribute to the growing threat of AMR. The study, which examined data from eight countries, found a severe shortfall in appropriate antibiotic treatments, with access ranging from just 0.2% in Kenya to 14.9% in Mexico and Egypt [126]. Experts argue that global strategies must balance antibiotic innovation with broader access, especially in poorer nations. Drawing parallels to HIV treatment efforts, researchers have called for treatment targets and global action to improve antibiotic accessibility. To address these disparities, it is essential to implement policies that promote both the development of new antimicrobial agents and the equitable distribution of existing ones. This includes investing in healthcare infrastructure, enhancing laboratory capacities, and ensuring that LMICs have the necessary resources to combat AMR effectively.

6.3. Enhancing Surveillance: Integrating Environmental Data

Integrated surveillance is essential to capture the full spectrum of AMR across human, animal, and environmental domains. However, most national and international surveillance systems still primarily focus on human and veterinary sectors, neglecting the critical role of environmental reservoirs. Environmental compartments—such as surface waters, wastewater, soils, and sediments—serve as hotspots for antibiotic residues and ARGs originating from healthcare facilities, pharmaceutical manufacturing, and agricultural runoff [11]. Recent studies have highlighted the global spread of ARGs through aquatic environments. For instance, Hendriksen et al. [32] demonstrated the use of metagenomics in wastewater-based surveillance across 60 countries, revealing diverse resistance genes including those for last-resort antibiotics such as colistin. This approach offers a scalable, cost-effective way to monitor resistance trends and complements clinical surveillance data. Xiong et al. [127] further emphasized that fecal contamination and poor sanitation infrastructure significantly exacerbate environmental AMR dissemination, especially in LMICs.

The Joint Programming Initiative on AMR (JPIAMR) and WHO have both called for harmonized environmental surveillance protocols to integrate AMR tracking across sectors [95,128]. While progress has been made in some regions—such as the inclusion of waterborne AMR in the European AMR surveillance network (EARS-Net)—broader implementation remains limited. In low-resource settings, environmental surveillance faces challenges including limited laboratory capacity, lack of standardized indicators, and fragmented data sharing mechanisms [129]. To close this gap, environmental AMR surveillance should be incorporated into National Action Plans (NAPs) and supported through international funding platforms such as the Fleming Fund. Countries like Sweden and the Netherlands have established successful wastewater AMR monitoring programs that could serve as models for global adaptation [79]. Ultimately, a One Health-aligned surveillance system must integrate human, animal, and environmental data to enable the timely detection of resistance hotspots, guide policy interventions, and promote sustainable antimicrobial use practices.

6.4. From Awareness to Behavior Change

Raising public and professional awareness about AMR is a foundational intervention, but awareness alone does not ensure behavioral change. Chukwu et al. [130] conducted a systematic review showing that while awareness campaigns often increase knowledge, they have limited impact on changing prescribing behaviors without structural support. Similarly, Haenssgen et al. [131] found that in Southeast Asia, increased antibiotic knowledge did not translate into reduced demand for antibiotics, highlighting the importance of context-specific behavioral interventions. Horwood et al. [132] emphasized the role of peer influence, benchmarking, and real-time prescribing feedback in improving clinician adherence to antimicrobial guidelines. In a randomized trial, In a randomized trial, Meeker et al. [133] demonstrated that integrating behavioral nudges into electronic health record systems led to measurable improvements in prescribing quality. Verma et al. [134] documented the positive impact of incorporating AMR education into undergraduate medical and veterinary curricula across India, noting a significant increase in knowledge retention and reported stewardship practices. Finally, Sirota et al. [135] concluded that AMR education programs grounded in social and behavioral science were more likely to result in sustained behavioral change compared to traditional lecture-based approaches.

6.5. Strengthening Global Policy Frameworks

Global agencies such as the WHO, FAO, and the WOAH have developed strategic frameworks to guide AMR mitigation efforts [67,91]. These frameworks promote multisectoral collaboration, capacity building, and integrated surveillance systems based on the One Health concept. However, national uptake of Codex guidelines and related global action plans is uneven [67,136]. The ACT project by FAO helps countries adapt global AMR standards to local contexts, offering technical assistance, training, and policy guidance. A notable example is Thailand's National Strategic Plan on AMR (2017–2021), which incorporated WHO guidance and established coordination among ministries of public health, agriculture, and environment. The plan resulted in measurable reductions in antibiotic use in humans and animals, supported by robust surveillance and community engagement.

Similarly, Sweden has demonstrated the effectiveness of sustained investment and cross-sectoral coordination in curbing AMR through the Swedish Strategic Programme Against Antibiotic Resistance (STRAMA) initiative. STRAMA has led to a significant decline in antibiotic prescriptions and improved public awareness through educational campaigns and strict prescribing regulations. Multilateral commitments, such as those made by the G7 and the United Nations, have elevated AMR on the global health agenda [137,138], but sustained political will and investment are required to operationalize these commitments at the country level. Integrating global recommendations into national action plans, ensuring multisectoral stakeholder buy-in, and establishing monitoring frameworks are critical steps for long-term policy effectiveness. To facilitate a clearer understanding of the conclusions drawn in this review, Table 3 presents a structured summary of the main thematic areas explored, the corresponding key findings, and strategic recommendations. This table underscores the multifaceted nature of foodborne AMR and provides a concise roadmap for researchers, policymakers, and public health professionals working to address this global threat through the One Health approach.

Table 3. Summary of Major Findings and Strategic Recommendations for Combating Foodborne AMR.

Thematic Area	Key Finding	Reference	Strategic Recommendation	Reference
Microbial Food Safety and AMR Threats	High prevalence of MDR Salmonella, *E. coli*, *L. monocytogenes*, and *Campylobacter* in foods, especially ready-to-eat products across the EU	[41]	Implement farm-to-fork microbiological monitoring programs targeting RTE and minimally processed foods	[139]
Antibiotic Use in Food Production	Non-therapeutic use of antimicrobials (e.g., growth promotion) is a major AMR driver in livestock	[30]	Ban prophylactic antimicrobial use, enforce veterinary oversight, and promote stewardship in animal sectors	[139]
Environmental Reservoirs	Sewage and agricultural runoff harbor diverse AMR genes, reflecting environmental contamination	[32]	Integrate environmental samples (wastewater, manure, soil) into AMR surveillance; enhance pollution control	[139]
One Health Implementation	Fragmented governance hinders multisectoral coordination in AMR control	[122]	Establish inter-ministerial One Health coordination platforms and capacity building in risk analysis	[122]
Surveillance Gaps	Underrepresentation of animal and environmental data in national AMR systems	[41]	Broaden AMR surveillance to include metagenomic and wastewater approaches; support LMIC implementation	[32]
Innovation and Access Inequalities	Aquaculture AMU is rising (~93,000 t in 2017, projected +11.5% by 2030); LMIC diagnostics remain limited	[30,140]	Invest in affordable diagnostics in LMICs; support regional labs and rapid molecular tools	[140]
Behavior Change and Public Awareness	AMR awareness campaigns alone often fail to change prescribing behavior	[141]	Integrate behavioral nudges, peer comparison, and electronic prescribing feedback in stewardship	[133]
Policy and Governance Challenges	Global AMR strategies have variable national implementation; funding remains inadequate	[142]	Support local adaptation; embed AMR plans into national health and development agendas; secure sustainable financing	[141]

7. Conclusions

The convergence of microbial food safety and AMR presents a formidable and evolving threat to global health, food security, and sustainable development. This review has elucidated the multifaceted nature of foodborne AMR, highlighting critical points of emergence, transmission, and amplification across human, animal, and environmental domains. The widespread use and misuse of antibiotics in agriculture, aquaculture, and human medicine have facilitated the persistence and dissemination of MDR pathogens such as *Salmonella* spp., *E. coli*, *L. monocytogenes*, and *Campylobacter* spp. throughout the food chain, especially in RTE products. Tackling this crisis requires a robust One Health approach that integrates antimicrobial stewardship, surveillance, regulation, innovation, and public education. Advances in molecular diagnostics, whole-genome sequencing, and resistome profiling have been improving our ability to detect, monitor, and predict resistance patterns, while alternative interventions such as phage therapy, bacteriocins, and probiotics offer promising tools for reducing antibiotic dependence. However, innovation must be matched by equitable access—particularly in LMICs where laboratory capacity, surveillance infrastructure, and treatment availability remain limited. Sustained political commitment, global policy harmonization, and integrated action plans are essential for operationalizing the One Health framework. Embedding environmental monitoring, regulating antibiotic use in food systems, and fostering behavioral change through education will be key to slowing the tide of AMR. As the food system continues to globalize, a coordinated international response is critical to safeguard both human health and food safety in the face of rising resistance. These findings underscore the need for integrated, farm-to-fork interventions, robust surveillance systems, and policy frameworks that address AMR risks in foods. In conclusion, tackling AMR in food systems is integral to safeguarding food security and requires harmonized One Health strategies spanning agriculture, food safety, veterinary care, and public health.

Author Contributions: Conceptualization, A.E. and E.M.; writing—original draft, A.E., E.M., A.A. (Adil Abalkhail), H.M.E., A.T.E., A.M.A., M.I., A.A. (Abdulrahman Almujaidel), F.A., A.A. (Abdullah Alqrni) and A.A.-O.; writing—review and editing, A.E., E.M., A.A. (Adil Abalkhail), H.M.E., A.T.E., A.M.A., M.I., A.A. (Abdulrahman Almujaidel), F.A., A.A. (Abdullah Alqrni) and A.A.-O.; visualization, A.E., E.M., A.A. (Adil Abalkhail), H.M.E., A.T.E., A.M.A., M.I., A.A. (Abdulrahman Almujaidel), F.A., A.A. (Abdullah Alqrni) and A.A.-O.; supervision, A.E. and E.M.; project administration, A.E. and A.A.; funding acquisition, A.E., E.M., A.A. (Adil Abalkhail), H.M.E., A.T.E., A.M.A., M.I., A.A. (Abdulrahman Almujaidel), F.A., A.A. (Abdullah Alqrni) and A.A.-O. All authors have read and agreed to the published version of the manuscript.

Funding: This research received no external funding.

Institutional Review Board Statement: Not applicable.

Informed Consent Statement: Not applicable.

Data Availability Statement: No new data were created or analyzed in this study. Data sharing is not applicable to this article.

Acknowledgments: The researchers would like to thank the Deanship of Graduate Studies and Scientific Research at Qassim University for financial support (QU-APC-2025).

Conflicts of Interest: The authors declare no conflicts of interest.

References

Verraes, C.; Van Boxstael, S.; Van Meervenne, E.; Van Coillie, E.; Butaye, P.; Catry, B.; De Schaetzen, M.-A.; Van Huffel, X.; Imberechts, H.; Dierick, K. Antimicrobial resistance in the food chain: A review. *Int. J. Environ. Res. Public Health* **2013**, *10*, 2643–2669. [CrossRef]

Kumar, A.; Kumar, V.; Arsenov, D.; Thakur, M.; Kumar, A.; Khokhar, A.; Seth, C.S.; Kumar, R. The science of food safety and their health impacts. *J. Geochem. Explor.* **2024**, *267*, 107596. [CrossRef]

Marshall, B.M.; Levy, S.B. Food animals and antimicrobials: Impacts on human health. *Clin. Microbiol. Rev.* **2011**, *24*, 718–733. [CrossRef]

Silbergeld, E.K.; Graham, J.; Price, L.B. Industrial food animal production, antimicrobial resistance, and human health. *Annu. Rev. Public Health* **2008**, *29*, 151–169. [CrossRef] [PubMed]

5. Landers, T.F.; Cohen, B.; Wittum, T.E.; Larson, E.L. A review of antibiotic use in food animals: Perspective, policy, and potential. *Public Health Rep.* **2012**, *127*, 4–22. [CrossRef] [PubMed]
6. Manyi-Loh, C.; Mamphweli, S.; Meyer, E.; Okoh, A. Antibiotic use in agriculture and its consequential resistance in environmental sources: Potential public health implications. *Molecules* **2018**, *23*, 795. [CrossRef] [PubMed]
7. Nobrega, D.B.; Tang, K.L.; Caffrey, N.P.; De Buck, J.; Cork, S.C.; Ronksley, P.E.; Polachek, A.J.; Ganshorn, H.; Sharma, N.; Kastelic, J.P. Prevalence of antimicrobial resistance genes and its association with restricted antimicrobial use in food-producing animals: A systematic review and meta-analysis. *J. Antimicrob. Chemother.* **2021**, *76*, 561–575. [CrossRef]
8. Allel, K.; Day, L.; Hamilton, A.; Lin, L.; Furuya-Kanamori, L.; Moore, C.E.; Van Boeckel, T.; Laxminarayan, R.; Yakob, L. Global antimicrobial-resistance drivers: An ecological country-level study at the human–animal interface. *Lancet Planet. Health* **2023**, *7*, e291–e303. [CrossRef]
9. European Food Safety Authority (EFSA); European Centre for Disease Prevention and Control (ECDC). The European Union summary report on antimicrobial resistance in zoonotic and indicator bacteria from humans, animals and food in 2016. *EFSA J.* **2018**, *16*, e05182.
10. Robinson, T.P.; Bu, D.; Carrique-Mas, J.; Fèvre, E.M.; Gilbert, M.; Grace, D.; Hay, S.I.; Jiwakanon, J.; Kakkar, M.; Kariuki, S. Antibiotic resistance is the quintessential One Health issue. *Trans. R. Soc. Trop. Med. Hyg.* **2016**, *110*, 377–380. [CrossRef]
11. Berendonk, T.U.; Manaia, C.M.; Merlin, C.; Fatta-Kassinos, D.; Cytryn, E.; Walsh, F.; Bürgmann, H.; Sørum, H.; Norström, M.; Pons, M.-N. Tackling antibiotic resistance: The environmental framework. *Nat. Rev. Microbiol.* **2015**, *13*, 310–317. [CrossRef]
12. Nethathe, B.; Matsheketsheke, P.A.; Mashau, M.E.; Ramashia, S.E. Microbial safety of ready-to-eat food sold by retailers in Thohoyandou, Limpopo province, South Africa. *Cogent Food Agric.* **2023**, *9*, 2185965. [CrossRef]
13. Organization, W.H. GLASS Manual for Antimicrobial Resistance Surveillance in Common Bacteria Causing Human Infection. 2023. Available online: https://www.who.int/news-room/fact-sheets/detail/antimicrobial-resistance (accessed on 23 May 2025).
14. O'Neill, J. Tackling Drug-Resistant Infections Globally: Final Report and Recommendations. 2016. Available online: https://www.cabidigitallibrary.org/doi/full/10.5555/20173071720 (accessed on 23 May 2025).
15. Woolhouse, M.; Ward, M.; Van Bunnik, B.; Farrar, J. Antimicrobial resistance in humans, livestock and the wider environment. *Philos. Trans. R. Soc. B Biol. Sci.* **2015**, *370*, 20140083. [CrossRef] [PubMed]
16. Martinez, J.L. Environmental pollution by antibiotics and by antibiotic resistance determinants. *Environ. Pollut.* **2009**, *157*, 2893–2902. [CrossRef] [PubMed]
17. Lammie, S.L.; Hughes, J.M. Antimicrobial resistance, food safety, and one health: The need for convergence. *Annu. Rev. Food Sci. Technol.* **2016**, *7*, 287–312. [CrossRef]
18. Zhou, Y.; Zhang, A.; van Klinken, R.D.; Jones, D.; Wang, J. Consumers' perspectives on antibiotic use and antibiotic resistance in food animals: A systematic review. *npj Sci. Food* **2025**, *9*, 29. [CrossRef]
19. Lee, A.W.-T.; Ng, I.C.-F.; Wong, E.Y.-K.; Wong, I.T.-F.; Sze, R.P.-P.; Chan, K.-Y.; So, T.-Y.; Zhang, Z.; Fung, S.K.-Y.; Wong, S.C.-Y. Comprehensive identification of pathogenic microbes and antimicrobial resistance genes in food products using nanopore sequencing-based metagenomics. *Food Microbiol.* **2024**, *121*, 104493. [CrossRef]
20. Tang, K.L.; Caffrey, N.P.; Nóbrega, D.B.; Cork, S.C.; Ronksley, P.E.; Barkema, H.W.; Polachek, A.J.; Ganshorn, H.; Sharma, N.; Kellner, J.D. Restricting the use of antibiotics in food-producing animals and its associations with antibiotic resistance in food-producing animals and human beings: A systematic review and meta-analysis. *Lancet Planet. Health* **2017**, *1*, e316–e327. [CrossRef]
21. FAO. The FAO Action Plan on Antimicrobial Resistance 2016–2020. 2016. Available online: https://openknowledge.fao.org/server/api/core/bitstreams/70dd55f7-3348-4895-b651-76308101cb45/content (accessed on 23 May 2025).
22. CDC. *Antibiotic Resistance Threats in the United States*; US Department of Health and Human Services: Washington, DC, USA, 2019; Volume 1, pp. 67–100.
23. Hollis, A.; Ahmed, Z. The path of least resistance: Paying for antibiotics in non-human uses. *Health Policy* **2014**, *118*, 264–270. [CrossRef]
24. Regan, Á.; Sweeney, S.; McKernan, C.; Benson, T.; Dean, M. Consumer perception and understanding of the risks of antibiotic use and antimicrobial resistance in farming. *Agric. Hum. Values* **2023**, *40*, 989–1001. [CrossRef]
25. Kasimanickam, V.; Kasimanickam, M.; Kasimanickam, R. Antibiotics use in food animal production: Escalation of antimicrobial resistance: Where are we now in combating AMR? *Med. Sci.* **2021**, *9*, 14. [CrossRef] [PubMed]
26. Adebisi, Y.A. Balancing the risks and benefits of antibiotic use in a globalized world: The ethics of antimicrobial resistance. *Glob. Health* **2023**, *19*, 27. [CrossRef]
27. Murray, C.J.; Ikuta, K.S.; Sharara, F.; Swetschinski, L.; Aguilar, G.R.; Gray, A.; Han, C.; Bisignano, C.; Rao, P.; Wool, E. Global burden of bacterial antimicrobial resistance in 2019: A systematic analysis. *Lancet* **2022**, *399*, 629–655. [CrossRef] [PubMed]
28. Havelaar, A.H.; Kirk, M.D.; Torgerson, P.R.; Gibb, H.J.; Hald, T.; Lake, R.J.; Praet, N.; Bellinger, D.C.; De Silva, N.R.; Gargouri, N. World Health Organization global estimates and regional comparisons of the burden of foodborne disease in 2010. *PLoS Med.* **2015**, *12*, e1001923. [CrossRef]

29. Laxminarayan, R.; Duse, A.; Wattal, C.; Zaidi, A.K.; Wertheim, H.F.; Sumpradit, N.; Vlieghe, E.; Hara, G.L.; Gould, I.M.; Goossens, H. Antibiotic resistance—The need for global solutions. *Lancet Infect. Dis.* **2013**, *13*, 1057–1098. [CrossRef]
30. Van Boeckel, T.P.; Brower, C.; Gilbert, M.; Grenfell, B.T.; Levin, S.A.; Robinson, T.P.; Teillant, A.; Laxminarayan, R. Global trends in antimicrobial use in food animals. *Proc. Natl. Acad. Sci. USA* **2015**, *112*, 5649–5654. [CrossRef]
31. Kujat Choy, S.; Neumann, E.-M.; Romero-Barrios, P.; Tamber, S. Contribution of Food to the Human Health Burden of Antimicrobial Resistance. *Foodborne Pathog. Dis.* **2024**, *21*, 71–82. [CrossRef]
32. Hendriksen, R.S.; Munk, P.; Njage, P.; Van Bunnik, B.; McNally, L.; Lukjancenko, O.; Röder, T.; Nieuwenhuijse, D.; Pedersen, S.K.; Kjeldgaard, J. Global monitoring of antimicrobial resistance based on metagenomics analyses of urban sewage. *Nat. Commun.* **2019**, *10*, 1124. [CrossRef]
33. Deng, Y.; Tan, A.; Zhao, F.; Wang, F.; Gong, H.; Lai, Y.; Huang, Z. Global distribution of antimicrobial resistance genes in aquaculture. *One Health Adv.* **2025**, *3*, 6. [CrossRef]
34. Yuan, X.; Lv, Z.; Zhang, Z.; Han, Y.; Liu, Z.; Zhang, H. A review of antibiotics, antibiotic resistant bacteria, and resistance genes in aquaculture: Occurrence, contamination, and transmission. *Toxics* **2023**, *11*, 420. [CrossRef]
35. Hossain, A.; Habibullah-Al-Mamun, M.; Nagano, I.; Masunaga, S.; Kitazawa, D.; Matsuda, H. Antibiotics, antibiotic-resistant bacteria, and resistance genes in aquaculture: Risks, current concern, and future thinking. *Environ. Sci. Pollut. Res.* **2022**, *29*, 11054–11075. [CrossRef]
36. Wilson, A.; Gray, J.; Chandry, P.S.; Fox, E.M. Phenotypic and genotypic analysis of antimicrobial resistance among Listeria monocytogenes isolated from Australian food production chains. *Genes* **2018**, *9*, 80. [CrossRef] [PubMed]
37. Unicomb, L.E.; Fullerton, K.E.; Kirk, M.D.; Stafford, R.J. Outbreaks of campylobacteriosis in Australia, 2001 to 2006. *Foodborne Pathog. Dis.* **2009**, *6*, 1241–1250. [CrossRef] [PubMed]
38. Wieczorek, K.; Osek, J. Antimicrobial resistance mechanisms among Campylobacter. *BioMed Res. Int.* **2013**, *2013*, 340605. [CrossRef]
39. European Food Safety Authority (EFSA); European Centre for Disease Prevention and Control (ECDC). The European Union Summary Report on Antimicrobial Resistance in zoonotic and indicator bacteria from humans, animals and food in 2020/2021. *EFSA J.* **2023**, *21*, e07867.
40. Kuehn, B.M. CDC establishes global networks to combat antimicrobial resistance. *JAMA* **2022**, *327*, 315. [CrossRef]
41. European Food Safety Authority (EFSA); European Centre for Disease Prevention and Control (ECDC). The European Union Summary Report on Antimicrobial Resistance in zoonotic and indicator bacteria from humans, animals and food in 2018/2019. *EFSA J.* **2021**, *19*, e06490.
42. Zhang, S.; Huang, Y.; Yang, G.; Wu, Q.; Zhang, J.; Wang, J.; Ding, Y.; Su, Y.; Ye, Q.; Wu, S. High prevalence of multidrug-resistant Escherichia coli in retail aquatic products in China and the first report of mcr-1-positive extended-spectrum β-lactamase-producing E. coli ST2705 and ST10 in fish. *Int. J. Food Microbiol.* **2024**, *408*, 110449. [CrossRef]
43. Roussel, C.; De Paepe, K.; Galia, W.; De Bodt, J.; Chalancon, S.; Denis, S.; Leriche, F.; Vandekerkove, P.; Ballet, N.; Blanquet-Diot, S. Multi-targeted properties of the probiotic saccharomyces cerevisiae CNCM I-3856 against enterotoxigenic escherichia coli (ETEC) H10407 pathogenesis across human gut models. *Gut Microbes* **2021**, *13*, 1953246. [CrossRef]
44. Shafiq, M.; Rahman, S.U.; Bilal, H.; Ullah, A.; Noman, S.M.; Zeng, M.; Yuan, Y.; Xie, Q.; Li, X.; Jiao, X. Incidence and molecular characterization of ESBL-producing and colistin-resistant Escherichia coli isolates recovered from healthy food-producing animals in Pakistan. *J. Appl. Microbiol.* **2022**, *133*, 1169–1182. [CrossRef]
45. Zhang, Y.-J.; Hu, H.-W.; Chen, Q.-L.; Singh, B.K.; Yan, H.; Chen, D.; He, J.-Z. Transfer of antibiotic resistance from manure-amended soils to vegetable microbiomes. *Environ. Int.* **2019**, *130*, 104912. [CrossRef]
46. Zhu, Y.-G.; Zhao, Y.; Li, B.; Huang, C.-L.; Zhang, S.-Y.; Yu, S.; Chen, Y.-S.; Zhang, T.; Gillings, M.R.; Su, J.-Q. Continental-scale pollution of estuaries with antibiotic resistance genes. *Nat. Microbiol.* **2017**, *2*, 16270. [CrossRef] [PubMed]
47. Keet, R.; Rip, D. Listeria monocytogenes isolates from Western Cape, South Africa exhibit resistance to multiple antibiotics and contradicts certain global resistance patterns. *AIMS Microbiol.* **2021**, *7*, 40. [CrossRef] [PubMed]
48. Petrovic Fabijan, A.; Lin, R.C.; Ho, J.; Maddocks, S.; Ben Zakour, N.L.; Iredell, J.R.; Westmead Bacteriophage Therapy Team. Safety of bacteriophage therapy in severe *Staphylococcus aureus* infection. *Nat. Microbiol.* **2020**, *5*, 465–472. [CrossRef] [PubMed]
49. Gradisteanu Pircalabioru, G.; Popa, L.I.; Marutescu, L.; Gheorghe, I.; Popa, M.; Czobor Barbu, I.; Cristescu, R.; Chifiriuc, M.-C. Bacteriocins in the era of antibiotic resistance: Rising to the challenge. *Pharmaceutics* **2021**, *13*, 196. [CrossRef]
50. Sijbom, M.; Büchner, F.L.; Saadah, N.H.; Numans, M.E.; De Boer, M.G. Determinants of inappropriate antibiotic prescription in primary care in developed countries with general practitioners as gatekeepers: A systematic review and construction of a framework. *BMJ Open* **2023**, *13*, e065006. [CrossRef]
51. Browne, A.J.; Chipeta, M.G.; Haines-Woodhouse, G.; Kumaran, E.P.; Hamadani, B.H.K.; Zaraa, S.; Henry, N.J.; Deshpande, A.; Reiner, R.C.; Day, N.P. Global antibiotic consumption and usage in humans, 2000–2018: A spatial modelling study. *Lancet Planet. Health* **2021**, *5*, e893–e904. [CrossRef]

52. Fleming-Dutra, K.E.; Hersh, A.L.; Shapiro, D.J.; Bartoces, M.; Enns, E.A.; File, T.M.; Finkelstein, J.A.; Gerber, J.S.; Hyun, D.Y.; Linder, J.A. Prevalence of inappropriate antibiotic prescriptions among US ambulatory care visits, 2010–2011. *JAMA* **2016**, *315*, 1864–1873. [CrossRef]
53. Pollack, L.A.; van Santen, K.L.; Weiner, L.M.; Dudeck, M.A.; Edwards, J.R.; Srinivasan, A. Antibiotic stewardship programs in US acute care hospitals: Findings from the 2014 National Healthcare Safety Network Annual Hospital Survey. *Rev. Infect. Dis.* **2016**, *63*, 443–449. [CrossRef]
54. CDC. Core Elements of Hospital Antibiotic Stewardship Programs. 2024. Available online: https://www.cdc.gov/antibiotic-use/hcp/core-elements/hospital.html (accessed on 23 May 2025).
55. Ardillon, A.; Ramblière, L.; Kermorvant-Duchemin, E.; Sok, T.; Zo, A.Z.; Diouf, J.-B.; Long, P.; Lach, S.; Sarr, F.D.; Borand, L. Inappropriate antibiotic prescribing and its determinants among outpatient children in 3 low-and middle-income countries: A multicentric community-based cohort study. *PLoS Med.* **2023**, *20*, e1004211. [CrossRef]
56. Van Boeckel, T.P.; Pires, J.; Silvester, R.; Zhao, C.; Song, J.; Criscuolo, N.G.; Gilbert, M.; Bonhoeffer, S.; Laxminarayan, R. Global trends in antimicrobial resistance in animals in low-and middle-income countries. *Science* **2019**, *365*, eaaw1944. [CrossRef]
57. Acosta, A.; Tirkaso, W.; Nicolli, F.; Van Boeckel, T.P.; Cinardi, G.; Song, J. The future of antibiotic use in livestock. *Nat. Commun.* **2025**, *16*, 2469. [CrossRef] [PubMed]
58. Citaristi, I. United Nations Environment Programme—UNEP. In *The Europa Directory of International Organizations 2022*; Routledge: London, UK, 2022; pp. 193–199.
59. Cheung, C.; Naughton, P.J.; Dooley, J.S.; Corcionivoschi, N.; Brooks, C. The spread of antimicrobial resistance in the aquatic environment from faecal pollution: A scoping review of a multifaceted issue. *Environ. Monit. Assess.* **2025**, *197*, 467. [CrossRef] [PubMed]
60. Fernández Salgueiro, M.; Cernuda Martínez, J.A.; Gan, R.K.; Arcos González, P. Climate change and antibiotic resistance: A scoping review. *Environ. Microbiol. Rep.* **2024**, *16*, e70008. [CrossRef] [PubMed]
61. Larsson, D.J.; Flach, C.-F. Antibiotic resistance in the environment. *Nat. Rev. Microbiol.* **2022**, *20*, 257–269. [CrossRef]
62. WHO. Global Antimicrobial Resistance and Use Surveillance System (GLASS) Report 2022. 2022. Available online: https://www.who.int/publications/i/item/9789240062702 (accessed on 24 May 2025).
63. ECDC. Antimicrobial Resistance Surveillance in Europe 2022—EARS-Net Annual Report. 2022. Available online: https://www.ecdc.europa.eu/en/publications-data/surveillance-antimicrobial-resistance-europe-2022 (accessed on 24 May 2025).
64. European Centre for Disease Prevention and Control (ECDC); European Food Safety Authority (EFSA); European Medicines Agency (EMA). Antimicrobial consumption and resistance in bacteria from humans and food-producing animals: Fourth joint inter-agency report on integrated analysis of antimicrobial agent consumption and occurrence of antimicrobial resistance in bacteria from humans and food-producing animals in the EU/EEA JIACRA IV–2019−2021. *EFSA J.* **2024**, *22*, e8589.
65. CDC. National Antimicrobial Resistance Monitoring System (NARMS): About NARMS. 2022. Available online: https://www.cdc.gov/narms/about/index.html (accessed on 24 May 2025).
66. The Fleming Fund. How the Fleming Fund Has Worked to Build Surveillance Capacity in LMICs Across Human and Animal Health Sectors. 2023. Available online: https://www.flemingfund.org/publications/how-the-fleming-fund-has-worked-to-build-surveillance-capacity-in-lmics-across-human-and-animal-health-sectors/ (accessed on 24 May 2025).
67. FAO. *FAO Assessment Tool for Laboratories and AMR Surveillance Systems (ATLASS)*; FAO: Rome, Italy, 2023; Available online: https://www.fao.org/antimicrobial-resistance/resources/tools/fao-atlass/en/ (accessed on 24 May 2025).
68. Argimón, S.; Masim, M.A.; Gayeta, J.M.; Lagrada, M.L.; Macaranas, P.K.; Cohen, V.; Limas, M.T.; Espiritu, H.O.; Palarca, J.C.; Chilam, J. Integrating whole-genome sequencing within the National Antimicrobial Resistance Surveillance Program in the Philippines. *Nat. Commun.* **2020**, *11*, 2719. [CrossRef]
69. Dyar, O.J.; Huttner, B.; Schouten, J.; Pulcini, C. What is antimicrobial stewardship? *Clin. Microbiol. Infect.* **2017**, *23*, 793–798. [CrossRef]
70. Baur, D.; Gladstone, B.P.; Burkert, F.; Carrara, E.; Foschi, F.; Döbele, S.; Tacconelli, E. Effect of antibiotic stewardship on the incidence of infection and colonisation with antibiotic-resistant bacteria and Clostridium difficile infection: A systematic review and meta-analysis. *Lancet Infect. Dis.* **2017**, *17*, 990–1001. [CrossRef]
71. Singh, S.; Menon, V.P.; Mohamed, Z.U.; Kumar, V.A.; Nampoothiri, V.; Sudhir, S.; Moni, M.; Dipu, T.; Dutt, A.; Edathadathil, F. Implementation and impact of an antimicrobial stewardship program at a tertiary care center in South India. *Open Forum Infect. Dis.* **2019**, *6*, ofy290. [CrossRef]
72. Debnath, F.; De, R.G.; Chakraborty, D.; Majumdar, A.; Mukhopadhyay, S.; Sarkar, M.D.; Singh, T.; Patra, S.K.; Saha, S.; Rehman, J. Antimicrobial stewardship implementation in primary and secondary tier hospitals in India: Interim findings from a need assessment study using mixed method design. *Sci. Rep.* **2024**, *14*, 28068. [CrossRef] [PubMed]
73. Sahni, A.; Bahl, A.; Martolia, R.; Jain, S.K.; Singh, S.K. Implementation of antimicrobial stewardship activities in India. *Indian J. Med. Spec.* **2020**, *11*, 5–9.

4. Aarestrup, F.M. The livestock reservoir for antimicrobial resistance: A personal view on changing patterns of risks, effects of interventions and the way forward. *Philos. Trans. R. Soc. B Biol. Sci.* **2015**, *370*, 20140085. [CrossRef]
5. FDA. Summary Report on Antimicrobials Sold or Distributed for Use in Food Producing Animals. 2023. Available online: https://www.fda.gov/animal-veterinary/antimicrobial-resistance/2022-summary-report-antimicrobials-sold-or-distributed-use-food-producing-animals (accessed on 24 May 2025).
6. Liu, Y.-Y.; Wang, Y.; Walsh, T.R.; Yi, L.-X.; Zhang, R.; Spencer, J.; Doi, Y.; Tian, G.; Dong, B.; Huang, X. Emergence of plasmid-mediated colistin resistance mechanism MCR-1 in animals and human beings in China: A microbiological and molecular biological study. *Lancet Infect. Dis.* **2016**, *16*, 161–168. [CrossRef] [PubMed]
7. Dar, O.A.; Hasan, R.; Schlundt, J.; Harbarth, S.; Caleo, G.; Dar, F.K.; Littmann, J.; Rweyemamu, M.; Buckley, E.J.; Shahid, M. Exploring the evidence base for national and regional policy interventions to combat resistance. *Lancet* **2016**, *387*, 285–295. [CrossRef] [PubMed]
8. Rizzo, L.; Manaia, C.; Merlin, C.; Schwartz, T.; Dagot, C.; Ploy, M.-C.; Michael, I.; Fatta-Kassinos, D. Urban wastewater treatment plants as hotspots for antibiotic resistant bacteria and genes spread into the environment: A review. *Sci. Total Environ.* **2013**, *447*, 345–360. [CrossRef]
9. Larsson, D.J.; Andremont, A.; Bengtsson-Palme, J.; Brandt, K.K.; de Roda Husman, A.M.; Fagerstedt, P.; Fick, J.; Flach, C.-F.; Gaze, W.H.; Kuroda, M. Critical knowledge gaps and research needs related to the environmental dimensions of antibiotic resistance. *Environ. Int.* **2018**, *117*, 132–138. [CrossRef]
0. Abedon, S.T.; García, P.; Mullany, P.; Aminov, R. Phage therapy: Past, present and future. *Front. Microbiol.* **2017**, *8*, 981. [CrossRef]
1. Laxminarayan, R.; Sridhar, D.; Blaser, M.; Wang, M.; Woolhouse, M. Achieving global targets for antimicrobial resistance. *Science* **2016**, *353*, 874–875. [CrossRef]
2. Gomaa, A.A.; Klumpe, H.E.; Luo, M.L.; Selle, K.; Barrangou, R.; Beisel, C.L. Programmable removal of bacterial strains by use of genome-targeting CRISPR-Cas systems. *MBio* **2014**, *5*, e00928-13. [CrossRef]
3. Rodrigues, M.; McBride, S.W.; Hullahalli, K.; Palmer, K.L.; Duerkop, B.A. Conjugative delivery of CRISPR-Cas9 for the selective depletion of antibiotic-resistant enterococci. *Antimicrob. Agents Chemother.* **2019**, *63*. [CrossRef] [PubMed]
4. Zolti, O.; Suganthan, B.; Ramasamy, R.P. Lab-on-a-chip electrochemical biosensors for foodborne pathogen detection: A review of common standards and recent progress. *Biosensors* **2023**, *13*, 215. [CrossRef]
5. Mahato, K.; Chandra, P. based miniaturized immunosensor for naked eye ALP detection based on digital image colorimetry integrated with smartphone. *Biosens. Bioelectron.* **2019**, *128*, 9–16. [CrossRef] [PubMed]
6. Omar, R.-L.E.; Natasha, N.T.; Edward, G.D. Regulatory Landscape and the Potential of Bacteriophage Applications in the United States' Food Industry. *J. Food Prot.* **2025**, *88*, 100510.
7. Pérez de la Lastra, J.M.; González-Acosta, S.; Otazo-Pérez, A.; Asensio-Calavia, P.; Rodríguez-Borges, V.M. Antimicrobial Peptides for Food Protection: Leveraging Edible Mushrooms and Nano-Innovation. *Dietetics* **2025**, *4*, 9. [CrossRef]
8. McCullough, A.; Parekh, S.; Rathbone, J.; Del Mar, C.; Hoffmann, T. A systematic review of the public's knowledge and beliefs about antibiotic resistance. *J. Antimicrob. Chemother.* **2016**, *71*, 27–33. [CrossRef]
9. Huttner, B.; Goossens, H.; Verheij, T.; Harbarth, S. Characteristics and outcomes of public campaigns aimed at improving the use of antibiotics in outpatients in high-income countries. *Lancet Infect. Dis.* **2010**, *10*, 17–31. [CrossRef]
0. Chapot, L.; Sarker, M.S.; Begum, R.; Hossain, D.; Akter, R.; Hasan, M.M.; Bupasha, Z.B.; Bayzid, M.; Salauddin, M.; Parvej, M.S. Knowledge, attitudes and practices regarding antibiotic use and resistance among veterinary students in Bangladesh. *Antibiotics* **2021**, *10*, 332. [CrossRef]
1. WHO. World Antimicrobial Awareness Week. 2024. Available online: https://www.who.int/campaigns/world-amr-awareness-week/2024 (accessed on 24 May 2025).
2. WHO. World Antimicrobial Awareness Week. 2015. Available online: https://www.who.int/news-room/events/detail/2015/11/16/default-calendar/world-antibiotic-awareness-week (accessed on 24 May 2025).
3. Codex Alimentarius Commission. *Guidelines on Integrated Monitoring and Surveillance of Foodborne Antimicrobial Resistance (CXG 94-2021)*; FAO and WHO: Rome, Italy, 2021; Available online: https://www.fao.org/fao-who-codexalimentarius/sh-proxy/de/?lnk=1&url=https://workspace.fao.org/sites/codex/Standards/CXG+94-2021/CXG_94e.pdf (accessed on 24 May 2025).
4. CARB-X. Collaboration. 2024. Available online: https://globalamrhub.org/collaboration/ (accessed on 24 May 2025).
5. WHO. Global Antimicrobial Resistance and Use Surveillance System (GLASS) Report 2023. 2023. Available online: https://www.who.int/publications/i/item/9789240070424 (accessed on 24 May 2025).
6. Severino, N.; Reyes, C.; Fernandez, Y.; Azevedo, V.; Francisco, L.E.D.; Ramos, R.T.; Maroto-Martín, L.O.; Franco, E.F. Bacterial Foodborne Diseases in Central America and the Caribbean: A Systematic Review. *Microbiol. Res.* **2025**, *16*, 78. [CrossRef]
7. Su, M.; Satola, S.W.; Read, T.D. Genome-based prediction of bacterial antibiotic resistance. *J. Clin. Microbiol.* **2019**, *57*, e01405-18. [CrossRef]
8. Grace, D. Food safety in low and middle income countries. *Int. J. Environ. Res. Public Health* **2015**, *12*, 10490–10507. [CrossRef] [PubMed]

99. Tang, B.; Chang, J.; Luo, Y.; Jiang, H.; Liu, C.; Xiao, X.; Ji, X.; Yang, H. Prevalence and characteristics of the mcr-1 gene in retail meat samples in Zhejiang Province, China. *J. Microbiol.* **2022**, *60*, 610–619. [CrossRef]
100. Nhung, N.T.; Van, N.T.B.; Van Cuong, N.; Duong, T.T.Q.; Nhat, T.T.; Hang, T.T.T.; Nhi, N.T.H.; Kiet, B.T.; Hien, V.B.; Ngoc, P.T. Antimicrobial residues and resistance against critically important antimicrobials in non-typhoidal Salmonella from meat sold at wet markets and supermarkets in Vietnam. *Int. J. Food Microbiol.* **2018**, *266*, 301–309. [CrossRef] [PubMed]
101. Phung, T.; Tran, T.; Pham, D.; To, A.; Le, H. Occurrence and molecular characteristics of Listeria monocytogenes isolated from ready-to-eat meats in Hanoi, Vietnam. *Ital. J. Food Saf.* **2020**, *9*, 8772. [CrossRef] [PubMed]
102. Kayode, A.J.; Okoh, A.I. Antimicrobial-resistant Listeria monocytogenes in ready-to-eat foods: Implications for food safety and risk assessment. *Foods* **2023**, *12*, 1346. [CrossRef]
103. Adel, W.A.; Ahmed, A.M.; Hegazy, Y.; Torky, H.A.; Shimamoto, T. High prevalence of ESBL and plasmid-mediated quinolone resistance genes in Salmonella enterica isolated from retail meats and slaughterhouses in Egypt. *Antibiotics* **2021**, *10*, 881 [CrossRef]
104. Bort, B.; Martí, P.; Mormeneo, S.; Mormeneo, M.; Iranzo, M. Prevalence and antimicrobial resistance of Campylobacter spp. isolated from broilers throughout the supply chain in Valencia, Spain. *Foodborne Pathog. Dis.* **2022**, *19*, 717–724. [CrossRef]
105. Tasho, R.P.; Cho, J.Y. Veterinary antibiotics in animal waste, its distribution in soil and uptake by plants: A review. *Sci. Total Environ.* **2016**, *563*, 366–376. [CrossRef]
106. Silva, V.; Araújo, S.; Monteiro, A.; Eira, J.; Pereira, J.E.; Maltez, L.; Igrejas, G.; Lemsaddek, T.S.; Poeta, P. *Staphylococcus aureus* and MRSA in livestock: Antimicrobial resistance and genetic lineages. *Microorganisms* **2023**, *11*, 124. [CrossRef]
107. Hazards, E.P.o.B.; Koutsoumanis, K.; Allende, A.; Alvarez-Ordóñez, A.; Bolton, D.; Bover-Cid, S.; Chemaly, M.; Davies, R.; De Cesare, A.; Hilbert, F. Update of the list of QPS-recommended biological agents intentionally added to food or feed as notified to EFSA 14: Suitability of taxonomic units notified to EFSA until March 2021. *EFSA J.* **2021**, *19*, e06689.
108. Murray, A.K.; Zhang, L.; Yin, X.; Zhang, T.; Buckling, A.; Snape, J.; Gaze, W.H. Novel insights into selection for antibiotic resistance in complex microbial communities. *MBio* **2018**, *9*, e00969-18. [CrossRef] [PubMed]
109. Price, E.D.; Dassanayake, R.P.; Bearson, S.M. Increasing antimicrobial susceptibility of MDR Salmonella with the efflux pump inhibitor 1-(1-Naphthylmethyl)-piperazine. *Biochem. Biophys. Res. Commun.* **2023**, *668*, 49–54. [CrossRef]
110. Alkhalaf, A.A.; El-Ghareeb, W.R.; Raheem, S.M.; Seliem, M.M.; Shosha, A.M.; Elzawahry, R.R. A study on the prevalence of multidrug resistant food poisoning Salmonella spp. in camel meat and offal with a reduction trial using organic acids. *J. Adv. Vet Res.* **2024**, *14*, 526–530.
111. Jacoby, G.A.; Strahilevitz, J.; Hooper, D.C. Plasmid-mediated quinolone resistance. *Plasmids Biol. Impact Biotechnol. Discov.* **2015**, *30*, 475–503.
112. Eaves, D.J.; Randall, L.; Gray, D.T.; Buckley, A.; Woodward, M.J.; White, A.P.; Piddock, L.J. Prevalence of mutations within the quinolone resistance-determining region of gyrA, gyrB, parC, and parE and association with antibiotic resistance in quinolone resistant Salmonella enterica. *Antimicrob. Agents Chemother.* **2004**, *48*, 4012–4015. [CrossRef]
113. Schumann, A.; Gaballa, A.; Yang, H.; Yu, D.; Ernst, R.K.; Wiedmann, M. Site-selective modifications by lipid A phosphoethanolamine transferases linked to colistin resistance and bacterial fitness. *mSphere* **2024**, *9*, e00731-24. [CrossRef] [PubMed]
114. Racewicz, P.; Majewski, M.; Madeja, Z.E.; Łukomska, A.; Kubiak, M. Role of integrons in the proliferation of multiple drug resistance in selected bacteria occurring in poultry production. *Br. Poult. Sci.* **2020**, *61*, 122–131. [CrossRef]
115. McEwen, S.A.; Collignon, P.J. Antimicrobial resistance: A one health perspective. In *Antimicrobial Resistance in Bacteria from Livestock and Companion Animals*; Wiley: New York, NY, USA, 2018; pp. 521–547.
116. Schmidt, V.T.; Smith, K.F.; Melvin, D.W.; Amaral-Zettler, L.A. Community assembly of a euryhaline fish microbiome during salinity acclimation. *Mol. Ecol.* **2015**, *24*, 2537–2550. [CrossRef]
117. Moura, A.; Tourdjman, M.; Leclercq, A.; Hamelin, E.; Laurent, E.; Fredriksen, N.; Van Cauteren, D.; Bracq-Dieye, H.; Thouvenot P.; Vales, G. Real-time whole-genome sequencing for surveillance of Listeria monocytogenes, France. *Emerg. Infect. Dis.* **2017**, *23*, 1462. [CrossRef]
118. Zhang, A.-N.; Gaston, J.M.; Dai, C.L.; Zhao, S.; Poyet, M.; Groussin, M.; Yin, X.; Li, L.-G.; van Loosdrecht, M.C.; Topp, E. An omics-based framework for assessing the health risk of antimicrobial resistance genes. *Nat. Commun.* **2021**, *12*, 4765. [CrossRef]
119. Xu, M.; Wang, R.; Li, Y. Electrochemical biosensors for rapid detection of Escherichia coli O157: H7. *Talanta* **2017**, *162*, 511–522 [CrossRef] [PubMed]
120. Cassini, A.; Högberg, L.D.; Plachouras, D.; Quattrocchi, A.; Hoxha, A.; Simonsen, G.S.; Colomb-Cotinat, M.; Kretzschmar, M.E. Devleesschauwer, B.; Cecchini, M. Attributable deaths and disability-adjusted life-years caused by infections with antibiotic resistant bacteria in the EU and the European Economic Area in 2015: A population-level modelling analysis. *Lancet Infect. Dis* **2019**, *19*, 56–66. [CrossRef]
121. Schar, D.; Klein, E.Y.; Laxminarayan, R.; Gilbert, M.; Van Boeckel, T.P. Global trends in antimicrobial use in aquaculture. *Sci. Rep* **2020**, *10*, 21878. [CrossRef] [PubMed]

122. Cella, E.; Giovanetti, M.; Benedetti, F.; Scarpa, F.; Johnston, C.; Borsetti, A.; Ceccarelli, G.; Azarian, T.; Zella, D.; Ciccozzi, M. Joining forces against antibiotic resistance: The one health solution. *Pathogens* **2023**, *12*, 1074. [CrossRef] [PubMed]
123. Delpy, L.; Astbury, C.C.; Aenishaenslin, C.; Ruckert, A.; Penney, T.L.; Wiktorowicz, M.; Ciss, M.; Benko, R.; Bordier, M. Integrated surveillance systems for antibiotic resistance in a One Health context: A scoping review. *BMC Public Health* **2024**, *24*, 1717. [CrossRef]
124. Global AMR R&D Hub. *Situation Analysis: Antibiotic Use and Resistance Monitoring in LMICs—Gaps and Opportunities*; Global AMR R&D Hub: Berlin, Germany, 2023; Available online: https://globalamrhub.org (accessed on 24 May 2025).
125. World Health Organization. *Regional Strategy on Antimicrobial Resistance 2023–2027*; WHO SEARO: New Delhi, India, 2023. Available online: https://cdn.who.int/media/docs/default-source/searo/india/antimicrobial-resistance/rev-sea-hlm-407.pdf?sfvrsn=986ba503_2 (accessed on 24 May 2025).
126. Wirtz, V.J.; Hogerzeil, H.V.; Gray, A.L.; Bigdeli, M.; de Joncheere, C.P.; Ewen, M.A.; Gyansa-Lutterodt, M.; Jing, S.; Luiza, V.L.; Mbindyo, R.M. Essential medicines for universal health coverage. *Lancet* **2017**, *389*, 403–476. [CrossRef]
127. Xiong, Q.; Hu, L.-X.; Liu, Y.-S.; Zhao, J.-L.; He, L.-Y.; Ying, G.-G. Microalgae-based technology for antibiotics removal: From mechanisms to application of innovational hybrid systems. *Environ. Int.* **2021**, *155*, 106594. [CrossRef]
128. JPIAMR. Environmental Dimensions of Antimicrobial Resistance: Current Surveillance Approaches and Research Needs. 2022. Available online: https://www.jpiamr.eu/environment (accessed on 24 May 2025).
129. Leonard, A.F.; Zhang, L.; Balfour, A.J.; Garside, R.; Gaze, W.H. Human recreational exposure to antibiotic resistant bacteria in coastal bathing waters. *Environ. Int.* **2015**, *82*, 92–100. [CrossRef]
130. Chukwu, E.E.; Oladele, D.A.; Enwuru, C.A.; Gogwan, P.L.; Abuh, D.; Audu, R.A.; Ogunsola, F.T. Antimicrobial resistance awareness and antibiotic prescribing behavior among healthcare workers in Nigeria: A national survey. *BMC Infect. Dis.* **2021**, *21*, 22. [CrossRef]
131. Haenssgen, M.J.; Charoenboon, N.; Zanello, G.; Mayxay, M.; Reed-Tsochas, F.; Lubell, Y.; Wertheim, H.; Lienert, J.; Xayavong, T.; Zaw, Y.K. Antibiotic knowledge, attitudes and practices: New insights from cross-sectional rural health behaviour surveys in low-income and middle-income South-East Asia. *BMJ Open* **2019**, *9*, e028224. [CrossRef]
132. Horwood, J.; Cabral, C.; Hay, A.D.; Ingram, J. Primary care clinician antibiotic prescribing decisions in consultations for children with RTIs: A qualitative interview study. *Br. J. Gen. Pract.* **2016**, *66*, e207–e213. [CrossRef] [PubMed]
133. Meeker, D.; Knight, T.K.; Friedberg, M.W.; Linder, J.A.; Goldstein, N.J.; Fox, C.R.; Rothfeld, A.; Diaz, G.; Doctor, J.N. Nudging guideline-concordant antibiotic prescribing: A randomized clinical trial. *JAMA Intern. Med.* **2014**, *174*, 425–431. [CrossRef] [PubMed]
134. Verma, A.R.; Mathur, V.P. Anti-microbial resistance and dentistry. *Indian J. Dent. Res.* **2021**, *32*, 272–273. [CrossRef] [PubMed]
135. Sirota, M.; Habersaat, K.B.; Betsch, C.; Bonga, D.L.; Borek, A.; Buckel, A.; Butler, R.; Byrne-Davis, L.; Caudell, M.; Charani, E. We must harness the power of social and behavioural science against the growing pandemic of antimicrobial resistance. *Nat. Hum. Behav.* **2024**, *8*, 11–13. [CrossRef]
136. Kickbusch, I.; Szabo, M.M.C. A new governance space for health. *Glob. Health Action* **2014**, *7*, 23507. [CrossRef]
137. UNEP. Quadripartite Key Recommendations and Priorities for the 2024 UNGA High-Level Meeting on Antimicrobial Resistance (AMR)—A Policy Brief. 2024. Available online: https://www.unep.org/resources/policy-and-strategy/quadripartite-key-recommendations-and-priorities-2024-unga-high-level (accessed on 23 May 2025).
138. Amir, A.; Tellez, V.M. Leveraging the Antimicrobial Resistance Declarations of 2024 to Reduce the Burden of Drug-Resistant Infections. 2023. Available online: https://www.southcentre.int/policy-brief-137-14-march-2025/ (accessed on 24 May 2025).
139. WHO. Integrated Surveillance of Antimicrobial Resistance in Foodborne Bacteria Application of a One Health Approach. 2017. Available online: https://iris.who.int/bitstream/handle/10665/255747/9789241512411-eng.pdf (accessed on 29 June 2025).
140. Tiseo, K.; Huber, L.; Gilbert, M.; Robinson, T.P.; Van Boeckel, T.P. Global trends in antimicrobial use in food animals from 2017 to 2030. *Antibiotics* **2020**, *9*, 918. [CrossRef]
141. Chukwu, E.E.; Oladele, D.A.; Awoderu, O.B.; Afocha, E.E.; Lawal, R.G.; Abdus-Salam, I.; Ogunsola, F.T.; Audu, R.A. A national survey of public awareness of antimicrobial resistance in Nigeria. *Antimicrob. Resist. Infect. Control* **2020**, *9*, 72. [CrossRef]
142. WHO. Action Against Antimicrobial Resistance Requires a One Health Approach. 2024. Available online: https://www.who.int/europe/publications/i/item/WHO-EURO-2024-9510-49282-73655?utm_source=chatgpt.com (accessed on 29 June 2025).

microorganisms

MDPI

Review

Emerging Technologies and Integrated Strategies for Microbial Detection and Control in Fresh Produce

Ayman Elbehiry [1,*], Eman Marzouk [1], Feras Alzaben [2], Abdulaziz Almuaither [3], Banan Abead [4], Mohammed Alamri [5], Abdulaziz M. Almuzaini [6] and Akram Abu-Okail [7]

1 Department of Public Health, College of Applied Medical Sciences, Qassim University, P.O. Box 6666, Buraydah 51452, Saudi Arabia; e.marzouk@qu.edu.sa
2 Department of Food Service, King Fahad Armed Hospital, Jeddah 23311, Saudi Arabia
3 First Air Defense Group, Department of Nutrition, Riyadh 13624, Saudi Arabia
4 Support Service Department, King Fahad Armed Hospital, Jeddah 23311, Saudi Arabia
5 Family & Community Medicine Department, Prince Sultan Medical Military City, Riyadh 12233, Saudi Arabia
6 Department of Veterinary Preventive Medicine, College of Veterinary Medicine, Qassim University, Buraydah 51452, Saudi Arabia
7 Department of Pathology and Laboratory Diagnosis, College of Veterinary Medicine, Qassim University, Buraydah 51452, Saudi Arabia
* Correspondence: ar.elbehiry@qu.edu.sa

Abstract: The global consumption of fresh and ready-to-eat (RTE) fruits and vegetables has surged due to increasing awareness of their nutritional benefits. However, this trend has been accompanied by a rise in foodborne illness outbreaks linked to microbial contamination. This narrative review synthesizes current knowledge on the prevalence and diversity of foodborne pathogens in fresh produce, including bacterial, viral, and fungal agents. It critically evaluates both conventional and emerging detection methods, ranging from culture-based techniques and immunoassays to advanced molecular diagnostics, biosensors, flow cytometry (FC), and hyperspectral imaging (HSI). Additionally, this review discusses cutting-edge control strategies, such as natural antifungal agents, essential oils, biocontrol methods, and non-thermal technologies like cold plasma and UV-C treatment. Emphasis is placed on sampling methodologies, sustainability, One Health perspectives, and regulatory considerations. By highlighting recent technological advances and their limitations, this review aims to support the development of integrated, effective, and safe microbial control approaches for the fresh produce supply chain.

Keywords: foodborne pathogens; fresh produce; microbial safety; rapid detection; public health

1. Introduction

The global increase in consumption of fresh, raw, and RTE fruits and vegetables is largely driven by heightened public awareness of their health benefits. These foods are rich in essential vitamins, minerals, phytochemicals, and dietary fiber, and have been strongly associated with reduced risks of cardiovascular disease, type 2 diabetes, stroke, and all-cause mortality [1–4]. However, this positive dietary shift has coincided with a notable rise in foodborne illness outbreaks linked to fresh produce, primarily due to microbial contamination [5,6].

Unlike animal-based foods that typically undergo thermal processing, fruits and vegetables are often consumed raw, without any microbial inactivation steps, thereby increasing the likelihood of ingesting foodborne pathogens [7]. While bacterial pathogens such as

https://doi.org/10.3390/microorganisms13071447

Escherichia coli (*E. coli*) O157:H7, *Salmonella enterica* (*S. enterica*), and *Listeria monocytogenes* (*L. monocytogenes*) are most frequently implicated in produce-related outbreaks, non-bacterial pathogens also pose considerable risks [8,9]. Enteric viruses—particularly norovirus and hepatitis A virus—can survive on fresh produce surfaces and persist through common decontamination methods [10]. Moreover, protozoan parasites such as *Cryptosporidium* spp., *Giardia duodenalis*, and *Cyclospora cayetanensis* have been increasingly detected on fruits and vegetables, contributing to foodborne illness worldwide [11,12]. These pathogens may resist sanitizing treatments or internalize within produce tissues, complicating detection and elimination efforts. Consequently, safety strategies for fresh produce must consider a broader spectrum of pathogens—beyond bacteria—to effectively reduce foodborne disease risk.

Contamination can occur at various stages along the farm-to-fork continuum. In the preharvest phase, sources include untreated irrigation water, manure, contaminated soil, and animal intrusions [13]. During harvest and postharvest, factors such as unhygienic handling, equipment contamination, improper washing, and inadequate storage can introduce or amplify microbial risks [14,15]. Leafy greens are particularly vulnerable due to their morphology, which favors bacterial adhesion and biofilm formation [16]. The globalization of the food supply chain further complicates produce safety. Extended transport and storage durations increase the window for microbial proliferation and cross-border contamination [17]. *E. coli* and *L. monocytogenes* were responsible for additional outbreaks, including those with high morbidity and mortality [14,18].

Traditional culture-based methods, while reliable, are time-intensive and not ideal for perishable commodities. Consequently, advanced molecular and rapid detection methods have gained attention. Recent studies have increasingly evaluated techniques such as real-time PCR, multiplex PCR, and loop-mediated isothermal amplification (LAMP) for detecting foodborne pathogens in fresh produce, demonstrating high sensitivity, specificity, and potential for rapid on-site diagnostics [19,20]. Immunological assays like enzyme-linked immunosorbent assay (ELISA) and PCR–ELISA hybrids provide practical tools for routine testing [21]. Innovations in biosensor technology, including electrochemical platforms based on bacteriophage T7, offer real-time, on-site pathogen detection with high sensitivity [22]. HSI and FC provide complementary approaches for detecting viable and non-culturable pathogens on complex surfaces [23,24].

On the control side, non-thermal technologies such as cold plasma, UV-C irradiation, and antimicrobial edible coatings have shown promise for decontaminating produce without compromising sensory quality [25,26]. Plasma-activated water (PAW) has emerged as an effective and eco-friendly decontamination method [27,28]. Despite these advances, implementation barriers remain, including regulatory inconsistencies, high costs, limited infrastructure in low-income regions, and the need for specialized technical skills [29]. This review synthesizes current knowledge on microbial contamination, detection technologies, and control strategies for raw and RTE fruits and vegetables, aiming to inform future research directions and practical solutions for enhanced produce safety.

This narrative review is based on a structured search of peer-reviewed literature published between January 2000 and April 2024. Searches were conducted using scientific databases including PubMed, Web of Science, and Scopus. Keywords used in various combinations included "foodborne pathogens", "fresh produce", "detection methods", "decontamination technologies", "antimicrobial resistance", and "microbial safety". Articles were selected based on their relevance to microbial contamination, detection, and control strategies in raw and RTE fruits and vegetables. Additional sources were identified through manual cross-referencing of citations in key publications.

Building upon these foundations, this review offers a novel contribution by critically integrating both traditional and emerging detection and control strategies for foodborne pathogens in fresh produce with a broader and more interdisciplinary scope than previously published reviews. While Abdelshafy et al. [7] primarily focused on bacterial contamination, our work expands the discussion to include fungal spoilage, natural antifungal agents, and innovative preservation technologies. We also incorporate cutting-edge developments such as CRISPR-based diagnostics, HSI, and smart packaging solutions published between 2023 and 2025. Moreover, this review emphasizes sustainability, regulatory frameworks, and One Health perspectives, framing microbial safety not only as a technical challenge but also as an ecological and public health priority. These distinctions establish this review as a more comprehensive and forward-looking resource for addressing microbial safety in raw and RTE fruits and vegetables.

2. Epidemiology of Produce-Linked Foodborne Illnesses and Contamination Pathways

Unlike animal-derived foods that are typically subjected to cooking or thermal treatment, fresh produce is often consumed without any microbial inactivation step, significantly increasing the risk of exposure to pathogens. According to a review of CDC data from 2006 to 2023, *S. enterica* was responsible for 34 outbreaks linked to fresh produce in the United States, resulting in 7256 illnesses and 10 deaths. In the same period, *E. coli* O157:H7 caused 16 outbreaks with 998 illnesses and 10 deaths, while *L. monocytogenes* was implicated in 10 outbreaks that led to 303 cases and 54 deaths [7,14]. Recent studies have reported similar contamination trends in Saudi Arabia. A study by Kuddus et al. [30] analyzing RTE salads from restaurants in Hail city found significant microbial loads, including coliforms and intestinal parasites, even after multiple washings, indicating potential health risks associated with the consumption of such salads.

2.1. Epidemiological Trends and High-Risk Commodities

Several categories of produce are consistently linked to foodborne outbreaks. Aiyedun et al. [31] highlighted leafy greens, cucumbers, tomatoes, cantaloupes, and sprouts as the most commonly implicated commodities in both North America and Europe. Leafy greens are particularly prone to microbial contamination due to their irregular surfaces, large surface area, and ability to internalize bacteria through stomatal openings or damaged tissue [16]. Outbreak data show that produce-associated bacterial pathogens account for a significant portion of foodborne outbreaks globally. Havelaar et al. [18] estimated that 36% of all foodborne illnesses in the United States and 42% in the European Union could be attributed to contaminated fruits and vegetables. International surveillance studies further support this trend. For example, Wartha et al. [32] reported frequent detection of *L. monocytogenes*, *E. coli*, and *Staphylococcus aureus* (*S. aureus*) in RTE salads in Germany. Similarly, Bai et al. [33] found multidrug-resistant *Clostridium perfringens* in vegetable markets across China, while Oyedele et al. [34] detected toxigenic *E. coli* strains in raw produce in Nigeria. In a Canadian assessment by Zhang et al. [35], *L. monocytogenes* was found in 0.51% of fresh-cut fruits and 0.24% of vegetables, while *Salmonella* and *E. coli* O157:H7 were not detected. These results underscore the sporadic yet significant risk associated with opportunistic pathogens in produce.

Other produce types, including herbs like coriander and parsley, have also been linked to contamination. Elsafi et al. [36] identified various antibiotic-resistant bacteria, including *Enterobacter cloacae*, *Klebsiella* spp., and *Citrobacter freundii*, on leafy vegetables such as parsley and lettuce in Saudi Arabia's eastern region. Notably, some isolates exhibited resistance to ampicillin, highlighting the potential for dissemination of resistant strains through fresh produce. In addition to bacterial contamination, the presence of fungi such as *Aspergillus* and *Penicillium* spp. on vegetables has raised concerns about mycotoxin production. Pandey et al. [37] emphasized that these fungi are common in tropical and subtropical regions, where high humidity favors their growth. Mycotoxins like ochratoxin A, produced by these fungi, pose significant health risks, including nephrotoxicity and carcinogenicity.

2.2. Contamination Pathways: From Preharvest to Retail

Microbial contamination of fresh produce can occur at virtually any stage along the production and distribution chain. In the preharvest environment, key contamination sources include the use of untreated or inadequately treated irrigation water, contact with contaminated soil, improperly composted manure, animal feces, and intrusion by wildlife [13,38]. A notable example is the 2006 outbreak of *E. coli* O157:H7 associated with spinach, which was traced to irrigation water contaminated by cattle feces [14]. During harvesting and handling, contamination may be introduced through the use of dirty tools, containers, gloves, or worker hands. The reuse of equipment across different fields without sufficient sanitation greatly increases the potential for cross-contamination [38,39]. Recent investigations have also highlighted the role of irrigation water quality in pathogen transfer. For example, Allende and Monaghan [40] reviewed multiple cases where *Salmonella*, *Listeria*, and pathogenic *E. coli* were introduced through water sources, particularly in regions lacking microbial water quality monitoring.

In the postharvest phase, including processing, storage, and retail display, additional microbial risks are introduced. Improper washing and cutting, unsanitary packaging conditions, and inadequate temperature control may facilitate bacterial survival and proliferation. Psychrotrophic pathogens such as *L. monocytogenes* are particularly concerning, as they are capable of growing at refrigeration temperatures [32]. Moreover, contact surfaces like conveyor belts and packaging materials can harbor bacteria and serve as long-term reservoirs if not regularly sanitized [35]. According to the findings of Olaimat and Holley [6], poorly maintained equipment and wet processing environments offer ideal conditions for biofilm formation, allowing pathogens to persist despite routine cleaning. These biofilms, especially in niches like blade assemblies and drains, can intermittently release viable cells into produce batches.

At the retail and consumer levels, produce may be stored near raw meat, subjected to improper refrigeration, or handled without adequate hygiene, further increasing the risk of contamination and cross-contact with harmful microorganisms [41,42]. Moreover, consumer behavior plays a substantial role. As shown by Redmond and Griffith [43], consumers often fail to wash produce adequately or separate it from raw animal products, creating additional routes for pathogen transfer within household kitchens.

2.3. Emerging Risk Factors: Climate Change, Globalization, and Antimicrobial Resistance

Emerging risk factors are reshaping the landscape of microbial safety in fresh produce. Climate change—through its effects on temperature, rainfall, and drought cycles—can influence pathogen persistence in agricultural environments [44,45]. For example, increased rainfall can result in runoff carrying fecal contaminants into irrigation water sources [17,46,47]. Simultaneously, urbanization and water scarcity have driven the reuse of untreated or poorly treated wastewater for irrigation in many developing countries, creating direct pathways for contamination [13]. A comprehensive review by Lake and Barker [48] emphasized that while precise predictions are challenging, climate-related factors such as rising temperatures and altered precipitation patterns are expected to increase the prevalence and survival of foodborne pathogens in the environment. The study called for improved modeling and surveillance systems to anticipate food safety risks under future climate conditions.

Globalization extends the distribution chain, often increasing the time between harvest and consumption. This added delay creates more opportunities for microbial proliferation and cross-border contamination. As produce travels across multiple countries and through various handlers, traceability becomes increasingly complex, and response times in outbreak scenarios are delayed [7,15,49,50]. Bintsis [51] highlighted that the expansion of global food trade has significantly elevated the risk of foodborne pathogen dissemination, including antibiotic-resistant strains, due to extended transport chains and variability in food safety standards across exporting countries.

A particularly alarming development is the growing threat of antimicrobial resistance (AMR). Multidrug-resistant *Campylobacter* spp., *L. monocytogenes*, and *S. aureus* have been isolated from vegetables in multiple countries [52,53]. These bacteria are not only more difficult to treat when infections occur, but they may also act as vectors for resistance gene transfer in the environment. Kim and Ahn [54] and Hashempour-Baltork et al. [55] both reported the presence of AMR genes in *E. coli* and *Enterococcus* strains isolated from fresh produce, highlighting the potential for produce to serve as a reservoir for resistance dissemination. Peng et al. [56] demonstrated through metagenomic analysis that mobile genetic elements carrying resistance genes—including those for beta-lactams, tetracyclines, and sulfonamides—are widespread across human, animal, and environmental sources in China. Their study revealed that fresh produce can serve as a bridge in this resistance-sharing network, underscoring the importance of a One Health approach to surveillance.

3. Detection Methods for Foodborne Pathogens in Fresh Produce

Ensuring the microbiological safety of fresh and RTE fruits and vegetables requires detection tools that are not only sensitive and specific but also rapid and adaptable to real-time monitoring. Effective pathogen detection also depends critically on sampling methodology, which forms the foundation for accurate analysis. Sampling of fresh produce must account for the heterogeneous distribution of pathogens across surfaces and internal tissues. For culture-based approaches, Uyttendaele et al. [57] recommend the aseptic collection of 25–50 g of produce, followed by homogenization and enrichment to optimize pathogen recovery. Swabbing and rinse techniques are commonly used for surface contamination assessment; comparative studies highlight that rinsing outperforms swabbing, especially for creviced produce [58]. In molecular methods such as qPCR and LAMP, effective removal of PCR inhibitors (e.g., polyphenols and polysaccharides) through optimized DNA extraction is essential. Miao et al. [59] demonstrated a one-step RT-qPCR capable of detecting *S. enterica* at 10 CFU/g in fresh-cut vegetables after a short enrichment, highlighting the importance of both sensitive amplification and efficient sample preparation.

For FC, Buzatu et al. [60] developed a high-sensitivity, real-time detection system and highlighted that rinsate pre-filtration and optimized staining greatly enhance accurate *L. monocytogenes* enumeration in produce matrices. HSI protocols require intact samples under controlled lighting and precise sample positioning; Lorente et al. [61] and ElMasry & Sun [62] emphasized that these parameters directly influence spoilage and microbial detection accuracy. Finally, maintaining cold-chain logistics and conducting prompt analysis (preferably within 24 h) are universally recommended to prevent microbial proliferation or die-off, as outlined by the Canadian Food Inspection Agency [63]. Appropriate sampling protocols are essential to ensure reproducibility across detection platforms and validity in assessing decontamination efficacy. Furthermore, the sensitivity and reproducibility of these methodologies vary. qPCR and RT-qPCR can detect pathogens at levels as low as 10–100 gene copies per reaction, while FC enables rapid quantification with high reproducibility in viable cell detection [60,64]. HSI shows strong spatial sensitivity but is more variable due to lighting conditions and sample geometry [61]. Clearly defined sampling protocols are therefore critical for ensuring data reliability across methods.

Figure 1 illustrates the diverse range of diagnostic methods currently employed to detect foodborne pathogens in fresh produce. These tools span from traditional culture-based techniques and immunological assays to advanced molecular and spectroscopic platforms. Methods such as PCR and qPCR offer rapid and specific DNA-based identification, while MALDI-TOF MS and HSI enable non-culture-based, high-resolution profiling of microbial presence. The inclusion of novel technologies like biosensors and CRISPR-based diagnostics reflects ongoing innovations aimed at enhancing detection sensitivity, speed, and field applicability. Together, these methods form a comprehensive toolkit that supports proactive microbial risk assessment and real-time safety monitoring across the fresh produce supply chain.

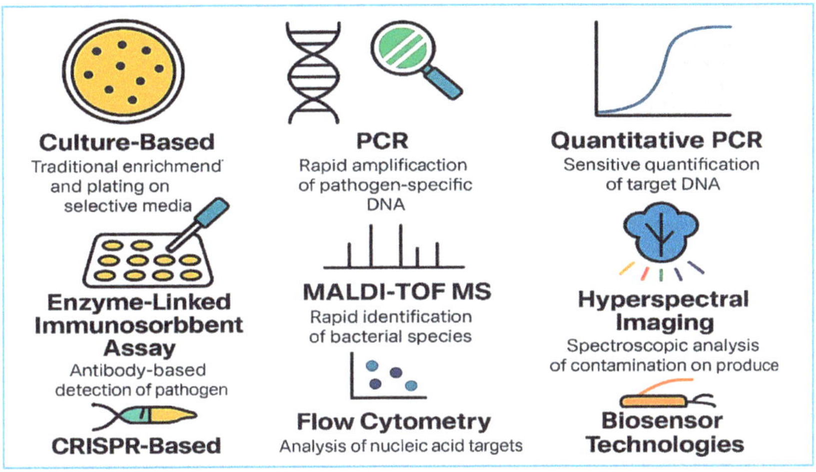

Figure 1. Detection methods for foodborne pathogens in fresh produce. This infographic illustrates eight major diagnostic approaches: culture-based methods, PCR, quantitative PCR (qPCR), ELISA, MALDI-TOF MS, HSI, FC, CRISPR-based detection, and biosensor technologies. Each method varies in its detection principle, sensitivity, and suitability for rapid or field-based application.

While traditional detection methods, such as culture-based techniques and immunological assays, have long served as the cornerstone of microbial diagnostics in fresh produce, they come with significant limitations. Culture-based methods, including selective enrichment and plating followed by biochemical or serological confirmation, remain the gold standard due to their high specificity and ability to recover live organisms for downstream analysis. They are widely accepted in regulatory frameworks and are cost-effective in terms of materials. However, these methods are labor-intensive, require skilled technicians, and are notably slow, often taking 5 to 10 days to yield actionable results. Moreover, they often fail to detect viable but non-culturable (VBNC) pathogens that may survive stress conditions such as refrigeration or sanitization [29,65].

Immunological assays, particularly enzyme-linked immunosorbent assays (ELISAs), offer quicker turnaround times and relatively simple operation. They are especially suitable for large-scale screening in routine quality control settings. Yet, their sensitivity can be compromised in complex food matrices, and they are prone to false positives or negatives due to cross-reactivity between structurally similar antigens [21]. These limitations restrict their utility when low levels of contamination must be accurately detected or when rapid results are critical.

To overcome these challenges, molecular and spectroscopic techniques have become increasingly prevalent. Quantitative PCR (qPCR) provides high sensitivity and specificity, enabling the detection of even low concentrations of pathogenic DNA in produce samples [66]. Its real-time amplification allows for quantification, which is essential for risk assessment. However, qPCR performance can be affected by inhibitors present in leafy greens or other matrix components, necessitating careful sample preparation. Droplet digital PCR (ddPCR) offers an even higher degree of accuracy and quantification by partitioning samples into thousands of micro-droplets, thus minimizing interference from inhibitors and improving reproducibility [67]. Despite their analytical advantages, both qPCR and ddPCR require high-cost instruments and reagents, as well as well-trained personnel, limiting their accessibility in low-resource settings.

MALDI-TOF MS represents a powerful proteomic tool for rapid species-level identification. It operates by generating mass spectra of bacterial proteins, which are then compared to reference libraries [68–71]. MALDI-TOF MS offers excellent throughput, low per-sample costs after setup, and minimal reagent consumption. However, its effectiveness is reduced when working directly with mixed microbial populations or environmental matrices, and spectral library limitations can impair accurate identification of novel or atypical strains.

Next-generation detection strategies include biosensor technologies, which offer portability, rapid analysis, and potential integration into automated systems. These devices employ biological recognition elements—such as antibodies, aptamers, or bacteriophages—combined with transducers that convert binding events into measurable signals. Electrochemical, magnetoelastic, and optical biosensors have shown success in detecting pathogens like *Salmonella*, *E. coli*, and *Listeria* directly from lettuce and spinach surfaces [22,72]. Still, widespread deployment remains limited by sensor robustness, matrix effects, and the need for calibration and validation.

CRISPR-based diagnostics are among the most promising innovations in molecular detection. Systems such as SHERLOCK and DETECTR leverage CRISPR-associated proteins (e.g., Cas12, Cas13) for highly specific nucleic acid detection, sometimes down to attomolar concentrations. These platforms have been adapted for food safety monitoring using lateral flow formats or fluorescence readouts [72,73]. Their advantages include programmability, speed, and compatibility with minimal infrastructure. Nonetheless, commercial adoption is still nascent due to regulatory and scalability challenges.

HSI, an advanced non-invasive technique, combines spatial and spectral data to detect microbial contamination and spoilage by analyzing unique spectral fingerprints. HSI can detect microbial growth, even in its early stages, on complex surfaces such as spinach, strawberries, and tomatoes [74–77]. However, implementation is limited by high equipment costs, the need for sophisticated data interpretation algorithms, and sensitivity to environmental lighting conditions. In parallel, the development of intelligent biosensors integrated with mobile technologies is reshaping real-time decision-making in food safety systems. Chen et al. [78] demonstrated a next-generation biosensor platform incorporating aptamer recognition, smartphone-based readout, and cloud connectivity, enabling rapid, decentralized pathogen detection. These systems align with the growing demand for on-site testing tools that are both accessible and scalable.

In summary, no single detection method is universally superior. Traditional culture-based and immunoassays remain essential for baseline and confirmatory testing, especially in regulatory and export settings. However, the increasing complexity of global food supply chains and the need for rapid, high-throughput, and field-deployable solutions underscore the importance of adopting newer molecular and spectroscopic tools. A comparative assessment of sensitivity, specificity, time-to-result, cost, technical complexity, and field applicability is crucial for selecting the appropriate platform for a given context. Blended strategies—combining traditional methods with modern technologies—may offer the most robust framework for safeguarding the microbial safety of fresh produce.

4. Control Strategies for Microbial Contamination in Fresh Produce

Traditional disinfection treatments such as chlorinated water washes, peracetic acid (PAA), and hydrogen peroxide rinses have historically formed the backbone of microbial control in the fresh produce industry. Chlorine-based sanitizers remain the most widely used due to their low cost, broad-spectrum efficacy, and ease of integration into large-scale washing systems [79,80]. However, they also present several critical limitations. The efficacy of chlorine is significantly reduced in the presence of organic matter, and its use can lead to the formation of potentially harmful chlorinated by-products such as trihalomethanes, raising safety and environmental concerns [81,82]. Additionally, increasing consumer and regulatory scrutiny is focused on chemical residues left on fresh produce surfaces [83].

PAA and hydrogen peroxide are viewed as residue-free alternatives with potent oxidizing capabilities. However, they may be less effective against biofilm-embedded or internalized pathogens, and improper concentrations can damage delicate fruits and vegetables, negatively impacting texture and shelf life [84,85]. From a regulatory standpoint, these sanitizers must comply with maximum residue limits and often require post-treatment rinsing, adding to operational complexity. In response to these challenges, emerging non-thermal decontamination technologies are being increasingly explored as promising alternatives that reduce or eliminate chemical use while preserving sensory and nutritional quality. These include cold plasma, UV-C, pulsed-light treatment, ozone-based sanitation, and electrolyzed water, all of which have shown substantial efficacy in inactivating a wide range of foodborne pathogens without compromising produce quality [86,87].

Additionally, natural antimicrobial agents—such as essential oils, pomegranate peel extracts, and chitosan-based edible coatings—offer eco-friendly solutions with significant antimicrobial activity, although their effectiveness may vary depending on concentration, food matrix, and storage conditions [88,89]. Smart packaging systems, including oxygen scavengers and antimicrobial films, further support microbial suppression and extend product shelf life [90]. A visual overview of these principal control strategies is illustrated in Figure 2. These approaches collectively represent a multifaceted strategy to enhance microbial safety and quality in both raw and RTE fruits and vegetables, aligning with sustainability and consumer health priorities.

The following subsections provide an in-depth evaluation of these control strategies, including both emerging and established interventions. Each method is examined in terms of its mechanism of action, efficacy against foodborne pathogens, impact on produce quality, and feasibility for commercial implementation. By dissecting the strengths and limitations of individual approaches—ranging from nonthermal technologies like cold plasma and UV-C to bio-based treatments such as essential oils (EOs) and smart packaging—this review aims to offer a comprehensive, evidence-based roadmap for enhancing microbial safety in raw and RTE fruits and vegetables.

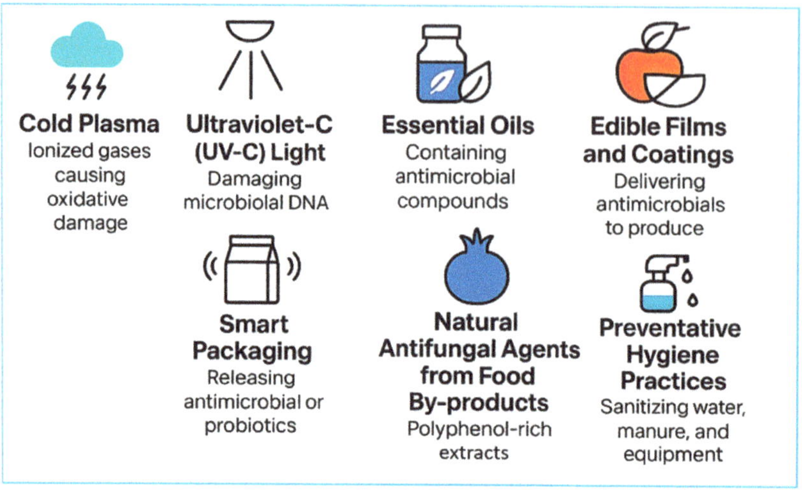

Figure 2. Strategies for microbial contamination control in fresh produce. This infographic highlights eight integrated approaches, including nonthermal decontamination (cold plasma and UV-C light), natural antimicrobials (essential oils and antifungal agents from pomegranate peels), advanced delivery systems (edible coatings and smart packaging), and preventative hygiene measures. These technologies work individually or synergistically to reduce microbial load while preserving produce quality.

4.1. Cold Plasma (CP)

CP is a promising nonthermal decontamination technique utilizing ionized gases rich in reactive oxygen and nitrogen species. These species interact with bacterial cell membranes, leading to irreversible oxidative damage. The application of PAW has shown considerable efficacy in reducing microbial populations in fresh produce. For example, de Araújo Bezerra et al. [91] demonstrated that CP treatments reduced total aerobic bacteria on fresh-cut pears below the detection limit after treatment at 8 kV for 2 min. Additional studies confirmed the synergistic effects of combining CP with hydrogen peroxide, improving microbial inactivation on leafy greens and apples [25]. Mahnot et al. [92] reported that CP treatment significantly reduced microbial load on fresh-cut strawberries without compromising sensory attributes. Misra et al. [93] investigated the effects of CP generated within a sealed package on the physical quality parameters and respiration rates of strawberries. Their findings suggest that CP treatment can effectively decontaminate strawberries while retaining their quality.

4.2. Ultraviolet-C (UV-C) Light

UV-C light is effective against surface-bound pathogens by damaging microbial DNA. Its efficacy increases when paired with Modified Atmosphere Packaging (MAP), a technique that alters the gaseous environment inside packaging—typically by reducing oxygen and increasing carbon dioxide—to suppress microbial growth and extend shelf life. Li et al. [94] reported that UV-C + MAP reduced microbial counts in fresh-cut carrots by over 1 log CFU/g compared to control samples. In raspberries, electron beam irradiation decreased mesophilic bacterial counts by 2 log CFU/g over seven days of storage [95]. Choi et al. [96] demonstrated that combining UV-C irradiation with modified MAP significantly enhanced microbial inactivation on cherry tomatoes during cold storage. The treatment achieved a >2 log CFU/g reduction in *S. enterica* serovar Typhimurium populations and extended the product's shelf life without compromising its sensory quality. This dual approach highlights the synergistic potential of UV-C and MAP in improving microbial safety and maintaining the freshness of fresh produce. Wang et al. [97] highlighted the potential of UV-C LED systems in reducing microbial loads on blueberries, emphasizing the importance of system design for effective surface decontamination.

4.3. Pulsed-Light (PL) Technology

Pulsed-light (PL) technology is a promising non-thermal decontamination technique that delivers short-duration, high-intensity pulses of broad-spectrum light (ranging from 200 to 1100 nm), including UV, visible, and infrared light. This intervention is particularly effective at inactivating surface microorganisms through mechanisms involving DNA damage, protein denaturation, and disruption of cell membranes. Its application in the fresh produce sector has gained momentum due to its minimal impact on food quality and its alignment with clean-label processing trends. Multiple studies have confirmed the efficacy of PL in reducing microbial contamination on fruits and vegetables. Bialka and Demirci [98] demonstrated that water-assisted PL treatment achieved up to 3.9 log CFU/g reductions in *E. coli* O157:H7 and *S. enterica* on raspberries without damaging the fruit. Similarly, Gómez-López et al. [99] reported a 4-log reduction of Listeria monocytogenes on packaged lettuce, while maintaining product quality.

Moreover, Ramos-Villarroel et al. [100] found that PL effectively reduced microbial counts on blueberries and strawberries when combined with water-assisted delivery, achieving significant decontamination with minimal sensory alteration. These findings are supported by Moreira et al. [101], who observed a >3 log CFU/g reduction in *Listeria innocua* on fresh-cut apples using PL at a fluence of 1.4 J/cm^2, while maintaining color and texture stability during storage. Further reinforcing these results, Krishnamurthy et al. [102] evaluated the microbial inactivation kinetics of *L. monocytogenes* and *Salmonella* on tomatoes and mangoes and found that PL treatments achieved >4 log CFU/g reductions without any significant deterioration in visual appearance. More recently, Woldemariam et al. [103] explored PL applications in spice and dry produce safety, demonstrating its effectiveness in reducing both microbial loads and mycotoxins in red pepper powder, which underscores its broad-spectrum utility beyond conventional applications.

The effectiveness of PL treatment depends on several parameters, including fluence (total energy delivered), number of pulses, treatment distance, produce surface geometry, and light absorption characteristics. Water-assisted PL systems have proven particularly effective in mitigating surface shadowing and improving uniform exposure. The advantages of PL include minimal heat generation, rapid processing time (typically seconds), chemical-free sanitation, and compatibility with automation in postharvest processing lines. Nevertheless, challenges such as equipment cost, surface irregularities, shielding microbes, and potential minor discoloration with overexposure must be addressed to optimize commercial implementation. Despite these limitations, PL stands out as a sustainable and consumer-friendly technology for enhancing microbial safety in fresh and minimally processed produce.

4.4. Essential Oils (EOs)

EOs such as rosemary, eucalyptus, cinnamon, thyme, and oregano possess bioactive compounds—most notably thymol and carvacrol—that exhibit broad-spectrum antimicrobial activity against foodborne pathogens. These oils can be applied through multiple delivery methods, including direct incorporation into food matrices, coating or dipping of produce, incorporation into edible films, or use in vapor-phase treatments. For example, vapor-phase application has shown promising results in fresh produce preservation. Xylia et al. [104] reported that eucalyptus and rosemary EO vapors significantly inhibited bacterial growth and preserved cucumber quality during cold storage, suggesting this method is effective for surface decontamination without altering taste or texture [105].

In meat products, EOs are often incorporated directly at specific concentrations. Vidaković Knežević et al. [106] observed strong antilisterial effects of thyme and oregano EOs in both biofilm models and minced pork meat systems, with minimum inhibitory concentrations ranging from 0.09 to 1.78 μL/mL. Similarly, Nedić et al. [107] found that adding 0.6% oregano or thyme EO to fermented sausages effectively eliminated *L. monocytogenes* by day 14 of ripening. Notably, a lower concentration (0.3%) maintained antimicrobial action while minimizing changes to sensory quality.

To improve EO efficacy and stability, recent studies have explored synergistic applications with non-thermal technologies. When combined with ultrasound or cold plasma, the antimicrobial action of EOs is enhanced due to increased permeability of microbial membranes and better distribution within the food matrix [105]. These approaches not only reduce the required concentration of EOs but also mitigate potential impacts on flavor and aroma, making them suitable for industrial-scale use in RTE foods.

4.5. Edible Films and Coatings

Edible films and coatings are gaining increased attention as sustainable and efficient carriers of antimicrobial agents for fresh produce preservation. Typically derived from biopolymers such as chitosan, alginate, pectin, and starch, these biodegradable matrices are applied directly to the surface of fruits and vegetables to create semi-permeable barriers that regulate gas exchange, reduce moisture loss, and act as carriers for bioactive compounds. Kostić et al. [26] demonstrated that chitosan-based films enriched with oregano essential oil significantly inhibited *E. coli*, *Bacillus subtilis*, and *Staphylococcus epidermidis* on coated fruit surfaces, showing broad-spectrum antimicrobial efficacy. Innovations in formulation have also explored polyphenol-enhanced coatings; for instance, tannic acid/chitosan-citric acid coatings extended the shelf life of strawberries by more than 100 h and achieved over 90% bacterial growth inhibition [108]. Yang et al. [109] developed edible films composed of chitosan enriched with turmeric and green tea extracts, which were applied to strawberries. These coatings effectively suppressed the growth of *Botrytis cinerea*, delayed ripening, and preserved total phenolic content and antioxidant activity during cold storage at 4 °C, demonstrating functional and nutritional benefits.

Similarly, Rojas-Graü et al. [110] showed that alginate and gellan-based coatings supplemented with antibrowning agents maintained the visual and sensory quality of fresh-cut Fuji apples. The coatings significantly reduced enzymatic browning, preserved firmness, and slowed microbial spoilage, supporting their utility in minimally processed fruit applications. Furthermore, edible coatings are increasingly integrated with emerging preservation technologies. For example, the combination of edible coatings with pulsed-light treatment or ozonation has been reported to enhance decontamination efficiency while minimizing chemical usage and preserving organoleptic properties. These synergistic approaches are being explored to meet consumer demands for chemical-free, safe, and high-quality fresh produce. Collectively, these findings reinforce the potential of edible films and coatings as environmentally friendly, multifunctional strategies to enhance the microbial safety, quality, and shelf life of raw and RTE fruits and vegetables.

4.6. Electrolyzed Water

Electrolyzed water (EW), especially slightly acidic electrolyzed water (SAEW) and neutral electrolyzed oxidizing water, has emerged as an environmentally friendly and highly effective method for sanitizing fresh produce. These fluids exhibit strong antimicrobial action due to hypochlorous acid and other active chlorine species, which damage microbial membranes, proteins, and DNA without leaving harmful residues or byproducts. A proteomic study by Chang et al. [111] (equivalent here to a recently conducted SAEW proteomic analysis) revealed that SAEW treatment induced oxidative stress and denatured ribosomal proteins in *L. monocytogenes*, subsequently leading to VBNC states—an important consideration for food safety monitoring.

In terms of practical decontamination, Guentzel et al. [112] (similar to early EW-spinach washing trials) found that using neutral EW (approx. 100 ppm free chlorine) with mild agitation reduced *E. coli*, *Salmonella enterica*, and *L. monocytogenes* by 4–5 log CFU/g on spinach and lettuce in a water bath. Another impactful study by Afari et al. [113] reported that automated produce washing using neutral-pH EW (155 mg/L free chlorine, pH 7.5) achieved reductions of 4.2–5.9 log CFU/g for *E. coli* O157:H7 and 4.6–6.0 log CFU/g for *Salmonella typhimurium* across romaine lettuce, iceberg lettuce, and tomatoes.

Moreover, Hamidi et al. [114] demonstrated that combining neutral EW with PAA produced synergistic effects, achieving >2.5 log reductions in *E. coli*, *Salmonella*, and *L. monocytogenes* on mixed produce and meat samples. Beyond planktonic cells, SAEW is also effective against biofilms: Chang et al. [111], in a recent ultrasonication-assisted SAEW study, showed >1.5 log CFU/mL reduction of Listeria and *E. coli* biofilms on stainless steel surfaces, illustrating the benefit of combining EW with physical agitation.

Despite its advantages, EW's effectiveness can be influenced by factors such as organic matter load, free chlorine concentration, and pH. Over-dilution or inappropriate application may lead to insufficient microbial inactivation or induction of viable-but-nonculturable states. Therefore, optimizing operational parameters and integrating EW with other physical or chemical treatments (e.g., ultrasonication or agitation) can improve its efficacy. In conclusion, EW represents a safe, effective, and sustainable alternative to traditional sanitizers for fresh produce, with broad applicability in both small- and large-scale operations.

4.7. Ozone Treatment

Ozone treatment has become an effective non-thermal method for microbial decontamination of fresh produce, applied in either gaseous or aqueous forms. It operates through the strong oxidative capacity of ozone, which damages microbial cell membranes, proteins, and nucleic acids, ultimately leading to cell death. One of the advantages of ozone is that it rapidly decomposes into oxygen, leaving no harmful residues, making it particularly suitable for organic and sustainable food processing systems. Aqueous ozone (ozonated water) has been widely studied for its use in postharvest washing of fruits and vegetables. A comprehensive meta-analysis by Hong et al. [115] evaluated multiple ozone application methods and reported that sparging or bubbling ozone through water could achieve 2–4 log CFU/g reductions in *Salmonella*, *L. monocytogenes*, and *E. coli* on produce like spinach, lettuce, and berries. Their review emphasized that treatment efficiency depends on factors such as water temperature, pH, ozone concentration, and contact time.

Gibson et al. [116] demonstrated that ozonated water at 13 mg/L significantly reduced *Salmonella enterica*, *Listeria innocua*, and *E. coli* on tomatoes and leafy greens by more than 4 log CFU/g, without altering texture or flavor. Their study also highlighted that ozone exposure times between 5 and 10 min are generally sufficient for effective microbial reduction. Gaseous ozone is also utilized, particularly in packaging and storage settings. Macías-Gallardo et al. [117] investigated the use of continuous low-dose gaseous ozone (0.3–1.0 ppm) during cold storage of strawberries. They found a significant reduction in fungal growth and maintenance of fruit firmness and color at 0.3 ppm, although higher concentrations led to mild oxidative changes. This illustrates the need for precise control to balance microbial safety and product quality.

Ozone treatment has demonstrated antifungal efficacy against key spoilage organisms such as *Botrytis cinerea* and *Penicillium expansum*. Luo et al. [118] showed that daily exposure to 79 ppm gaseous ozone significantly reduced fungal growth on kiwifruit while preserving quality. Similarly, Ozkan et al. [119] reported effective suppression of fungal decay on table grapes using 1.5 ppm ozone gas during storage. When combined with interventions like MAP or ultrasonication, ozone further enhances microbial inactivation. Its residue-free nature and broad-spectrum activity make ozone an eco-friendly option for decontaminating minimally processed and organic produce, provided application parameters are carefully optimized.

4.8. Hurdle Technology

Combining multiple treatments such as UV-C light, cold plasma, essential oils, and antimicrobial coatings can offer synergistic benefits. Abdelshafy et al. [120] reported that hurdle technologies significantly enhanced disinfection performance while maintaining sensory quality, especially for pathogens like *Salmonella typhimurium* and *L. monocytogenes* on leafy greens. Zhou et al. [121] demonstrated the application of power ultrasound combined with chemical sanitizers (chlorine and peroxyacetic acid) on grape tomatoes, romaine lettuce, and spinach. This combination achieved a microbial reduction of up to 3.99 log CFU/g for *L. monocytogenes* and 3.62 log CFU/g for *Salmonella* Newport, significantly outperforming the use of sanitizers or ultrasound alone.

Liu et al. [122] applied a novel physical hurdle approach by combining a low-voltage electrostatic field with MAP to extend the shelf life of button mushrooms. This combination reduced microbial spoilage while preserving sensory and nutritional quality throughout cold storage. In another study, Melhem et al. [123] conducted a meta-analysis assessing the synergy of high-pressure processing and natural antimicrobials, such as bacteriocins. While primarily applied to cured meats, the findings provide strong evidence for the cross-applicability of this hurdle strategy to fresh produce, especially for the control of *L. monocytogenes* and other resilient pathogens. Together, these studies underscore the value of combining complementary interventions to overcome microbial resistance mechanisms and preserve product quality, aligning with consumer demand for safe, minimally processed produce.

4.9. Smart Packaging and Responsive Films

Smart packaging technologies represent an innovative approach to improving the microbial safety, shelf life, and consumer confidence in fresh and minimally processed produce. These systems utilize responsive materials, biosensors, or visual indicators to monitor spoilage-related changes such as gas emission, pH fluctuation, or microbial growth. Mahović Poljaček et al. [124] demonstrated the development of starch-based films enhanced with bacterial nanocellulose and natural dyes that change color with pH shifts, offering both antimicrobial activity and freshness indicators. Such smart films are promising for future applications in produce monitoring and consumer safety.

Wu et al. [125] developed high-sensitivity intelligent packaging films incorporating rose anthocyanins that provide clear, colorimetric changes in response to pH variation. These films, tested on shrimp, effectively indicated spoilage progression and demonstrated potential adaptability to monitor the freshness of produce through pH-sensitive reactions. Zhang et al. [126] introduced biopolymer-based intelligent packaging integrated with natural colorimetric sensors for real-time food quality assessment. Their packaging system, based on biodegradable matrices, successfully detected spoilage-related VOCs and pH changes in perishable foods, aligning with sustainability goals in food safety monitoring. Additionally, Palanisamy et al. [127] reviewed recent advances in intelligent food packaging systems that combine biodegradable polymer matrices with natural sensors for detecting spoilage-related changes such as gas emission or temperature shifts. Their study emphasized how these smart films reduce food waste and improve food safety across the produce supply chain.

In a related innovation, Douaki et al. [128] developed battery-free, stretchable, and autonomous smart packaging that enables real-time monitoring of freshness through closed-loop sensing and release mechanisms. These systems operate without external power sources, making them suitable for wireless freshness tracking of high-value produce in cold chain logistics. Together, these findings highlight the growing utility of smart packaging as an effective strategy to reduce spoilage, monitor quality, and deliver intelligent, eco-friendly solutions tailored to fresh produce safety.

4.10. Application of Natural Antifungal Agents

Fungal spoilage is a major concern in the preservation of fresh produce, contributing to significant postharvest losses and reduced consumer safety. The application of natural antifungal agents, particularly those derived from plant extracts, EOs, and probiotic metabolites, has gained momentum as a safer alternative to synthetic fungicides. Nawaz et al. [129] demonstrated that polyphenol-rich extracts from pomegranate peels exhibited potent antifungal activity against a broad spectrum of postharvest pathogens, including *Aspergillus niger*, *Penicillium expansum*, and *Colletotrichum gloeosporioides*. The n-hexane fraction, containing bioactive compounds like nobiletin and salidroside, was found to be the most effective. Boubrik et al. [130] further supported the antifungal efficacy of EOs by demonstrating that Cinnamomum cassia essential oil inhibited the growth of spoilage yeasts and molds such as *Saccharomyces cerevisiae* and *Acremonium* spp. Their study revealed minimum inhibitory concentrations of 6.25% and 3.125%, respectively, underscoring the oil's potential as a natural preservative in fruit-based applications.

Similarly, Aljuhani et al. [131] assessed the antifungal activity of methanolic extracts of *Carica papaya* fruits against *Microsporum canis*. The extract produced a 37 mm zone of inhibition and an MIC of 1000 μg/mL. Phytochemical analysis identified compounds like xanthosine and decanoic acid as contributors to its antifungal action, highlighting the potential use of papaya extracts in antifungal formulations for produce protection. Additionally, Vasundaradevi et al. [132] isolated a strain of *Lactiplantibacillus argentoratensis* from tropical fruits that exhibited potent antifungal activity against *Fusarium oxysporum*. The cell-free supernatant from this probiotic strain inhibited spore germination and reduced fungal biomass by 94%, with citric, lactic, and malic acids identified as key antifungal metabolites. These findings support the development of bio-based fungistatic treatments for postharvest disease management.

Despite their natural origin, some plant-derived compounds and EOs may pose risks of cytotoxicity, allergenicity, or organoleptic alterations at higher concentrations. For example, certain phenolic constituents—such as cinnamaldehyde and eugenol—have been associated with skin sensitization or irritation in sensitive individuals [133]. Moreover, the Generally Recognized As Safe status does not preclude the need for dose optimization and regulatory evaluation in food applications [134]. Therefore, while these natural agents offer promising residue-free antifungal solutions, safety profiling and controlled application are essential for consumer protection. Collectively, these studies demonstrate that natural antifungal agents—ranging from botanical extracts to microbial metabolites—offer sustainable, residue-free alternatives for controlling fungal contamination in fresh and minimally processed produce.

4.11. Preventative Measures

Preventing bacterial and fungal contamination in fresh produce requires an integrated approach that begins at the farm level and continues throughout postharvest handling, packaging, and distribution. This "farm-to-fork" strategy is widely recognized as essential for reducing microbial risks across the fresh produce supply chain [135]. At the preharvest stage, adherence to Good Agricultural Practices (GAPs)—such as using clean irrigation water, proper composting of manure, crop rotation, and selecting disease-resistant cultivars—is critical for minimizing the introduction of pathogens and fungal colonization [136]. For instance, irrigation timing and water quality have been shown to significantly influence contamination levels, with studies reporting higher prevalence of *L. monocytogenes* in fields irrigated shortly before harvest [137]. Public outreach resources, such as those provided by the University of Nevada, Reno Extension, emphasize that GAPs tailored to local environmental conditions and farming practices are essential to prevent contamination at the source [138].

In postharvest handling, microbial contamination may arise from unsanitary equipment, inadequate worker hygiene, or improper washing and cooling procedures. While chlorine-based sanitizers are widely used, concerns over chemical residues and by-product formation have led to the adoption of alternatives such as EW and ozone treatments. These eco-friendly methods reduce microbial load without compromising the safety or sensory quality of fresh produce. Biopreservation methods are gaining prominence, particularly the use of lactic acid bacteria and their metabolic by-products, such as organic acids, hydrogen peroxide, and bacteriocins. These compounds exert antifungal effects by lowering pH, damaging fungal membranes, and inhibiting spore germination. Recent studies have demonstrated that lactic acid bacteria exhibit strong antifungal activity, effectively extending the shelf life of fresh fruits and vegetables while preserving sensory quality [139,140].

In addition, biological control agents (BCAs) such as antagonistic yeasts and beneficial bacteria offer a natural alternative to chemical fungicides. These BCAs function through mechanisms like competition for nutrients and space, production of antifungal enzymes, and interference with fungal cell signaling. As reviewed by Nunes [141], the application of BCAs has shown considerable efficacy in controlling postharvest diseases in citrus, grapes, apples, and other high-value produce. Finally, innovative packaging solutions such as MAP and active packaging can create hostile environments for fungal proliferation. These systems regulate internal oxygen and moisture levels, often incorporating antimicrobial films or coatings that further suppress microbial activity. Together, these preventative measures form a robust and sustainable framework for microbial risk reduction in fresh produce supply chains, contributing to safer, longer-lasting products from farm to consumer.

5. Implementation Challenges and Future Directions for Pathogen Control in Fresh Produce

Despite significant advances in pathogen detection and control technologies for fresh and RTE produce, practical implementation remains limited by multifaceted challenges (Table 1). Barriers to adoption include economic constraints, regulatory inconsistencies, technical complexity, and infrastructural deficiencies, particularly in low- and middle-income countries (LMICs). High equipment and maintenance costs, along with limited scalability, hinder the widespread use of advanced diagnostic platforms (Table 1). Technologies such as real-time PCR, ddPCR, MALDI-TOF MS, and HSI offer superior sensitivity and specificity; however, they require expensive instrumentation, specialized consumables, and technically skilled personnel. Ferone et al. [29] emphasized that these tools are largely confined to high-resource laboratory environments due to their operational demands. Abdelshafy et al. [7] similarly observed that implementation challenges are especially acute in LMICs, where laboratory infrastructure is insufficient despite frequent foodborne disease outbreaks.

Table 1. Key implementation challenges and strategic solutions for pathogen control in fresh produce.

Challenge Category	Specific Barriers	Strategic Solutions/Future Directions
Economic Constraints	High cost of advanced detection technologies (e.g., PCR, MALDI-TOF MS)	Investment in low-cost portable diagnostics (e.g., lateral flow assays, portable PCR)
Technical Limitations	Need for skilled personnel; complex data interpretation	Capacity-building through training programs; simplified, user-friendly diagnostic platforms
Regulatory Gaps	Fragmented or outdated food safety regulations, especially in LMICs	Policy harmonization guided by One Health approach; enhanced enforcement and international cooperation
Infrastructure Deficiencies	Lack of clean water, refrigeration, hygienic packaging, and cold chains	Development of basic infrastructure in rural/agricultural zones; support for smallholder supply chains
Laboratory–Field Disconnect	Technologies remain confined to labs; limited real-world application	Promote public–private partnerships; validation of tools in field settings; scale-up of innovations
Environmental Conditions	Climate variability, poor electricity access, and tropical humidity exacerbate risks	Deploy solar-powered cooling, weather-resilient logistics, and robust packaging systems
Workforce Shortage	Limited availability of microbiologists, bioinformaticians, and food safety experts	Establish food safety certification programs; invest in STEM education and international knowledge transfer

A further obstacle is the lack of harmonized and enforceable regulatory frameworks for produce safety. In the United States, for instance, the FDA encourages voluntary adherence to GAPs and Good Handling Practices (GHPs), but enforcement remains inconsistent and often decentralized [142]. In many developing countries, food safety regulations are outdated, poorly enforced, or entirely absent, while food surveillance programs suffer from chronic underfunding. In response, the WHO [143] has advocated for global policy coordination, emphasizing the strengthening of surveillance systems and AMR monitoring in foodborne pathogens.

Infrastructure limitations in rural and peri-urban regions—such as lack of access to clean water, refrigerated transportation, sanitary packaging environments, and reliable cold storage—further compound microbial risks. These conditions are particularly problematic for high-risk commodities like leafy greens and soft-skinned fruits, which possess intricate surface morphologies that favor microbial adhesion and biofilm formation [16]. De Araújo Bezerra et al. [91] also highlighted that the absence of continuous electricity and cold chain systems in tropical climates increases postharvest vulnerability to contamination.

Another key challenge is the workforce skill gap. The effective use of advanced technologies such as MALDI-TOF MS, next-generation sequencing, and biosensor systems requires trained microbiologists, molecular biologists, and bioinformaticians. However, many laboratories—especially in resource-limited settings—lack personnel with the expertise to process and interpret complex datasets. Ferone et al. [29] noted that this deficit significantly impedes the deployment of rapid diagnostic tools at the point of need. Moreover, there remains a substantial disconnect between laboratory innovation and commercial application. Promising technologies, such as CRISPR-based diagnostics and magnetoelastic biosensors, have shown strong performance in controlled trials but face hurdles in cost-effective validation, regulatory approval, and integration into existing food safety frameworks [144]. As Abdelshafy et al. [7] stressed, bridging this translational gap will require robust public–private partnerships and sustained investment.

To overcome these systemic challenges, a shift toward integrated, science-driven strategies is essential. Investment in mobile, low-cost diagnostic tools—such as portable PCR units and lateral flow assays—should be paired with infrastructural development, particularly in rural agricultural hubs. Concurrently, national and regional capacity-building initiatives, including workforce training programs and food safety certification schemes, can address skill shortages and promote regulatory compliance. Furthermore, policy harmonization guided by the One Health approach will be pivotal in addressing interconnected risks across human, animal, and environmental health domains.

A paradigm shift is urgently needed—one that prioritizes proactive, systems-based prevention over reactive responses. This transition should be grounded in scientific evidence, sustainability principles, and cross-sectoral collaboration. Ultimately, global cooperation, inclusive policies, and strategic investment will be indispensable to ensure the microbiological safety of fresh produce across all socioeconomic contexts.

6. Recommendations and Future Directions for Enhancing Microbial Safety of Fresh Produce

Enhancing the microbial safety of fresh and RTE fruits and vegetables requires a coordinated, systems-based approach that integrates regulatory, technological, and educational interventions (Table 2). Key priorities include the standardization of preventive practices, the advancement of accessible technologies, the harmonization of regulatory oversight, and capacity-building across the supply chain.

Table 2. Strategic actions to enhance microbial safety of fresh produce.

Strategic Focus Area	Action Item	Expected Outcome
Preventive Frameworks	Enforce GAPs, GHPs, and sanitation standards with third-party audits	Improved baseline hygiene and reduced contamination risk
Technology Innovation	Develop and deploy cost-effective tools (e.g., PAW, UV-C, qPCR, biosensors)	Real-time pathogen detection and non-thermal decontamination
Regulatory Harmonization	Adopt Codex-based global standards supported by WHO/FAO/OIE	Consistent safety benchmarks and enhanced AMR surveillance
Capacity Building	Train personnel in sanitation, diagnostics, and AMR control	Workforce readiness and effective tool deployment
Translational Research	Fund prototype development and field validation	Accelerated adoption of innovations
Public-Private Partnerships	Support scale-up of coatings, HSI, and smart packaging	Commercialization of pilot technologies
Smart Surveillance	Integrate AI, IoT, blockchain, and CRISPR-based tools	Real-time traceability and rapid outbreak response

A critical first step is the widespread enforcement and harmonization of GAPs, GHPs, and hygiene protocols. While agencies like the U.S. FDA and USDA provide guidance for voluntary adoption, uptake remains uneven, especially in low-resource contexts [145–147]. Integrating third-party audits and routine monitoring—focused on irrigation water quality, manure management, field sanitation, and postharvest infrastructure—is essential for ensuring compliance and reducing pathogen risk [57].

Targeted investment in cost-effective, scalable technologies is vital for low- and middle-income countries. Non-thermal decontamination methods such as PAW, UV-C light, and essential oil-based nanoemulsions offer effective microbial control while preserving produce quality [91,105]. Meanwhile, field-deployable diagnostic tools—such as multiplex qPCR and biosensors—should be prioritized for rapid, on-site detection of foodborne pathogens [148,149]. To support global coordination, a unified regulatory framework aligned with Codex Alimentarius and backed by the WHO, FAO, and OIE is needed. Integrating whole-genome sequencing and resistome surveillance into national monitoring programs will strengthen AMR detection and response as part of a One Health strategy [143].

Human capital development is equally critical. Training programs for food handlers, laboratory technicians, and regulatory staff should focus on sanitation, diagnostics, and AMR mitigation. Competency-based curricula and applied technical workshops—delivered through partnerships with academic institutions and industry—can enhance readiness to deploy advanced tools such as MALDI-TOF MS and CRISPR-based diagnostics [55]. To accelerate translation of research into practice, funding should support prototype development, field validation, and regulatory approval of emerging technologies. Public–private partnerships can scale promising solutions, such as antimicrobial edible coatings and HSI-integrated screening systems [7].

Looking ahead, the future of produce safety will be shaped by smart surveillance networks. Artificial intelligence, IoT-enabled sensors, and blockchain platforms will enhance traceability and enable real-time risk assessments [150,151]. Emerging efforts are already demonstrating the value of integrating CRISPR diagnostics with portable sequencing platforms for strain-level detection in field settings [152]. These technologies—combined with mobile tracking apps and blockchain-verified distribution chains—are being piloted in agricultural hubs for real-time monitoring of leafy greens and other high-risk produce. In sum, ensuring the microbial safety of fresh produce demands cross-sectoral collaboration, innovation, and sustained investment. Only through integrated strategies that unite science, technology, policy, and education can we build resilient, safe food systems for the future.

7. Conclusions

The microbial safety of fresh and RTE fruits and vegetables is a complex challenge that spans preharvest practices, postharvest handling, detection technologies, and intervention strategies. While traditional microbiological methods remain important, the integration of rapid, sensitive, and field-deployable tools—such as CRISPR-based diagnostics, biosensors, and MALDI-TOF MS—is critical to modern surveillance. Natural antifungal agents and non-thermal treatments show great promise as eco-friendly control strategies but require further standardization and safety assessments, particularly regarding toxicity and allergenicity. A major advancement presented in this review is the inclusion of comprehensive sampling methodologies and the critical evaluation of method sensitivity, specificity, and reproducibility. Future research must address scalability, cost-effectiveness, and policy harmonization under the One Health framework. A multifaceted and interdisciplinary approach is essential to safeguard public health and meet the growing demand for safe, minimally processed produce.

Author Contributions: Conceptualization, A.E., and E.M.; writing—original draft preparation, A.E., E.M., F.A., A.A., B.A., M.A., A.M.A., and A.A.-O.; writing—review and editing, A.E., E.M., F.A., A.A., B.A., M.A., A.M.A., and A.A.-O.; visualization, A.E., E.M., F.A., A.A., B.A., M.A., A.M.A., and A.A.-O.; supervision, A.E., and E.M.; project administration, A.E., and A.A.; funding acquisition, A.E., E.M., F.A., A.A., B.A., M.A., A.M.A., and A.A.-O. All authors have read and agreed to the published version of the manuscript.

Funding: This research received no external funding.

Institutional Review Board Statement: Not applicable.

Informed Consent Statement: Not applicable.

Data Availability Statement: No new data were created or analyzed in this study. Data sharing is not applicable to this article.

Acknowledgments: The researchers would like to thank the Deanship of Graduate Studies and Scientific Research at Qassim University for financial support (QU-APC-2025).

Conflicts of Interest: The authors declare no conflicts of interest.

References

1. Wang, X.; Ouyang, Y.; Liu, J.; Zhu, M.; Zhao, G.; Bao, W.; Hu, F.B. Fruit and vegetable consumption and mortality from all causes, cardiovascular disease, and cancer: Systematic review and dose-response meta-analysis of prospective cohort studies. *BMJ* **2014**, *349*, g4490. [CrossRef] [PubMed]
2. Aune, D.; Giovannucci, E.; Boffetta, P.; Fadnes, L.T.; Keum, N.; Norat, T.; Greenwood, D.C.; Riboli, E.; Vatten, L.J.; Tonstad, S. Fruit and vegetable intake and the risk of cardiovascular disease, total cancer and all-cause mortality—A systematic review and dose-response meta-analysis of prospective studies. *Int. J. Epidemiol.* **2017**, *46*, 1029–1056. [CrossRef] [PubMed]
3. Onni, A.T.; Balakrishna, R.; Perillo, M.; Amato, M.; Arjmand, E.J.; Thomassen, L.M.; Lorenzini, A.; Fadnes, L.T. Umbrella review of systematic reviews and meta-analyses on consumption of different food groups and risk of all-cause mortality. *Adv. Nutr.* **2025**, *16*, 100393. [CrossRef]
4. Ndanuko, R.N.; Tapsell, L.C.; Charlton, K.E.; Neale, E.P.; Batterham, M.J. Dietary patterns and blood pressure in adults: A systematic review and meta-analysis of randomized controlled trials. *Adv. Nutr.* **2016**, *7*, 76–89. [CrossRef]
5. Bennett, S.; Sodha, S.; Ayers, T.; Lynch, M.; Gould, L.; Tauxe, R. Produce-associated foodborne disease outbreaks, USA, 1998–2013. *Epidemiol. Infect.* **2018**, *146*, 1397–1406. [CrossRef]
6. Olaimat, A.N.; Holley, R.A. Factors influencing the microbial safety of fresh produce: A review. *Food Microbiol.* **2012**, *32*, 1–19. [CrossRef]
7. Abdelshafy, A.M.; Younis, H.A.; Osman, A.I.; Hussein, S.M.; El-Ela, A.S.A.; Mahmoud, E.A.; Elsherbiny, O.; Rashwan, A.K. Recent Advances in Detection and Control Strategies for Foodborne Bacteria in Raw and Ready-to-Eat Fruits and Vegetables. *Food Front.* **2025**, *6*, 605–629. [CrossRef]
8. Polkowska, A.; Räsänen, S.; Nuorti, P.; Maunula, L.; Jalava, K. Assessment of food and waterborne viral outbreaks by using field epidemiologic, modern laboratory and statistical methods—Lessons learnt from seven major norovirus outbreaks in Finland. *Pathogens* **2021**, *10*, 1624. [CrossRef] [PubMed]
9. Almeria, S.; Chacin-Bonilla, L.; Maloney, J.G.; Santin, M. Cyclospora cayetanensis: A perspective (2020–2023) with emphasis on epidemiology and detection methods. *Microorganisms* **2023**, *11*, 2171. [CrossRef]
10. Dawson, D.; Paish, A.; Staffell, L.; Seymour, I.; Appleton, H. Survival of viruses on fresh produce, using MS2 as a surrogate for norovirus. *J. Appl. Microbiol.* **2005**, *98*, 203–209. [CrossRef]
11. Li, J.; Wang, Z.; Karim, M.R.; Zhang, L. Detection of human intestinal protozoan parasites in vegetables and fruits: A review. *Parasites Vectors* **2020**, *13*, 380. [CrossRef]
12. Eslahi, A.V.; Mamedova, S.; Nassiba, R.; Karanis, P. Unveiling risks in healthy food: Vegetables and fruits are linked to the distribution chain of protozoan parasites. *Food Microbiol.* **2024**, *123*, 104592.
13. Nassarawa, S.S.; Luo, Z.; Lu, Y. Conventional and emerging techniques for detection of foodborne pathogens in horticulture crops: A leap to food safety. *Food Bioprocess Technol.* **2022**, *15*, 1248–1267. [CrossRef]
14. Thomas, G.A.; Gil, T.P.; Müller, C.T.; Rogers, H.J.; Berger, C.N. From field to plate: How do bacterial enteric pathogens interact with ready-to-eat fruit and vegetables, causing disease outbreaks? *Food Microbiol.* **2024**, *117*, 104389. [CrossRef] [PubMed]
15. Ölmez, H. Foodborne pathogenic bacteria in fresh-cut vegetables and fruits. In *Food Hygiene and Toxicology in Ready-to-Eat Foods*; Elsevier: Amsterdam, The Netherlands, 2016; pp. 151–166.
16. Azimirad, M.; Nadalian, B.; Alavifard, H.; Panirani, S.N.; Bonab, S.M.V.; Azimirad, F.; Gholami, F.; Jabbari, P.; Yadegar, A.; Busani, L. Microbiological survey and occurrence of bacterial foodborne pathogens in raw and ready-to-eat green leafy vegetables marketed in Tehran, Iran. *Int. J. Hyg. Environ. Health* **2021**, *237*, 113824. [CrossRef]
17. Rashwan, A.K.; Bai, H.; Osman, A.I.; Eltohamy, K.M.; Chen, Z.; Younis, H.A.; Al-Fatesh, A.; Rooney, D.W.; Yap, P.-S. Recycling food and agriculture by-products to mitigate climate change: A review. *Environ. Chem. Lett.* **2023**, *21*, 3351–3375. [CrossRef]
18. Havelaar, A.H.; Kirk, M.D.; Torgerson, P.R.; Gibb, H.J.; Hald, T.; Lake, R.J.; Praet, N.; Bellinger, D.C.; De Silva, N.R.; Gargouri, N. World Health Organization global estimates and regional comparisons of the burden of foodborne disease in 2010. *PLoS Med.* **2015**, *12*, e1001923. [CrossRef]
19. Chen, Z.-D.; Kang, H.-J.; Chai, A.-L.; Shi, Y.-X.; Xie, X.-W.; Li, L.; Li, B.-J. Development of a loop-mediated isothermal amplification (LAMP) assay for rapid detection of Pseudomonas syringae pv. tomato in planta. *Eur. J. Plant Pathol.* **2020**, *156*, 739–750. [CrossRef]
20. Zheng, T.; Li, X.; Xie, Y.-N.; Yang, B.; Wu, P. Dual-gene isothermal amplification coupled with lateral flow strip for on-site accurate detection of E. coli O157: H7 in food samples. *Anal. Chem.* **2023**, *95*, 6053–6060. [CrossRef]

21. Hu, J.; Huang, R.; Wang, Y.; Wei, X.; Wang, Z.; Geng, Y.; Jing, J.; Gao, H.; Sun, X.; Dong, C. Development of duplex PCR-ELISA for simultaneous detection of Salmonella spp. and Escherichia coli O157: H7 in food. *J. Microbiol. Methods* **2018**, *154*, 127–133. [CrossRef]
22. El-Moghazy, A.Y.; Wisuthiphaet, N.; Yang, X.; Sun, G.; Nitin, N. Electrochemical biosensor based on genetically engineered bacteriophage T7 for rapid detection of Escherichia coli on fresh produce. *Food Control* **2022**, *135*, 108811. [CrossRef]
23. Johnson, N.; Kniel, K.; Bais, H.; Ragone, A. Deep ultraviolet fluorescence sensing with multispectral imaging to detect and monitor food-borne pathogens on the leafy green phyllosphere. *J. Food Saf.* **2023**, *43*, e13056. [CrossRef]
24. Zand, E.; Froehling, A.; Schoenher, C.; Zunabovic-Pichler, M.; Schlueter, O.; Jaeger, H. Potential of flow cytometric approaches for rapid microbial detection and characterization in the food industry—A review. *Foods* **2021**, *10*, 3112. [CrossRef]
25. Lee, H.W.; Oh, Y.J.; Min, S.C. Microbial inhibition in mixed vegetables packaged in plastic containers using combined treatment with hydrogen peroxide and cold plasma. *Food Control* **2024**, *161*, 110107. [CrossRef]
26. Kostić, M.; Bajac, B.; Janjušević, L.; Bajac, J.; Antov, M. Edible coatings based on plant components for active packaging of fresh/fresh-cut fruits. *S. Afr. J. Bot.* **2023**, *161*, 395–403. [CrossRef]
27. Rahman, M.; Hasan, M.S.; Islam, R.; Rana, R.; Sayem, A.; Sad, M.A.A.; Matin, A.; Raposo, A.; Zandonadi, R.P.; Han, H. Plasma-activated water for food safety and quality: A review of recent developments. *Int. J. Environ. Res. Public Health* **2022**, *19*, 6630. [CrossRef] [PubMed]
28. Gao, Y.; Francis, K.; Zhang, X. Review on formation of cold plasma activated water (PAW) and the applications in food and agriculture. *Food Res. Int.* **2022**, *157*, 111246. [CrossRef]
29. Ferone, M.; Gowen, A.; Fanning, S.; Scannell, A.G. Microbial detection and identification methods: Bench top assays to omics approaches. *Compr. Rev. Food Sci. Food Saf.* **2020**, *19*, 3106–3129. [CrossRef] [PubMed]
30. Kuddus, M.; Shahid, S.M.; Kausar, M.A.; Alzayed, F.S.; Aldhamadi, H.F.; Aljameel, O.S. Microbial analysis of vegetable salad served in restaurants of Hail City, Saudi Arabia. *Biochem. Cell Arch.* **2017**, *17*, 153–158.
31. Aiyedun, S.O.; Onarinde, B.A.; Swainson, M.; Dixon, R.A. Foodborne outbreaks of microbial infection from fresh produce in Europe and North America: A systematic review of data from this millennium. *Int. J. Food Sci. Technol.* **2021**, *56*, 2215–2223. [CrossRef]
32. Wartha, S.; Huber, S.; Kraemer, I.; Alter, T.; Messelhäußer, U. Presence of Listeria at primary production and processing of food of non-animal origin (FNAO) in Bavaria, Germany. *J. Food Prot.* **2023**, *86*, 100015. [CrossRef]
33. Bai, Y.; Li, J.; Huang, M.; Yan, S.; Li, F.; Xu, J.; Peng, Z.; Wang, X.; Ma, J.; Sun, J. Prevalence and characterization of foodborne pathogens isolated from fresh-cut fruits and vegetables in Beijing, China. *Int. J. Food Microbiol.* **2024**, *421*, 110804. [CrossRef]
34. Oyedele, O.A.; Kuzamani, K.Y.; Adetunji, M.C.; Osopale, B.A.; Makinde, O.M.; Onyebuenyi, O.E.; Ogunmola, O.M.; Mozea, O.C.; Ayeni, K.I.; Ezeokoli, O.T. Bacteriological assessment of tropical retail fresh-cut, ready-to-eat fruits in south-western Nigeria. *Sci. Afr.* **2020**, *9*, e00505. [CrossRef]
35. Zhang, H.; Yamamoto, E.; Murphy, J.; Locas, A. Microbiological safety of ready-to-eat fresh-cut fruits and vegetables sold on the Canadian retail market. *Int. J. Food Microbiol.* **2020**, *335*, 108855. [CrossRef]
36. Elsafi, S.H.; Al Zahrani, E.M.; Al Zaid, R.F.; Alshagifi, S.A.; Farghal, T.A.; Alshamuse, K.B.; Albalawi, A.S.; Alkhalaf, F.; Sumaily, A.A.; Almusabi, S. Antibiotic-resistant bacteria contaminating leafy vegetables in Saudi Arabia's eastern region. *BMC Microbiol.* **2024**, *24*, 303. [CrossRef] [PubMed]
37. Pandey, A.K.; Samota, M.K.; Kumar, A.; Silva, A.S.; Dubey, N.K. Fungal mycotoxins in food commodities: Present status and future concerns. *Front. Sustain. Food Syst.* **2023**, *7*, 1162595. [CrossRef]
38. Iwu, C.D.; Okoh, A.I. Preharvest transmission routes of fresh produce associated bacterial pathogens with outbreak potentials: A review. *Int. J. Environ. Res. Public Health* **2019**, *16*, 4407. [CrossRef] [PubMed]
39. Mahunu, G.; Osei-Kwarteng, M.; Ogwu, M.C.; Afoakwah, N.A. Safe food handling techniques to prevent microbial contamination. In *Food Safety and Quality in the Global South*; Springer: Berlin/Heidelberg, Germany, 2024; pp. 427–461.
40. Allende, A.; Monaghan, J. Irrigation water quality for leafy crops: A perspective of risks and potential solutions. *Int. J. Environ. Res. Public Health* **2015**, *12*, 7457–7477. [CrossRef] [PubMed]
41. Alimi, B.A.; Lawal, R.; Odetunde, O.N. Food safety and microbiological hazards associated with retail meat at butchery outlets in north-central Nigeria. *Food Control* **2022**, *139*, 109061. [CrossRef]
42. Foods, N.A.C.o.M.C.f. Microbiological safety evaluations and recommendations on fresh produce. *Food Control* **1999**, *10*, 117–143.
43. Redmond, E.C.; Griffith, C.J. The importance of hygiene in the domestic kitchen: Implications for preparation and storage of food and infant formula. *Perspect. Public Health* **2009**, *129*, 69–76. [CrossRef]
44. Rose, J.B.; Epstein, P.R.; Lipp, E.K.; Sherman, B.H.; Bernard, S.M.; Patz, J.A. Climate variability and change in the United States: Potential impacts on water-and foodborne diseases caused by microbiologic agents. *Environ. Health Perspect.* **2001**, *109*, 211–221.
45. Hellberg, R.S.; Chu, E. Effects of climate change on the persistence and dispersal of foodborne bacterial pathogens in the outdoor environment: A review. *Crit. Rev. Microbiol.* **2016**, *42*, 548–572. [CrossRef] [PubMed]

46. Tryland, I.; Robertson, L.; Blankenberg, A.G.B.; Lindholm, M.; Rohrlack, T.; Liltved, H. Impact of rainfall on microbial contamination of surface water. *Int. J. Clim. Change Strateg. Manag.* **2011**, *3*, 361–373. [CrossRef]
47. Kostyla, C.; Bain, R.; Cronk, R.; Bartram, J. Seasonal variation of fecal contamination in drinking water sources in developing countries: A systematic review. *Sci. Total Environ.* **2015**, *514*, 333–343. [CrossRef] [PubMed]
48. Lake, I.; Barker, G. Climate change, foodborne pathogens and illness in higher-income countries. *Curr. Environ. Health Rep.* **2018**, *5*, 187–196. [CrossRef] [PubMed]
49. Bhutta, M.N.M.; Ahmad, M. Secure identification, traceability and real-time tracking of agricultural food supply during transportation using internet of things. *IEEE Access* **2021**, *9*, 65660–65675. [CrossRef]
50. Irvin, K.; Viazis, S.; Fields, A.; Seelman, S.; Blickenstaff, K.; Gee, E.; Wise, M.E.; Marshall, K.E.; Gieraltowski, L.; Harris, S. An overview of traceback investigations and three case studies of recent outbreaks of Escherichia coli O157: H7 infections linked to romaine lettuce. *J. Food Prot.* **2021**, *84*, 1340–1356. [CrossRef]
51. Bintsis, T. Foodborne pathogens. *AIMS Microbiol.* **2017**, *3*, 529. [CrossRef]
52. Grudlewska-Buda, K.; Bauza-Kaszewska, J.; Wiktorczyk-Kapischke, N.; Budzyńska, A.; Gospodarek-Komkowska, E.; Skowron, K. Antibiotic Resistance in selected emerging bacterial foodborne Pathogens—An issue of concern? *Antibiotics* **2023**, *12*, 880. [CrossRef]
53. Rajaei, M.; Moosavy, M.-H.; Gharajalar, S.N.; Khatibi, S.A. Antibiotic resistance in the pathogenic foodborne bacteria isolated from raw kebab and hamburger: Phenotypic and genotypic study. *BMC Microbiol.* **2021**, *21*, 272. [CrossRef] [PubMed]
54. Kim, J.; Ahn, J. Emergence and spread of antibiotic-resistant foodborne pathogens from farm to table. *Food Sci. Biotechnol.* **2022**, *31*, 1481–1499. [CrossRef]
55. Hashempour-Baltork, F.; Hosseini, H.; Shojaee-Aliabadi, S.; Torbati, M.; Alizadeh, A.M.; Alizadeh, M. Drug resistance and the prevention strategies in food borne bacteria: An update review. *Adv. Pharm. Bull.* **2019**, *9*, 335. [CrossRef] [PubMed]
56. Peng, S.; Luo, M.; Long, D.; Liu, Z.; Tan, Q.; Huang, P.; Shen, J.; Pu, S. Full-length 16S rRNA gene sequencing and machine learning reveal the bacterial composition of inhalable particles from two different breeding stages in a piggery. *Ecotoxicol. Environ. Saf.* **2023**, *253*, 114712. [CrossRef] [PubMed]
57. Uyttendaele, M.; Jaykus, L.A.; Amoah, P.; Chiodini, A.; Cunliffe, D.; Jacxsens, L.; Holvoet, K.; Korsten, L.; Lau, M.; McClure, P. Microbial hazards in irrigation water: Standards, norms, and testing to manage use of water in fresh produce primary production. *Compr. Rev. Food Sci. Food Saf.* **2015**, *14*, 336–356. [CrossRef]
58. Gravning, G.E.N.; Røtterud, O.-J.; Bjørkøy, S.; Forseth, M.; Skjerve, E.; Llarena, A.-K.; Lian, A.; Johannessen, G.S.; Hauge, S.J. Comparison of four sampling methods for microbiological quantification on broiler carcasses. *Food Control* **2021**, *121*, 107589. [CrossRef]
59. Miao, Y.; Xiong, G.; Bai, M.; Ge, Y.; Wu, Z. Detection of live Salmonella enterica in fresh-cut vegetables by a TaqMan-based one-step reverse transcription real-time PCR. *Lett. Appl. Microbiol.* **2018**, *66*, 447–454. [CrossRef]
60. Buzatu, D.A.; Moskal, T.J.; Williams, A.J.; Cooper, W.M.; Mattes, W.B.; Wilkes, J.G. An integrated flow cytometry-based system for real-time, high sensitivity bacterial detection and identification. *PLoS ONE* **2014**, *9*, e94254. [CrossRef]
61. Lorente, D.; Aleixos, N.; Gómez-Sanchis, J.; Cubero, S.; García-Navarrete, O.L.; Blasco, J. Recent advances and applications of hyperspectral imaging for fruit and vegetable quality assessment. *Food Bioprocess Technol.* **2012**, *5*, 1121–1142. [CrossRef]
62. ElMasry, G.; Sun, D.-W. Principles of hyperspectral imaging technology. In *Hyperspectral Imaging for Food Quality Analysis and Control*; Elsevier: Amsterdam, The Netherlands, 2010; pp. 3–43.
63. Canadian Food Inspection Agency. Sampling Procedures for Fresh Produce and Ready-to-Eat Products. 2023. Available online: https://inspection.canada.ca (accessed on 16 June 2025).
64. Postollec, F.; Falentin, H.; Pavan, S.; Combrisson, J.; Sohier, D. Recent advances in quantitative PCR (qPCR) applications in food microbiology. *Food Microbiol.* **2011**, *28*, 848–861. [CrossRef]
65. Foddai, A.C.; Grant, I.R. Methods for detection of viable foodborne pathogens: Current state-of-art and future prospects. *Appl. Microbiol. Biotechnol.* **2020**, *104*, 4281–4288. [CrossRef]
66. Kim, J.-H.; Oh, S.-W. Rapid and sensitive detection of E. coli O157: H7 and S. Typhimurium in iceberg lettuce and cabbage using filtration, DNA concentration, and qPCR without enrichment. *Food Chem.* **2020**, *327*, 127036. [CrossRef] [PubMed]
67. Fang, Z.; Zhou, X.; Wang, X.; Shi, X. Development of a 3-plex droplet digital PCR for identification and absolute quantification of Salmonella and its two important serovars in various food samples. *Food Control* **2023**, *145*, 109465. [CrossRef]
68. Xu, X.; Liu, G.; Huang, X.; Li, L.; Lin, H.; Xu, D. MALDI-TOF MS-based identification of bacteria and a survey of fresh vegetables with pathogenic bacteria in Beijing, China. *Food Biosci.* **2021**, *41*, 100746. [CrossRef]
69. Zeng, X.; Wang, Y.; Shen, X.; Wang, H.; Xu, Z.-L. Application of Matrix-Assisted Laser Desorption Ionization Time-of-Flight Mass Spectrometry for Identification of Foodborne Pathogens: Current Developments and Future Trends. *J. Agric. Food Chem.* **2024**, *72*, 22001–22014. [CrossRef]
70. Chai, Z.; Soko, W.C.; Xie, J.; Bi, H. Microchip coupled with MALDI-TOF MS for the investigation of bacterial contamination of fish muscle products. *Food Chem.* **2022**, *396*, 133658. [CrossRef] [PubMed]

1. Elbehiry, A.; Marzouk, E.; Hamada, M.; Al-Dubaib, M.; Alyamani, E.; Moussa, I.M.; AlRowaidhan, A.; Hemeg, H.A. Application of MALDI-TOF MS fingerprinting as a quick tool for identification and clustering of foodborne pathogens isolated from food products. *New Microbiol.* **2017**, *40*, 269–278. [PubMed]
2. Gootenberg, J.S.; Abudayyeh, O.O.; Kellner, M.J.; Joung, J.; Collins, J.J.; Zhang, F. Multiplexed and portable nucleic acid detection platform with Cas13, Cas12a, and Csm6. *Science* **2018**, *360*, 439–444. [CrossRef]
3. Zhang, X. Development of CRISPR-mediated nucleic acid detection technologies and their applications in the livestock industry. *Genes* **2022**, *13*, 2007. [CrossRef]
4. Tang, T.; Zhang, M.; Mujumdar, A.S. Intelligent detection for fresh-cut fruit and vegetable processing: Imaging technology. *Compr. Rev. Food Sci. Food Saf.* **2022**, *21*, 5171–5198. [CrossRef]
5. Vignati, S.; Tugnolo, A.; Giovenzana, V.; Pampuri, A.; Casson, A.; Guidetti, R.; Beghi, R. Hyperspectral imaging for fresh-cut fruit and vegetable quality assessment: Basic concepts and applications. *Appl. Sci.* **2023**, *13*, 9740. [CrossRef]
6. Soni, A.; Dixit, Y.; Reis, M.M.; Brightwell, G. Hyperspectral imaging and machine learning in food microbiology: Developments and challenges in detection of bacterial, fungal, and viral contaminants. *Compr. Rev. Food Sci. Food Saf.* **2022**, *21*, 3717–3745. [CrossRef] [PubMed]
7. Manthou, E.; Karnavas, A.; Fengou, L.-C.; Bakali, A.; Lianou, A.; Tsakanikas, P.; Nychas, G.-J.E. Spectroscopy and imaging technologies coupled with machine learning for the assessment of the microbiological spoilage associated to ready-to-eat leafy vegetables. *Int. J. Food Microbiol.* **2022**, *361*, 109458. [CrossRef] [PubMed]
8. Chen, Y.; Wang, Y.; Zhang, Y.; Wang, X.; Zhang, C.; Cheng, N. Intelligent biosensors promise smarter solutions in food safety 4.0. *Foods* **2024**, *13*, 235. [CrossRef] [PubMed]
9. Gil, M.I.; Selma, M.V.; López-Gálvez, F.; Allende, A. Fresh-cut product sanitation and wash water disinfection: Problems and solutions. *Int. J. Food Microbiol.* **2009**, *134*, 37–45. [CrossRef]
0. López-Gálvez, F.; Tudela, J.A.; Allende, A.; Gil, M.I. Microbial and chemical characterization of commercial washing lines of fresh produce highlights the need for process water control. *Innov. Food Sci. Emerg. Technol.* **2019**, *51*, 211–219. [CrossRef]
1. Rodgers, S.L.; Cash, J.N.; Siddiq, M.; Ryser, E.T. A comparison of different chemical sanitizers for inactivating Escherichia coli O157: H7 and Listeria monocytogenes in solution and on apples, lettuce, strawberries, and cantaloupe. *J. Food Prot.* **2004**, *67*, 721–731. [CrossRef]
2. Qi, H.; Wang, L.; Huang, Q.; Hung, Y.-C. Effect of organic load on the efficacy of activated persulfate in inactivating Escherichia coli O157: H7 and the production of halogenated by-products. *Food Control* **2020**, *114*, 107218. [CrossRef]
3. Van Haute, S.; Sampers, I.; Holvoet, K.; Uyttendaele, M. Physicochemical quality and chemical safety of chlorine as a reconditioning agent and wash water disinfectant for fresh-cut lettuce washing. *Appl. Environ. Microbiol.* **2013**, *79*, 2850–2861. [CrossRef]
4. Banach, J.L.; Sampers, I.; Van Haute, S.; Van der Fels-Klerx, H. Effect of disinfectants on preventing the cross-contamination of pathogens in fresh produce washing water. *Int. J. Environ. Res. Public Health* **2015**, *12*, 8658–8677. [CrossRef]
5. Gómez-López, V.M.; Rajkovic, A.; Ragaert, P.; Smigic, N.; Devlieghere, F. Chlorine dioxide for minimally processed produce preservation: A review. *Trends Food Sci. Technol.* **2009**, *20*, 17–26. [CrossRef]
6. Misra, N.; Tiwari, B.; Raghavarao, K.; Cullen, P. Nonthermal plasma inactivation of food-borne pathogens. *Food Eng. Rev.* **2011**, *3*, 159–170. [CrossRef]
7. Ricciardi, F.E.; Plazzotta, S.; Conte, A.; Manzocco, L. Effect of pulsed light on microbial inactivation, sensory properties and protein structure of fresh ricotta cheese. *LWT* **2021**, *139*, 110556. [CrossRef]
8. Nazzaro, F.; Fratianni, F.; De Martino, L.; Coppola, R.; De Feo, V. Effect of essential oils on pathogenic bacteria. *Pharmaceuticals* **2013**, *6*, 1451–1474. [CrossRef] [PubMed]
9. Rojas-Graü, M.A.; Soliva-Fortuny, R.; Martín-Belloso, O. Edible coatings to incorporate active ingredients to fresh-cut fruits: A review. *Trends Food Sci. Technol.* **2009**, *20*, 438–447. [CrossRef]
0. Li, S.; Ren, Y.; Hou, Y.; Zhan, Q.; Jin, P.; Zheng, Y.; Wu, Z. Polysaccharide-Based Composite Films: Promising Biodegradable Food Packaging Materials. *Foods* **2024**, *13*, 3674. [CrossRef]
1. de Araújo Bezerra, J.; Lamarão, C.V.; Sanches, E.A.; Rodrigues, S.; Fernandes, F.A.; Ramos, G.L.P.; Esmerino, E.A.; Cruz, A.G.; Campelo, P.H. Cold plasma as a pre-treatment for processing improvement in food: A review. *Food Res. Int.* **2023**, *167*, 112663. [CrossRef]
2. Mahnot, N.K.; Chakraborty, S.; Gupta, K.; Saikia, S. Cold Plasma Processing Methods: Impact of Decontamination on Food Quality. In *Advances in Food Process Engineering*; Apple Academic Press: Palm Bay, FL, USA, 2023; pp. 255–275.
3. Misra, N.; Patil, S.; Moiseev, T.; Bourke, P.; Mosnier, J.; Keener, K.; Cullen, P. In-package atmospheric pressure cold plasma treatment of strawberries. *J. Food Eng.* **2014**, *125*, 131–138. [CrossRef]
4. Li, L.; Li, C.; Sun, J.; Xin, M.; Yi, P.; He, X.; Sheng, J.; Zhou, Z.; Ling, D.; Zheng, F. Synergistic effects of ultraviolet light irradiation and high-oxygen modified atmosphere packaging on physiological quality, microbial growth and lignification metabolism of fresh-cut carrots. *Postharvest Biol. Technol.* **2021**, *173*, 111365. [CrossRef]

95. Elias, M.; Madureira, J.; Santos, P.; Carolino, M.; Margaça, F.; Verde, S.C. Preservation treatment of fresh raspberries by e-beam irradiation. *Innov. Food Sci. Emerg. Technol.* **2020**, *66*, 102487. [CrossRef]
96. Choi, D.S.; Park, S.H.; Choi, S.R.; Kim, J.S.; Chun, H.H. The combined effects of ultraviolet-C irradiation and modified atmosphere packaging for inactivating Salmonella enterica serovar Typhimurium and extending the shelf life of cherry tomatoes during cold storage. *Food Packag. Shelf Life* **2015**, *3*, 19–30. [CrossRef]
97. Wang, C.-P.; Liao, J.-Y. Effect of UV-C LED arrangement on the sterilization of Escherichia coli in planar water disinfection reactors. *J. Water Process Eng.* **2023**, *56*, 104399. [CrossRef]
98. Bialka, K.; Demirci, A. Efficacy of pulsed UV-light for the decontamination of Escherichia coli O157: H7 and Salmonella spp. on raspberries and strawberries. *J. Food Sci.* **2008**, *73*, M201–M207. [CrossRef]
99. Gómez-López, V.M.; Bhat, R.; Pellicer, J.A. Pulsed light. In *Electromagnetic Technologies in Food Science*; Wiley: Hoboken, NJ, USA, 2021; pp. 200–219.
100. Ramos-Villarroel, A.; Martín-Belloso, O.; Soliva-Fortuny, R. Pulsos de luz intensa: Inactivación microbiana en frutas y hortalizas. *CyTA-J. Food* **2013**, *11*, 234–242. [CrossRef]
101. Moreira, M.R.; Cassani, L.; Martín-Belloso, O.; Soliva-Fortuny, R. Effects of polysaccharide-based edible coatings enriched with dietary fiber on quality attributes of fresh-cut apples. *J. Food Sci. Technol.* **2015**, *52*, 7795–7805. [CrossRef] [PubMed]
102. Krishnamurthy, K.; Demirci, A.; Irudayaraj, J. Inactivation of Staphylococcus aureus by pulsed UV-light sterilization. *J. Food Prot.* **2004**, *67*, 1027–1030. [CrossRef]
103. Woldemariam, H.W.; Harmeling, H.; Emire, S.A.; Teshome, P.G.; Toepfl, S.; Aganovic, K. Pulsed light treatment reduces microorganisms and mycotoxins naturally present in red pepper (*Capsicum annuum* L.) powder. *J. Food Process Eng.* **2022**, *45*, e13948. [CrossRef]
104. Xylia, P.; Chrysargyris, A.; Shahwar, D.; Ahmed, Z.F.; Tzortzakis, N. Application of rosemary and eucalyptus essential oils on the preservation of cucumber fruit. *Horticulturae* **2022**, *8*, 774. [CrossRef]
105. Perumal, A.B.; Huang, L.; Nambiar, R.B.; He, Y.; Li, X.; Sellamuthu, P.S. Application of essential oils in packaging films for the preservation of fruits and vegetables: A review. *Food Chem.* **2022**, *375*, 131810. [CrossRef]
106. Vidaković Knežević, S.; Knežević, S.; Vranešević, J.; Kravić, S.Ž.; Lakićević, B.; Kocić-Tanackov, S.; Karabasil, N. Effects of selected essential oils on Listeria monocytogenes in biofilms and in a model food system. *Foods* **2023**, *12*, 1930. [CrossRef]
107. Nedic, D.; Grkovic, N.; Kalaba, V.; Golic, B.; Ilic, T.; Kasagic, D.; Suvajdzic, B.; Vejnovic, B.; Djuric, S.; Vasilev, D. Antimicrobial activity of thyme (*Thymus vulgaris*) and oregano (*Origanum vulgare*) essential oils against Listeria monocytogenes in fermented sausages. *IOP Conf. Ser. Earth Environ. Sci.* **2021**, *854*, 012064. [CrossRef]
108. Chang, L.; Xu, L.; Yang, Z.; Liu, L.; Qiu, D. Antibacterial and antioxidative biogenic films for room-temperature strawberry preservation. *Food Chem.* **2023**, *405*, 134893. [CrossRef] [PubMed]
109. Yang, C.; Lu, J.-H.; Xu, M.-T.; Shi, X.-C.; Song, Z.-W.; Chen, T.-M.; Herrera-Balandrano, D.D.; Zhang, Y.-J.; Laborda, P.; Shahriar, M. Evaluation of chitosan coatings enriched with turmeric and green tea extracts on postharvest preservation of strawberries. *LWT* **2022**, *163*, 113551. [CrossRef]
110. Rojas-Graü, M.; Tapia, M.; Rodríguez, F.; Carmona, A.; Martin-Belloso, O. Alginate and gellan-based edible coatings as carriers of antibrowning agents applied on fresh-cut Fuji apples. *Food Hydrocoll.* **2007**, *21*, 118–127. [CrossRef]
111. Chang, H.-Y.; Gui, C.-Y.; Huang, T.-C.; Hung, Y.-C.; Chen, T.-Y. Quantitative proteomic analysis on the slightly acidic electrolyzed water triggered viable but non-culturable Listeria monocytogenes. *Int. J. Mol. Sci.* **2023**, *24*, 10616. [CrossRef]
112. Guentzel, J.L.; Lam, K.L.; Callan, M.A.; Emmons, S.A.; Dunham, V.L. Reduction of bacteria on spinach, lettuce, and surfaces in food service areas using neutral electrolyzed oxidizing water. *Food Microbiol.* **2008**, *25*, 36–41. [CrossRef]
113. Afari, G.K.; Hung, Y.C.; King, C.H. Efficacy of neutral pH electrolyzed water in reducing Escherichia coli O157: H7 and Salmonella Typhimurium DT 104 on fresh produce items using an automated washer at simulated food service conditions. *J. Food Sci.* **2015**, *80*, M1815–M1822. [CrossRef]
114. Hamidi, R.M.; Shekarforoush, S.S.; Hosseinzadeh, S.; Basiri, S. Near Neutral Electrolyzed Water and Peroxyacetic Acid and Their Effect on the Survival of Escherichia coli, Salmonella Typhimurium, and Listeria monocytogenes Inoculated on Poultry Meat. *Food Prot. Trends* **2021**, *41*, 380–388. [CrossRef]
115. Hong, H.; Rizzi, M.F.; Wang, D.; McLandsborough, L.; Lu, J. A Meta-Analysis on the Antimicrobial Effectiveness of Ozonated Water Treatments for Fresh Produce Washing—Effect of Ozonation Methods. *Foods* **2024**, *13*, 3906. [CrossRef]
116. Gibson, K.E.; Almeida, G.; Jones, S.L.; Wright, K.; Lee, J.A. Inactivation of bacteria on fresh produce by batch wash ozone sanitation. *Food Control* **2019**, *106*, 106747. [CrossRef]
117. Macías-Gallardo, F.; Barajas-Díaz, C.G.-M.; Mireles-Arriaga, A.I.; Ozuna, C. Strawberry variety influences the effectiveness of postharvest treatment with gaseous ozone: Impact on the physicochemical, microbiological, and bioactive properties of the fruit. *Processes* **2023**, *11*, 346. [CrossRef]

18. Luo, A.; Bai, J.; Li, R.; Fang, Y.; Li, L.; Wang, D.; Zhang, L.; Liang, J.; Huang, T.; Kou, L. Effects of ozone treatment on the quality of kiwifruit during postharvest storage affected by Botrytis cinerea and Penicillium expansum. *J. Phytopathol.* **2019**, *167*, 470–478. [CrossRef]
19. Öztekin, S.; Zorlugenç, B.; Zorlugenç, F.K.l. Effects of ozone treatment on microflora of dried figs. *J. Food Eng.* **2006**, *75*, 396–399. [CrossRef]
20. Abdelshafy, A.M.; Luo, Z.; Belwal, T.; Ban, Z.; Li, L. A comprehensive review on preservation of shiitake mushroom (*Lentinus edodes*): Techniques, research advances and influence on quality traits. *Food Rev. Int.* **2023**, *39*, 2742–2775. [CrossRef]
21. Zhou, X.; Salazar, J.K.; Fay, M.L.; Zhang, W. Efficacy of power ultrasound-based hurdle technology on the reduction of bacterial pathogens on fresh produce. *Foods* **2023**, *12*, 2653. [CrossRef]
22. Liu, F.; Xu, Y.; Zeng, M.; Zhang, Y.; Pan, L.; Wang, J.; Huang, S. A novel physical hurdle technology by combining low voltage electrostatic field and modified atmosphere packaging for long-term stored button mushrooms (*Agaricus bisporus*). *Innov. Food Sci. Emerg. Technol.* **2023**, *90*, 103514. [CrossRef]
23. Melhem, L.C.d.M.; Do Rosario, D.K.A.; Monteiro, M.L.G.; Conte-Junior, C.A. High-Pressure Processing and Natural Antimicrobials Combined Treatments on Bacterial Inactivation in Cured Meat. *Sustainability* **2022**, *14*, 10503. [CrossRef]
24. Mahović Poljaček, S.; Tomašegović, T.; Strižić Jakovljević, M.; Jamnicki Hanzer, S.; Murković Steinberg, I.; Žuvić, I.; Leskovac, M.; Lavrič, G.; Kavčič, U.; Karlovits, I. Starch-Based Functional Films Enhanced with Bacterial Nanocellulose for Smart Packaging: Physicochemical Properties, pH Sensitivity and Colorimetric Response. *Polymers* **2024**, *16*, 2259. [CrossRef]
25. Wu, W.; Zheng, L.; Yu, J.; Liu, L.; Goksen, G.; Shao, P. High-sensitivity intelligent packaging films harnessing rose anthocyanins and hydrophilic silica aerogel for visual food freshness monitoring. *Food Qual. Saf.* **2024**, *8*, fyad051. [CrossRef]
26. Zhang, J.; Liu, S.; Xie, C.; Wang, C.; Zhong, Y.; Fan, K. Recent advances in pH-sensitive indicator films based on natural colorants for smart monitoring of food freshness: A review. *Crit. Rev. Food Sci. Nutr.* **2024**, *64*, 12800–12819. [CrossRef]
27. Palanisamy, Y.; Kadirvel, V.; Ganesan, N.D. Recent technological advances in food packaging: Sensors, automation, and application. *Sustain. Food Technol.* **2025**, *3*, 161–180. [CrossRef]
28. Douaki, A.; Ahmed, M.; Longo, E.; Windisch, G.; Riaz, R.; Inam, S.; Tran, T.N.; Papadopoulou, E.L.; Athanassiou, A.; Boselli, E. Battery-Free, Stretchable, and Autonomous Smart Packaging. *Adv. Sci.* **2024**, *12*, e2417539. [CrossRef]
29. Nawaz, M.; Pan, J.; Liu, H.; Umer, M.J.; Liu, J.; Yang, W.; Lv, Z.; Zhang, Q.; Jiao, Z. Integrated evaluation of antifungal activity of pomegranate peel polyphenols against a diverse range of postharvest fruit pathogens. *Bioresour. Bioprocess.* **2025**, *12*, 34. [CrossRef] [PubMed]
30. Boubrik, F.; Boubellouta, T.; Benyoucef, N.; Bellik, Y.; Gali, L.; Akdoğan, A.; Kucher, D.E.; Utkina, A.O.; Kucher, O.D.; Rebouh, N.Y. Investigating the chemical composition and antifungal effect of *Cinnamomum cassia* essential oil against *Saccharomyces cerevisiae* and *Acremonium* sp. *Sci. Rep.* **2025**, *15*, 10195.
31. Aljuhani, S.; Rizwana, H.; Aloufi, A.S.; Alkahtani, S.; Albasher, G.; Almasoud, H.; Elsayim, R. Antifungal activity of Carica papaya fruit extract against Microsporum canis: In vitro and in vivo study. *Front. Microbiol.* **2024**, *15*, 1399671. [CrossRef]
32. Vasundaradevi, R.; Sarvajith, M.; Divyashree, S.; Deepa, N.; Achar, P.N.; Sreenivasa, M. Tropical fruit-derived Lactiplantibacillus as potential probiotic and antifungal agents against Fusarium oxysporum. *Sci. Rep.* **2025**, *15*, 2144. [CrossRef]
33. Bakkali, F.; Averbeck, S.; Averbeck, D.; Idaomar, M. Biological effects of essential oils–A review. *Food Chem. Toxicol.* **2008**, *46*, 446–475. [CrossRef]
34. Raut, J.S.; Karuppayil, S.M. A status review on the medicinal properties of essential oils. *Ind. Crops Prod.* **2014**, *62*, 250–264. [CrossRef]
35. Alegbeleye, O.O.; Singleton, I.; Sant'Ana, A.S. Sources and contamination routes of microbial pathogens to fresh produce during field cultivation: A review. *Food Microbiol.* **2018**, *73*, 177–208. [CrossRef]
36. Weller, D.; Wiedmann, M.; Strawn, L.K. Irrigation is significantly associated with an increased prevalence of Listeria monocytogenes in produce production environments in New York State. *J. Food Prot.* **2015**, *78*, 1132–1141. [CrossRef]
37. Weller, D.; Wiedmann, M.; Strawn, L.K. Spatial and temporal factors associated with an increased prevalence of Listeria monocytogenes in spinach fields in New York State. *Appl. Environ. Microbiol.* **2015**, *81*, 6059–6069. [CrossRef]
38. University of Nevada, Reno Extension. Good Agricultural Practices (GAPs) and Good Handling Practices (GHPs). 2013. Available online: https://extension.unr.edu/publication.aspx?PubID=2168 (accessed on 22 May 2025).
39. Agriopoulou, S.; Stamatelopoulou, E.; Sachadyn-Król, M.; Varzakas, T. Lactic acid bacteria as antibacterial agents to extend the shelf life of fresh and minimally processed fruits and vegetables: Quality and safety aspects. *Microorganisms* **2020**, *8*, 952. [CrossRef]
40. Sathe, S.; Nawani, N.; Dhakephalkar, P.; Kapadnis, B. Antifungal lactic acid bacteria with potential to prolong shelf-life of fresh vegetables. *J. Appl. Microbiol.* **2007**, *103*, 2622–2628. [CrossRef]
41. Nunes, C.A. Biological control of postharvest diseases of fruit. *Eur. J. Plant Pathol.* **2012**, *133*, 181–196. [CrossRef]
42. Jagannathan, B.V.; Vijayakumar, P.P. The need for prevention-based food safety programs for fresh produce. *Food Prot. Trends* **2019**, *39*, 572–579.

143. WHO. One Health. 2023. Available online: https://www.who.int/health-topics/one-health#tab=tab_1 (accessed on 21 May 2025).
144. Suliman Maashi, M. CRISPR/CAS-based aptasensor as an innovative sensing approaches for food safety analysis: Recent progresses and new horizons. *Crit. Rev. Anal. Chem.* **2024**, *54*, 2599–2617. [CrossRef]
145. Olaniran, A.F.; Taiwo, A.E.; Iranloye, Y.M.; Okonkwo, C.E. The role of good agricultural practices (GAPs) and good manufacturing practices (GMPs) in food safety. In *Food Safety and Toxicology: Present and Future Perspectives*; De Gruyter: Berlin, Germany, 2023; pp. 417–432.
146. Okpala, C.O.R.; Korzeniowska, M. Understanding the relevance of quality management in agro-food product industry: From ethical considerations to assuring food hygiene quality safety standards and its associated processes. *Food Rev. Int.* **2023**, *39*, 1879–1952. [CrossRef]
147. Harris, D.; Lott, M.; Lakins, V.; Bowden, B.; Kimmons, J. Farm to institution: Creating access to healthy local and regional foods. *Adv. Nutr.* **2012**, *3*, 343–349. [CrossRef] [PubMed]
148. Yadav, A.; Yadav, K. Portable solutions for plant pathogen diagnostics: Development, usage, and future potential. *Front. Microbiol.* **2025**, *16*, 1516723. [CrossRef]
149. Verma, N.V.; Shukla, M.; Kulkarni, R.; Srivastava, K.; Claudic, B.; Savara, J.; Joseph Mathew, M.; Maurya, R.; Bhattacharjee, G.; Singh, V. Emerging extraction and diagnostic tools for detection of plant pathogens: Recent trends, challenges, and future scope. *ACS Agric. Sci. Technol.* **2022**, *2*, 858–881. [CrossRef]
150. Lewin, E.R. Applications of Next Generation Sequencing for the Assessment of Microbiological Safety of Fresh Produce. Ph.D. Thesis, Newcastle University, Newcastle, UK, 2020.
151. Kearns, E.A.; Gustafson, R.E.; Castillo, S.M.; Alnughaymishi, H.; Lim, D.V.; Ryser, E.T. Rapid large-volume concentration for increased detection of Escherichia coli O157: H7 and Listeria monocytogenes in lettuce wash water generated at commercial facilities. *Food Control* **2019**, *98*, 481–488. [CrossRef]
152. Hadi, J.; Rapp, D.; Dhawan, S.; Gupta, S.K.; Gupta, T.B.; Brightwell, G. Molecular detection and characterization of foodborne bacteria: Recent progresses and remaining challenges. *Compr. Rev. Food Sci. Food Saf.* **2023**, *22*, 2433–2464. [CrossRef] [PubMed]

microorganisms

MDPI

Review

A Review on Recent Trends in Bacteriophages for Post-Harvest Food Decontamination

Márcia Braz [1,2], Carla Pereira [1], Carmen S. R. Freire [2,*] and Adelaide Almeida [1,*]

[1] CESAM—Centre for Environmental and Marine Studies, Department of Biology, University of Aveiro, Campus Universitário de Santiago, 3810-193 Aveiro, Portugal; marciabraz96@ua.pt (M.B.); csgp@ua.pt (C.P.)
[2] CICECO—Aveiro Institute of Materials, Department of Chemistry, University of Aveiro, Campus Universitário de Santiago, 3810-193 Aveiro, Portugal
[*] Correspondence: cfreire@ua.pt (C.S.R.F.); aalmeida@ua.pt (A.A.)

Abstract: Infectious diseases resulting from unsafe food consumption are a global concern. Despite recent advances and control measures in the food industry aimed at fulfilling the growing consumer demand for high-quality and safe food products, infection outbreaks continue to occur. This review stands out by providing an overview of post-harvest food decontamination methods against some of the most important bacterial foodborne pathogens, with particular focus on the advantages and challenges of using phages, including their most recent post-harvest applications directly to food and integration into active food packaging systems, highlighting their potential in providing safer and healthier food products. The already approved commercial phage products and the numerous available studies demonstrate their antibacterial efficacy against some of the most problematic foodborne pathogens in different food products, reinforcing their possible use in the future as a current practice in the food industry for food decontamination. Moreover, the incorporation of phages into packaging materials holds particular promise, providing protection against harsh conditions and enabling their controlled and continuous release into the food matrix. The effectiveness of phage-added packaging materials in reducing the growth of pathogens in food systems has been well-demonstrated. However, there are still some challenges associated with the development of phage-based packaging systems that need to be addressed with future research.

Keywords: foodborne pathogens; post-harvest stage; phage biocontrol; active food packaging; food safety

1. Introduction

Infectious diseases caused by the consumption of contaminated food pose a significant public health threat worldwide. The recurrent use of antibiotics has contributed to the alarming rise in multidrug-resistant bacterial strains and resistance genes, exacerbating this problem. Despite several efforts of the food industry to reduce foodborne pathogens in their products, foodborne infections remain a leading cause of hospitalizations and fatalities on a global scale. According to the World Health Organization (WHO), unsafe food consumption leads to 600 million foodborne illnesses and 420,000 deaths annually, imposing an economic burden exceeding USD 110 billion [1].

Campylobacter spp., *Salmonella* spp., Shiga toxin-producing *Escherichia coli* (STEC), and *Listeria monocytogenes* are among the most reported foodborne pathogens in food outbreaks [2]. Other bacteria, such as *Vibrio* spp., *Cronobacter sakazakii*, and *Shigella* spp. have also been reported in a considerable number of cases. *Bacillus cereus*, *Clostridium*

 https://doi.org/10.3390/microorganisms13030515

botulinum, *Clostridium perfringens*, and *Staphylococcus aureus* are particularly important due to the production of bacterial toxins. In 2021, although the majority of foodborne outbreaks caused by toxigenic bacteria were related to *Bacillus cereus* toxins, the highest number of deaths and hospitalizations were mainly caused by *C. perfringens* and *S. aureus* toxins, respectively. The European Centre for Disease Prevention and Control (ECDC) has reported several outbreaks associated with contamination of different foods with these bacterial pathogens and toxins, emphasizing the importance of addressing contamination in the food supply [2].

Efforts to mitigate the microbial burden in food, particularly in raw products like fresh fruits and vegetables, involve an array of decontamination strategies, including thermal and non-thermal methods. However, these methods face inherent limitations [3–5]. Therefore, there is a pressing need for more efficient, safer, and environmentally friendly approaches to reduce food contamination.

Bacteriophages are increasingly recognized as biocontrol agents with enormous potential for the food industry. Bacteriophages, or simply phages, are viruses that specifically infect bacteria, the most abundant entities on the planet. Phages are obligate intracellular parasites that can reproduce only in the presence of a host bacterial cell. These viruses are innocuous to humans, animals, and plants [6], making them a safe alternative to conventional antibiotics and other treatments used in the food industry [7].

The interest in phages is clearly highlighted by several phage-based products for direct application in food, with Generally Recognized as Safe (GRAS) status, to control some of the leading foodborne bacterial pathogens that have already reached the market. Examples of commercial phage companies with Food and Drug Administration (FDA) approval for their food safety products include Micreos Food Safety, Intralytix, FINK TEC GmbH, Phagelux, and Passport Food Safety Solutions [8,9].

Several reviews have appraised the application of approved phages and also of other new phage suspensions directly in food against some of the most important foodborne pathogens [8,10–14] or specific bacteria, namely *E. coli* [15], *Salmonella* spp. [16], *L. monocytogenes* [17], and *Campylobacter* spp. [18]. Also, some review papers focus on the direct application of phages in a specific type of food, specifically on poultry meat against *Salmonella* spp. [16,19–21]. However, the successful application of phages in food still requires strategies to ensure their stability during food processing, transportation, and storage [9].

Many food environments subject phages to challenging physicochemical conditions, such as low pH levels on the surface of certain fruits and high storage temperatures during food transportation, which may lead to phage inactivation. The incorporation of phages into different materials are often employed in the context of active packaging systems designed to safeguard phages, maintain their stability, and facilitate controlled release, as recently reviewed by García-Anaya et al. (2023) [22]. In particular, this review addressed studies on phage incorporation in edible films and coatings and the associated challenges of applying them to foods.

The reviews available in the literature generally either discuss free phages or incorporated phages, some focus only on a specific bacteria or food, while others bring together pre- and post-harvest phage applications, with an integrative and detailed overview of phages for post-harvest food decontamination still missing in the literature. Thus, with this review, we attempted to fill this gap and gather here all the important information regarding food decontamination only at post-harvest level, focusing on the application of phages in their free form directly in food or incorporated into food packaging in a variety of different products and against the most important foodborne bacteria. This review examines some of the studies already outlined in other reviews but also includes the most recent articles in this field; since there is a growing interest in phages, this is an area in constant development and research. Regarding phage incorporation in food packaging, we tried to include all available relevant studies in the literature, considering that the majority of reviews only present a few studies on this topic.

In this context, this study offers an overview of available methods to reduce microbial contamination in food, with a particular focus on recent developments in the application of phages (Figure 1).

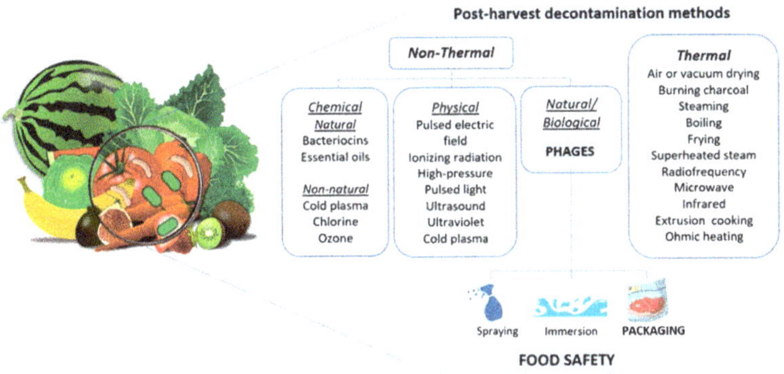

Figure 1. Post-harvest decontamination methods with a focus on phage application directly in food and phage incorporation into active food packaging.

Specifically, we delve into the advantages and challenges of phages in the food industry, including their direct post-harvest application in food, and examine in detail some of the most recent studies on this topic. This review also includes the available *in vitro* and in-food studies involving the incorporation of phages into active food packaging systems.

2. Available Methods to Reduce the Microbial Burden in Food

2.1. Thermal Methods

Heat processing is one of the most efficient strategies used in the food industry to improve the quality and prolong the shelf life of food and includes pasteurization, sterilization, convective hot air drying, vacuum drying, burning charcoal treatment, steaming, boiling, and frying [23–25]. In these methods, the heat energy is transferred into a product through conduction or convection.

Pasteurization and sterilization are commonly used conventional techniques for the inactivation of enzymes and microorganisms in foods [26]. The difference between these two methods is the induction temperature. While pasteurization is performed at a temperature in the range of 60–80 °C to eliminate pathogenic microorganisms from food products and degrade enzymes, sterilization works at a temperature above 100 °C to destroy spores or spore-forming microorganisms in food products [27]. Pasteurization is commonly applied at a specific temperature and for a specific period of time in order to decontaminate and preserve liquid foods (milk, fruit juices, beer, and other fermented drinks) and pre-packed food products due to alarming food safety concerns related to ready-to-eat (RTE) food [26]. However, pasteurization does not completely eliminate all pathogenic microorganisms or kill heat-resistant spore-forming bacteria. Thus, moderate heat treatments between 40 and 60 °C are preferable to their high-temperature counterparts to reduce quality loss in food [23,25].

Superheated steam has higher heat transfer coefficients, which reduce microorganisms on the surface of food products efficiently with low energy consumption levels, non-oxidative conditions, and low environmental impacts. Applied for several food products, such as vegetables, fruits, cereals, and meat, this approach can facilitate the development of products with the desired quality, reduce nutrient loss, and improve the physicochemical properties of foodstuffs, solving the problems faced by traditional thermal procedures. However, this approach requires high maintenance costs due to the complexity of the equipment [24,28].

Infrared heating uses electromagnetic waves to transfer energy from the infrared source to the product. The heat needed for microbial decontamination is generated by the rotational and vibrational state of molecules or atoms [29] and also from hot sources like a metal rod, quartz tube, or quartz lamp. The germicidal property of this technology has been reported in different types of foods [30,31], namely in fruits [32,33], shelled corn [34,35], eggs [36], cottage cheese [37], and milk [38]. However, in food products with a complex matrix, radiation cannot reach the inner part of the sample and food heating is not uniform [39]. The food processing sector also uses infrared heating for blanching, thawing, broiling, frying, roasting, baking, drying, dehydration, peeling, polyphenols and antioxidants recovery, sterilization of grains, baking bread, manufacture of juices, and cooking food [30,31].

Table 1 summarizes the advantages and disadvantages of these and other post-harvest food decontamination techniques, which are further discussed in the following paragraphs.

Table 1. Advantages and disadvantages of post-harvest food decontamination techniques.

Thermal Methods			
Type of Decontamination Method	**Advantages**	**Disadvantages**	**References**
Conventional thermal methods	Pasteurization, sterilization, air or vacuum drying, burning charcoal, steaming, boiling, frying • Effective reduction in the microbial load in food products	• High energy consumption • High environmental footprints • Poor food quality	[23–25]
Novel thermal methods	Superheated steam • Fast processing rate • High energy efficiency • Enhanced safety • Low environmental footprint • Water saving • Excellent product quality	• Can lead to poor product quality • High maintenance costs due to the complexity of the equipment	[4,24]
	Radiofrequency and microwave • Cost-effectiveness • Easy and quick operation • High energy efficiency • Non-toxicity • Suitability for heat-sensitive fluids	• Limited use to high-moisture, high-salt, and high-fat content food products • Non-uniformity heating that creates hot and cold spots inside the food product affecting its microbial safety • Microwave treatment can lead to moisture and nutrient loss	[4,23,40]

Table 1. *Cont.*

Thermal Methods			
Type of Decontamination Method	**Advantages**	**Disadvantages**	**References**
Novel thermal methods	Infrared • Precise and rapid heating • High heat-transfer efficiency and heating rate • High levels of time and temperature control • Easy operation • Low maintenance costs	• Relatively low penetrability • Sometimes causes color degradation • The equipment design and parameters must be optimized to avoid the overheating problem in industrial applications	[4,23,39]
	Extrusion cooking • Low cost • High productivity • Enhanced product quality by retaining heat-sensitive components of food	• Changes in the physical property of food products • Puffing up of the food products	[4,41]
	Ohmic heating • Rapid heating of the food • Overcooking can be avoided • Low heat losses • No residual heat transfer after the current shut off • Better impact on sensory properties of food • Low maintenance costs • Suitable for viscous products and pumpable foods containing particles	• Requires uniform conductivity within the food matrix to avoid cold spots • Might lead to considerable color differences	[40]
Non-thermal methods			
Chemical methods (natural)	Bacteriocins • More stable and effective at: - Acidic pH - Higher-than-normal temperatures or lower-than-normal temperatures • Degraded by proteases	• Lactic acid bacteria have been linked with sepsis, endocarditis, and bacteremia • May not be effective in satisfactorily reducing the microbial load • Research must determine if the lactic acid bacteria strains have long-term effects on bacterial reduction during shelf life • Limited range of activity	[42–44]

Table 1. *Cont.*

Thermal Methods			
Type of Decontamination Method	**Advantages**	**Disadvantages**	**References**
Chemical methods (natural)	Essential oils • Natural origin • Improve the flavor, odor, and color of food	• The elimination of a specific microorganism in food is still limited • Level required for inhibition may introduce too strong a flavor to the food	[42,45]
Chemical methods (non-natural)	Chlorine-based methods • Inexpensive • Easy to use	• Generation of toxic by-products as well as off-tastes and odors • May not be effective in satisfactorily reduce the microbial load on surfaces containing biofilm or in food matrices with high organic content	[42,46]
	Ozone • Can be used in both a gaseous or liquid form • Higher percentage of reduction in microorganisms compared to chlorine • Relatively cheap with low running costs • Environmentally sustainable and commercially feasible technology • Short contact time for disinfection compared to other conventional methods	• Oxidative spoilage of the food • Limited to surface treatment • High initial investment • Unstable nature • Highly toxic and corrosive compound with a pungent disagreeable odor • Effectiveness diminished by the presence of organic matter	[3,42,47]
	Cold Plasma • High percentage of reduction • Broad spectrum approach • Allows in package treatment • Eco-friendly benefits: - diminished water utilization - absence of chemical residues - utilization of environmental air as a working gas • Minimal changes in the food matrix	• Low penetration efficiency • High microbial loads reduce its efficiency • Efficiency dependent on humidity, gas composition, and flow rate • May result in lipid deterioration and the destruction of inherent antioxidants by the free radicals formed, leaving the food with an undesirable taste and aroma • High capital investment • Still at its lowest level (laboratory or pilot scale)	[3,48]

Table 1. *Cont.*

Thermal Methods			
Type of Decontamination Method	**Advantages**	**Disadvantages**	**References**
	Pulsed electric field • Retains nutrients and sensory characteristics, promotes durability, and ensures the safety of the foodstuffs • Environmentally friendly and highly energy-efficient technique • Decrease in energy costs, processing time and degradative effects of heat-sensitive food components	• Huge start-up costs • Some bacteria cells have developed resistance against this method • Not yet effective in treating solid foodstuffs, compared to partial solids or liquid foods	[3,40,46,47]
Physical methods	High-pressure processing • Commercially available technique for bulk quantities of samples, either solid or liquid • Retains the taste and nutrient composition of food and elongates its shelf-life • Eco-friendly approach	• Not suitable for dehydrated and porous foodstuffs • The treated food must be kept in cold/ refrigerated conditions • Only plastic materials appear the best fit as packing materials • High equipment costs • Need for appropriate skill and space to effectively operate	[3,40,47]
	Ultrasound • Commercially available technique for both solid and liquid food • Highly reduced treatment time for handling food • Uses a minimal amount of energy • Safe and environmentally friendly technology • Quite feasible, as it is simple and economically cheap • Useful for eliminating microbial entities that can hinder the food fermentation processes	• Free radicals may decline food product quality • Harmful effects in food characteristics such as sensory parameters as well as nutrient composition • Better reduction when applied in combination with other treatments • The sonication is more confined to liquid foods	[3,40,46,47, 49]

Table 1. *Cont.*

Thermal Methods			
Type of Decontamination Method	**Advantages**	**Disadvantages**	**References**
Physical methods	Ultraviolet radiation • Better nutrient preservation • Higher lethal effects against microbes when compared to some conventional chemical agents • Easy to use, cost-effective, and environmentally friendly approach • Minimal effects on the quality of food • Can be used for liquid and solid food • Prevents recontamination as it can be applied in already packed food products • Its processing time is described as shorter	• Lack of complete recognition and acceptability by the consumers • Effect of UV-C light application on liquid food is affected by their turbidity • Ineffective when applied to foodstuffs with indefinite shape and structure given its low penetration capacity • Huge investment requirements	[3,23,46,50]
	Pulsed light system • Rapid disinfection food processing technology • Ensure microbial inactivation and at the same time retain the sensory characteristics of foodstuffs • Outstandingly short time energy transmission compared to ultraviolet • Flexible and eco-friendly procedure • In-package treatment	• High investment costs • Not suitable for application in foods that are opaque and irregularly shaped • Possible heating of products due to extended periods of treatment	[3,51]
	Ionizing radiation • Minimum effect on quality, taste, appearance, and texture of food • Effective in destroying microbes but also insects, mites, and pests • Processing time is reasonably shorter comparatively to other technologies • Eco-friendly approach leaving no chemicals or residues • The dosage applied for food preservation is generally lower and not dangerous for humans after eating irradiated food	• If the dose is too high, the functional and the sensory properties of food can be affected • Ionizing radiation can be harmful to processors and workers • Huge investment costs • Problem of consumer acceptance	[3,40,52]

Table 1. *Cont.*

Type of Decontamination Method	Advantages	Disadvantages	References
	Thermal Methods		
Biological methods	Phage treatment • Natural • Low cost • Self-replicating • Commercialization timeframe less stringent than for human therapeutic application • High specificity for their target bacterial host with no significant impact on consumers' resident microbiota • No impact on the sensory and quality characteristics of food • Ability to infiltrate bacterial biofilms and infect biofilm-embedded bacteria	• Sometimes a high phage titer is required for the success of treatment • Phages can be inactivated by extreme environmental factors (extreme pH, temperature, UV irradiation, and low oxygen) • Some food matrices limit phage diffusion or interfere with phages, reducing their encounter with host bacteria • Food additives can impair phage activity • Bacteria can develop phage resistance following repeated exposure	[9,11,53–55]

Ohmic heating is based on the movement of an alternating electric current through a food matrix that is transformed into heat due to the electrical resistance of the food, allowing faster food heating when compared with conventional heating techniques. This process inactivates microorganisms mainly by thermal effects although other nonthermal inactivation mechanisms are also involved [40] and has been proposed as an alternative fast-heating method for fruit juices [56,57]. This method can also be effective for viscous products and pumpable food containing particles inside the food matrix due to the heat generation. Nevertheless, it requires uniform conductivity to avoid cold spots (Table 1) [40,58].

Microwave and radio-frequency heating treatments involve high-frequency alternating radio waves, microwave electromagnetic radiation, or electric fields [23].

Microwave treatment is applied in the food industry for thawing, blanching, precooking, cooking, tempering, baking, puffing, foaming, concentration, drying, pasteurization, and sterilization. The inactivation of microorganisms is caused by the denaturation of cell protein structure, leading to the extrusion of cellular matrix and death [29]. This technology has been studied in liquid foods, like grape juice, apple juice, apple cider, coconut water, and milk [23]. However, at high power levels, microwave heating can cause acrylamide formation (a neurotoxic and carcinogenic substance), unlike conventional food heat treatments. On the other hand, short exposure to microwaves (during blanching and thawing) at low power may even limit the formation of acrylamide during the final heat treatment (Table 1) [59].

Radio-frequency heating has been already used in the food industry for a considerable period post the baking of biscuits and cereals as well as the drying of food. Deeper penetration, compared to microwave treatment, makes this process more efficient. The heat causes inactivation and changes in the cell membrane of microorganisms and has been used to decontaminate different types of foods (meat products, salmon caviar, eggs, pasta products, peanut butter, crackers, sandwiches, fresh carrots, fruit juices, apple cider, whole milk, and soybean milk) [60]. Its main limitation is the inconsistent heating of the food matrix, which can impact food quality and safety (Table 1) [23,40,60].

Extrusion cooking combines heat transfer, mass transfer, pressure changes, and shear to produce different effects like cooking, kneading, shearing, shaping, and forming, among others [41]. This technique has gained popularity in the food and feed industries due to its low cost, high productivity, and enhanced product quality by retaining the heat-sensitive components of food (Table 1) since high temperatures are applied for short periods [41]. The extruded products have less moisture content, longer shelf life, and are considered microbiologically safe due to enzyme inactivation and microbial reduction. The success of extrusion cooking has been mainly reported in breakfast cereals, coextruded snacks, texturized vegetable proteins, and pet food; however, despite the advantages, this approach can cause changes in the physical properties of the treated food products [4,61].

2.2. Non-Thermal Methods

2.2.1. Chemical Methods

Non-Natural Approaches

Washing, sanitation, and use of oxidative and non-oxidative biocides are commonly employed for the disinfection of food. Washing is responsible for removing oil, dirt, physical matter, and debris but also helps to reduce the microbial populations on the surface of vegetables. The addition of sanitizers to the washing water can increase the efficiency of disinfection, mainly in leafy vegetables. The majority of common sanitizers are oxidizing agents and work by developing oxidation potential in water [46]. Several sanitizing agents or disinfectants (including chlorine-based products, hydrogen peroxide, organic acids, and ozone) have been approved in order to reduce bacterial populations on minimally processed food, such as fruits and vegetables [5,42].

Chlorine is a commonly used disinfectant in the fresh produce industry. However, studies indicate that chlorine concentrations traditionally used are not effective in reducing pathogen loads on fresh-cut produce [42]. In addition, the use of chlorine as a sanitizer, as well as the use of other chemical compounds for food decontamination, might result in the formation of undesirable by-products such as carcinogenic compounds (Table 1) [62], with their use forbidden in several countries, namely European countries like Belgium, Switzerland, and the Netherlands [46,63].

Ozone is used for the surface decontamination of fruits and vegetables, drinking water disinfection, meat, and seafood processing. This sanitizer is effective against a broad spectrum of microorganisms, including bacterial spores [64], by interfering with their cellular respiration and interacting with the unsaturated lipids of the cell membrane causing microbial death However, ozone is not suitable for the treatment of liquid food and products rich in unsaturated fats and soluble proteins, and, as a powerful oxidant, must be handled carefully to ensure the safety of employees and durability of equipment (Table 1) [3,64].

Cold (non-thermal) plasma is one of the more recent and promising food decontamination technologies. This method uses a partial ionized gas composed of positively and negatively charged ions, electrons and neutral atoms, molecules, and radicals to inactivate microorganisms in food and packaging materials, namely by membrane rupture, oxidation of proteins, and nucleic acids damage. Due to the low penetration levels, the inactivation of microorganisms only occurs on the surface of solid food (Table 1) [48]. However, Xiang et al. (2022) reported the efficiency of water and other plasma-activated liquids in disinfecting food products with complex matrices [65]. This method is suitable for the surface treatment of fresh meats and greens and for seafood preservation [3,40,48,65].

Natural Approaches

Lactic acid bacteria produce significant metabolites from which bacteriocins, antimicrobial peptides, are particularly important because of the applications that these peptides may have in the food area. Bacteriocins act mainly by forming pores in the cell membrane of bacteria, causing apoptosis [66,67]. Since nisin received GRAS status from FDA in 1988 [68], numerous bacteriocins have been discovered and reported on; some of these are also already commercially available [69,70]. Bacteriocins are used to preserve meat and dairy products and are also applied as a bioprotective technology in fruits and vegetables [44,66,71,72] to reduce the cell counts of problematic strains, such as *Salmonella enterica* subsp. *enterica* serovar Typhimurium, *E. coli*, and *L. monocytogenes* [42,66,73]. However, some challenges (Table 1), namely those associated with their limited range of activity, can impair their efficacy [42–44].

Essential oils are widely used as flavorings for food and beverages, but several articles have demonstrated their capability as preservatives in different types of food due to their ability to inhibit the growth of or to eliminate pathogens [45,74,75] such as *Salmonella* spp. [76–78], *E. coli* [76,77,79], and *L. monocytogenes* [76,77,80]. Salanță and Cropotova (2022) recently reviewed the current knowledge on the applicability of essential oils in food preservation, reporting that the levels of inactivation differ between the used oils [75]. Also, in some circumstances, to achieve the same *in vitro* inhibitory effect, a higher concentration is needed to treat fruits, leading to possible negative effects on sensory properties that may limit their application (Table 1) [74].

2.2.2. Physical Methods

Pulsed electric field is an environmentally friendly and highly energy-efficient technique that uses high-intensity short electrical pulses for food preservation with minimal loss in food quality and operates by two fundamental processes, namely electroporation and electrical breakdown [81]. This approach has been mostly applied for the treatment of orange, tomato, apple, and carrot juices [81]; apple sauce, pea soup; salad dressing; eggs [82]; and milk and milk products [83]. This technique increases the durability and safety of the products, retaining their sensory and nutritional characteristics. However, studies at the industrial scale are still very scarce [84]. The high start-up costs are one of the main barriers to the application of pulsed electric field on a large scale (Table 1) [40].

High-pressure processing is based on the treatment of liquid or solid foods at extreme pressures that can alter the anatomy of the bacterial cells as well as their enzymatic functions, resulting in their weakening and eventual death [3,40]. This procedure has been used commercially for several years, and its use is increasing continuously; it is employed in the treatment of foods whose water activity exceeds 0.8, such as fruits, meats, vegetables, milk and milk products, juices, beverages, seafood, and fish [3,40]. Despite the efficiency of this technology, and similarly to the pulsed electric field method, high-pressure processing also requires high initial costs associated with the establishment of appropriate facilities and the acquisition of the equipment (Table 1) [3,40]. In addition, this approach can influence sensory properties of foods [85].

Ultrasound is a safe, environmentally friendly, simple, and economically cheap (Table 1) technology that involves pressure waves with a frequency range of between 20 and 100 kHz, classified as minimum- or maximum-intensity sonication or, according to their frequency ranges, as power, high-frequency, or diagnostic ultrasound methods [86]. Many studies have demonstrated the potential of ultrasounds in eliminating microorganisms from meat, fruits, vegetables, dairy products, and eggs, with the preservation of physicochemical, bioactive, and nutritional components of the foods [3,40,46,49,86–90]. However, this technology is associated with negative effects on food characteristics, such as sensory parameters and nutrient composition (Table 1) [3,40,46,49].

Ultraviolet (UV) radiation is a non-ionizing radiation with germicidal properties at wavelengths in the range of 200–280 nm (UV-C). UV light can directly affect microbial pathogens by modifying their genetic material and/or damaging their proteins/lipids. Usually, different types of food products require specific doses of radiation to inactivate different types of pathogens. Apart from its lethality against all types of microorganisms, UV light also possesses other advantages, such as leaving no chemical residues, being easy to set up, and using low amounts of energy. Still, it also has certain drawbacks, for instance, it discolors some foods, only acts on the food's surface, and exposure is harmful to humans (Table 1) [91]. Several studies have examined the application of UV-C light in the food industry for the surface decontamination of fresh foods, increasing the shelf life of foods, the pasteurization of fruit juices, and drinking water disinfection [3,23,46,50].

Pulsed light rapidly inactivates pathogenic and food spoilage microorganisms through the application of a comprehensive array of short but highly energized pulses; the wide band of white light affects microbial structures and leads to their inactivation [92]. The potential of this method has already been reported for a variety of foods [51], such as fish [93], meat [94], fruits [95,96], vegetables [96,97], milk [98], fruit juices [92], and water [99]. Pulsed light has been shown as suitable for application in liquid and regularly shaped foods due to its low degree of penetration. Also, long treatment periods can result in a 'heating effect' in food products that can impact the effectiveness of microbial destruction (Table 1) [3,92].

Ionizing radiation involves the application of three major types of ionizing radiation: (i) gamma, (ii) electron beam, and (iii) X-rays. This technology inhibits the enzymatic activity of microorganisms through direct acid nucleic damage and the production of reactive molecules that can lead to metabolic pathway damage inside the cells, intracellular oxidation, and consequently, cell lysis [3,40,52,100]. Several studies showed that this approach is effective in eliminating microorganisms in fruits, vegetables [3,40], and fish and meat products [101], thus preserving the physicochemical properties of the foods (Table 1). However, in some cases, in order to achieve a substantial level of microorganism elimination, a higher dose of irradiation is required, leading to food deterioration and risking processors and workers [63]. The huge capital/investment requirements associated with the purchase of an ionizing radiation facility and concerns about consumer acceptance are further disadvantages of this technology (Table 1) [3].

2.2.3. Biological Methods

Bio-preservatives may originate from microorganisms, plants, and animals, being a potential alternative to the physical and chemical processing of foods [102]. Although bacteriocins and essential oils are considered natural chemical methods, as previously mentioned, they are also considered biological methods. Lactic acid bacteria, yeast, phages, bacteriocins, and endolysins are among the most common bio-preservatives effective in eliminating pathogens and preventing food spoilage [103]. These can be applied to food systems as additives, direct ingredients, or protective cultures. For instance, bacteriocins help in food bio-preservation, acting as antagonists, inhibitors, and antimicrobials against pathogens and spoilage microorganisms. In addition, they also provide various health benefits to humans, reinforcing the human immune system [104]. Yeasts are also known for their positive role in food fermentation, extending the shelf life of foods by creating an hostile environment to spoilage microorganisms [105].

Phages can be used as an eco-friendly approach to prevent and control pathogenic bacteria in food, with a substantial number of research reports describing the use of phages in various types of foods, ranging from meats [106–108] to fresh produce such as fruits [109,110] and vegetables [108–110], with promising results. Information regarding phages will be discussed in more detail in the following section (Section 3).

In addition to all the aforementioned food decontamination approaches, there is also the possibility of combining some of them, namely thermal and non-thermal methods, in order to take advantage of the benefits of each, resulting in a synergistic effect. Sometimes, these combinations can achieve more significant reductions in microorganisms than the individual treatments alone while retaining product quality. Information regarding these combinations can be seen in detail in different reviews [4,6,23,46,47,72,111].

3. Phages Against Foodborne Bacterial Pathogens

Phages were identified at the beginning of the twentieth century; however, a poor understanding of the nature of phage–host interactions and mechanisms underlying bacterial pathogenesis led to a succession of badly designed and executed experiments. Despite this, their potential as therapeutic antibacterial agents has been recognized. Phages were used to treat and prevent bacterial infectious diseases both in Eastern Europe and the former Soviet Union, but in the 1940s, with the advent of antibiotics, the application of phages as antibacterial agents diminished in much of the world [112]. However, the use of phage therapy has continued in regions of the former Soviet Union. Lately, with the emergence of pathogenic bacteria resistant to antibiotics, there has been a renewed interest in the use of phages as an alternative to conventional antimicrobials [12,54,113].

The differences in the infection process depend on the phage life cycle, which can be either lytic or lysogenic. In the cases of the lytic cycle (lytic phages), after the attachment and injection of phage genome into the host cell, the host DNA replication and protein synthesis machinery is exploited to produce new phage particles, which are released from the cell, culminating in the death of the bacterium. In the case of lysogenic cycle, the phages (designed as temperate phages) do not automatically start a lytic cycle, but instead the phage genome is integrated into the host cell genome (as a prophage) and may prevent the subsequent phage infection of the same host, thus delaying bacteriolysis. The induction, or excision of the prophage from the genome can occur spontaneously or in response to cellular internal or external triggers [114]. This results in the lysis and release of progeny phages that may progress and lyse or lysogenize other susceptible bacteria. The integration of the phage genome into the host genome may increase the virulence of the host. Temperate phages, in addition, may not cause the immediate death of the bacteria, but can be even favorable to bacterial evolution, particularly in regard to their pathogenicity and fitness. Through integration into the host genome, temperate phages can introduce new functional genes to the bacteria, such as virulence genes and antibiotic resistance genes, as reviewed by Chen et al. (2022) [115]. For these reasons, temperate phages should not be used in phage treatment unless they are genetically engineered.

Phage treatment uses lytic phages to destroy the host through a lytic life cycle [54,116–118].

Phages possess several benefits (Table 1) when compared to other post-harvest food treatment techniques that justify their use, including the following: (i) high specificity within a single genus or species or even within a subgroup of bacterial strains within a species, as opposed to some decontamination methods that indiscriminately kill bacteria that are naturally present and beneficial in food [119,120]; (ii) self-replication, which means that low or unique doses are enough for phage treatment as long as there are enough host cells to allow the phages to multiply [121]; (iii) no food safety issues related to the oral ingestion of phages in food, as phages are naturally present in high numbers in the environment and exist naturally on many food products and are thus often unintentionally ingested by humans [122]; (iv) phages do not affect the organoleptic properties, such as the sensory and quality characteristics, of food, unlike other post-harvest techniques such as thermal methods [12,122,123]; (v) self-limiting, since phages (in addition to being harmless) cannot persist for a long time in the environment without a host [124], unlike chemical compounds that have the capacity to leach into food produce and persist in their surroundings with a further risk of bacterial resistance; (vi) the purification, identification, and propagation of phages is relatively fast, simple, and production costs are considerably less expensive in comparison to other post-harvest decontamination methods [125–127]; and (vii) phages also have the ability to penetrate bacterial biofilms and infect cells contained within, which is unlike other decontamination methods that work effectively on planktonic cells [11,53–55].

Beyond the advantages, and like any other decontamination methods, phages also possess some disadvantages (Table 1): (i) phages may not completely kill or inhibit the pathogen and bacteria can naturally develop resistance to phages [128]; (ii) phages, mainly temperate phages, can transfer genetic material by horizontal gene transfer, which may increase the virulence of the host bacteria [129] and their antibiotic resistance [130]; (iii) phages are often exposed to, and inactivated by, certain factors like UV, pH, temperature, and osmotic pressure during different stages of food production, transportation, and storage [119,131–135]; (iv) the food matrix and the composition of the food surface (including food additives) can affect the efficacy of phages both by limiting phage diffusion or by specifically interfering with phages, thus reducing their encounter with host bacteria [123,136,137]; (v) release of pro-inflammatory compounds including endotoxins from lysed bacterial pathogens [121]; (vi) it is necessary to know which pathogen(s) are present to ensure that the correct phage or phage cocktail is applied [138]; (vii) the components of the phage product must have a host range broad enough to kill all members within the target pathogenic genus, species, or strains [9,11]; and (viii) in order to work, phages need to be applied in high enough quantities that allow phage particles to physically encounter all or most of the target bacterial cells, i.e., the initial phage titer applied to food samples needs to be high enough, and this is a determinant variable for the success of phage treatment [9,11].

Regarding the problem of phage-resistant bacteria, the combination of multiple phages (phage cocktails) [139,140] or the combination of phages with other antibacterial strategies in a multi-hurdle approach [141–143] may help to deal with the development of phage resistant-bacteria and thus inhibit pathogens, ensuring food safety. Komora and colleagues reported the synergistic effect of a multi-hurdle approach, combining mild high-pressure processing, phages, and bacteriocin against *L. monocytogenes* in milk [141] and in a fermented meat sausage model [142]. These two types of food matrices seem to be protective matrices that support phage application with mild high-pressure processing [143]. Promising results were also obtained with phages combined with essential oils and UV light against *L. monocytogenes* in beef, resulting in higher bacterial inactivation relative to each treatment individually [144].

Despite these strategies, bacterial resistance to phages is not as challenging as resistance towards antibiotics, since bacteria and phages co-evolve in the natural world, with phages evolving and adapting themselves to overcome bacterial resistance mechanisms [145]. Moreover, phage-resistant bacteria frequently exhibit a reduction in their fitness. Phage resistance frequently triggers several trade-offs in pathogenic bacteria, including a decrease in pathogenicity, a delay in bacterial growth, and even an increase in antibiotic susceptibility [115]. Phage engineering to increase bacterial host range is another strategy to deal with the emergence of phage-resistant bacteria [146].

Due to horizontal gene transfer, temperate phages must be avoided, and lytic phages must be characterized in terms of antimicrobial resistance, virulence, or integrase-coding genes to ensure their safety [9]. Despite these challenges, phage biocontrol is increasingly recognized as an attractive method to eliminate spoilage and pathogenic bacteria from food [42,117,122].

3.1. *Use of Phages in the Food Industry at the Post-Harvest Stage*

Since the regulatory acceptance of ListShield™ in 2006, the first phage-based product approved by the FDA for use in the control of *Listeria* in meat and poultry products, the number of new available phage-based products for foodborne pathogen control at the post-harvest stage has increased [147]. Phages can be used in the post-harvest stage directly into food, food processing equipment surfaces and incorporated into food packaging systems to control food contamination [9,147]; in this review, only the use of free phages directly in food or incorporated into packaging systems will be addressed (Figure 2).

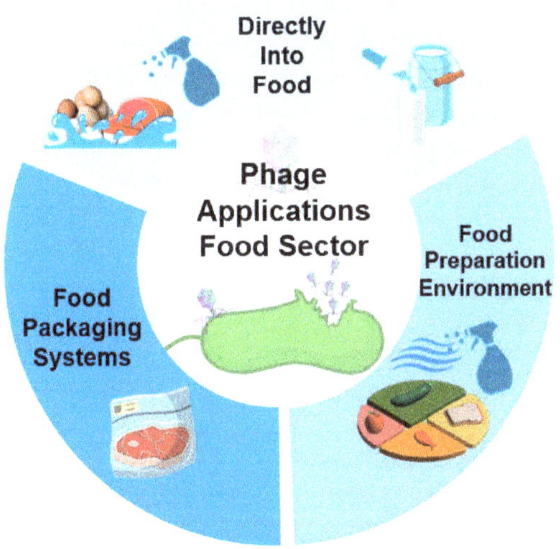

Figure 2. Phage applications in the food sector, highlighting the applications addressed in this review: direct phage application in food and their incorporation into food packaging systems.

3.1.1. Approved Phages for Direct Post-Harvest Applications in Food

Some private companies have demonstrated their intention to work on phage-based solutions to control or prevent foodborne pathogens in the food industry, but only a few products have been developed and are available in the market for direct post-harvest applications in food. (i) Micreos Food Safety (Wageningen, The Netherlands) developed a phage-based application to fight *L. monocytogenes* (PhageGuard Listex™) in RTE meat products; fish, fruits, vegetables, cheese, seafood, and pet food; *Salmonella* spp. (PhageGuard S™) in pork and poultry products, fruits, vegetables, and pet food; and *E. coli* 0157 (PhageGuard E™) in beef carcasses, subprimals, beef cuts, and trimmings intended for ground beef, poultry, fruits, vegetables, and pet food. (ii) Intralytix Inc. (Baltimore, MD, USA) has commercialize phage-based products for controlling *L. monocytogenes* (ListShield™) in RTE meat, fish (also smoked fish), and fresh and processed fruits and vegetables; *Salmonella enterica* (SalmoFresh™) in poultry, red meat, fish, shellfish, fresh and processed fruits, and vegetables; *Shigella* spp. (ShigaShield™) in cooked beef and chicken, smoked salmon, honeydew melon, lettuce, and yogurt; *E. coli* (EcoShield PX™) in meat, poultry, fruits, vegetables, dairy products, fish, and other seafood; *E. coli* O157:H7 (EcoShield™) in red meat parts and trim intended to be ground; and *Campylobacter* spp. (CampyShield^{TM}) in raw red meat (including whole carcasses, primals, subprimals, cuts, trimmings, and organs) and raw poultry. (iii) Phagelux Inc. (Montréal, QC, Canada) supply a phage cocktail called SalmoPro® for the biocontrol of *Salmonella enterica* on poultry, red meat, fresh and processed fruits, fresh and processed vegetables, eggs, fish and shellfish. (iv) FINK TEC GmbH (Hamm, Germany) developed a phage-based product called Secure Shield E1 for control of *E. coli* in beef carcasses [8,9,147,148].

3.1.2. Recent Studies on Direct Post-Harvest Phage Treatment in Food

Recent years have witnessed successful phage-based biocontrol measures against some of the most problematic foodborne bacteria. Some of the most recent studies on phage application (mainly by immersion, pipetting, and spraying) against *Salmonella* spp., *E. coli*, *L. monocytogenes*, and *Campylobacter* spp. using approved or new phages in different foods are discussed in the following sections.

Phages for Biocontrol of *Salmonella enterica*

Salmonella spp. are a common cause of foodborne disease outbreaks and are recognized as one of the top four global causes of diarrheal diseases [149]. Salmonellosis is the second most common zoonotic disease after campylobacteriosis in the European Union [2]. *Salmonella* colonizes a broad range of animals and can then be transmitted to humans through eating, in particular, through contaminated animal-based food. *Salmonella* species are most frequently found in meat, dairy products, and eggs [150,151] but also can also be detected in fruits, vegetables and cereals due to cross-contamination during harvest and post-harvest periods [152].

Of the most recent studies on the post-harvest application of phages in food, those concerning *Salmonella* spp., namely *Salmonella enterica* subsp. *enterica* serovar Enteritidis [153–157] and *S.* Typhimurium [154,158–161], are the most common.

Meat [153,154,158,160,162], vegetables [159,163–165], dairy products [153,163,165,166], and eggs [155,156,164,167] are the most discussed food products. Other food matrices, such as fruit or vegetable juices [153,168–170], fish [171,172], seafood [172,173], water [153,174], fruits [175,176], and beef broth [169] were also reported on, but in smaller numbers. In these studies, new or previous isolated phages were mostly applied as single phage treatments [153,155,158,159,166,177]. A smaller but still significant number of studies also examined the treatment of food products with phage cocktails [154,156,157,161,162,167]. A few studies also investigated commercially approved phage cocktails, including SalmoFresh^{TM} [178] and PhageGuard S^{TM} [179–181].

Recent studies reported a significant *Salmonella* reduction in different food matrices after phage application through pipetting [171], immersion [177], and spraying [154] with bacterial reductions ranging from 1.7 to 4.5 log CFU, depending on the incubation temperature, MOI, and food matrix. Information regarding the detailed results of these studies can be found in Supplementary Table S1.

Phages for the Biocontrol of *Escherichia coli*

E. coli is responsible for severe foodborne infectious diseases. In foodborne *E. coli*, STEC strains are the most reported in outbreaks, accounting for 20% of global foodborne illnesses [182]. Enterohemorrhagic *E. coli* (EHEC) strains, especially *E. coli* O157:H7 (a STEC strain), are major pathogens associated with a higher risk of severe bloody diarrhea and hemolytic uremic syndrome when compared with other pathotypes of *E. coli* [183]. Meat and dairy products are among the foods most frequently contaminated with this pathogen [184,185]. Nevertheless, fruits and vegetables have also been implicated in *E. coli*–related illness [5,186–188].

Of the studies on recent phage treatments against *E. coli* directly in food, those concerned with the *E. coli* O157:H7 strain were the most frequent [189–193]. Also, the majority of these studies were performed on meat [189,191,194–196] and vegetables [189,192,194,197–199], followed by dairy products [189,194,195,197,200]. The potential of phages against *E. coli* on chicken skin [197,201], fish [109,202], eggs [201], fruits [109], seafood [173], and cooked rice [194] was only addressed in a limited way. Most studies applied new or previously isolated phages separately [189,192,194,197,198,200] or combined in phage cocktails [190,195,199,203], with a reduced number of studies using approved phage products, namely PhageGuard S™ [191], PhageGuard E™ [108,204], and EcoShield™ [109,205].

Promising results on *E. coli* reduction in different types of food were obtained in recent studies after phage application by pipetting [189,206] and immersion [192]. In general, higher bacterial reduction was observed in liquid compared to solid matrices with reductions of 0.7 log CFU/piece in meat and 4.7 log CFU/mL in milk. For detailed results please consult Supplementary Table S1.

Phages for the Biocontrol of *Listeria monocytogenes*

Listeria monocytogenes is a foodborne pathogen responsible for causing listeriosis. Listeriosis is the third leading cause of death from foodborne illnesses, with about 1600 people suffering from listeriosis each year and about 260 dying [207]. Seafood, soft (raw milk) cheese, unpasteurized milk, and meat spreads are regarded as moderate- to high-risk foods for *L. monocytogenes* contamination in retail. Also, fresh or minimally processed fruits, such as apples and lemons, and vegetables are often associated with infection with this bacterium [208–210]. The ability of *L. monocytogenes* to replicate at refrigeration temperatures makes it one of the most important pathogens in RTE food [211,212].

Despite the reduced number of post-harvest studies using phages against *L. monocytogenes*, compared to *E. coli* and *Salmonella*, phage potential against this bacterium has been evaluated on different food matrices, including meat [142,213–218], vegetables [215,219–223], fish [224–226], dairy products [141,216,222,227,228], and fruit juice [226]. The majority of studies focus on the use of the approved phage product PhageGuard Listex™, containing a single *Listeria* phage P100 [141,142,216,217,220,221,223–225]. The approved phage cocktail ListShield™ was also applied in a few studies [218,227]. Other works report the use of new or previously isolated single phages [222,226,228] or phage cocktail [214,215,219].

Recent studies reported a significant decrease in *L. monocytogenes* levels after phage addition, by pipetting [227] and immersion [215,219], to different foods. Despite the differences in the tested matrices, in general, a maximum reduction of around 2 log CFU/g was obtained. The detailed results of these studies can be found in Supplementary Table S1.

Phages for Biocontrol of *Campylobacter* spp.

Campylobacter spp. infections are the most common bacterial cause of human gastroenteritis in the world [229]. Within the *Campylobacter* genus, the species *Campylobacter jejuni* and *Campylobacter coli* are considered the most important human pathogens, involved in most cases of campylobacteriosis. Campylobacteriosis was the most reported zoonosis in the European Union in 2021 with poultry meat and milk as the main sources of *Campylobacter* infection in humans [2]. Several control strategies for *Campylobacter* spp. have been developed with most of them focused on the reduction in Campylobacter colonization at the farm level. Strict biosecurity measures, good manufacturing practice, hazard analysis and critical control points, vaccines, phages, bacteriocins, probiotics, and phytochemicals are among these strategies [18,230–232].

From the beginning of 2018 until now, there have only been two available studies on phage application in food which are concerned with the use of a new single phage against

C. jejuni [233] and a previously isolated phage against *C. coli* [234], both at a post-harvest level on meat. The reduced number of *Campylobacter* studies is probably a result of the difficulty of working with *Campylobacter* phages [235].

Despite the reduced number of studies against *Campylobacter*, promising results were reported after phage application by spraying [233] and pipetting [234] in meat, with reductions between 1 and 2 log CFU/g. More detailed results can be found in Supplementary Table S1.

3.1.3. Factors Affecting the Effectiveness of Phage Treatment Directly in Food

Despite the numerous studies and promising results, there are some challenges associated with phage application in food. One of the main limitations is phage vulnerability towards various conditions [236,237], and it is difficult to maintain a constant phage titer due to the complexity of some foods [236,237]. Whether applied as a prophylactic or as a disinfectant, phage stability is very variable and depends on certain parameters, such as phage type, food composition, temperature, pH, and the presence of antiphage compounds [238]. Phages are often exposed to, and inactivated by, extreme environmental factors and physicochemical conditions in the food matrices. Extreme pH (common in fruits and fruit juice) [133], high temperature [133,239,240], salinity [133,240], UV irradiation [132,240], desiccation [132], and antiphage food compounds, such as milk caseins and organic acids and tannins from fruits and vegetables [236,237], have all been demonstrated to significantly reduce the concentration of viable phages and their activity. These factors can lead to damage in structural protein elements of a phage (capsid, sheath, tail) as well as lipid loss and/or can promote DNA or RNA structural changes [241]; therefore, these must be considered during food processing, transportation, and storage in order to maintain phage viability and ensure the effectiveness of the treatment [9].

The protein structure of phage particles determines its stability [242]. Tailed phages are known as the most stable in adverse conditions, and phages with larger capsids have higher survivability [133,241]. Temperature and pH are the main factors limiting phage activity [238]. Low and high pH values and high temperatures tend to inactivate phages [133].

Temperature is a fundamental factor for phage stability and activity [240,243–245], playing an important role in the attachment, invasion, and replication of phages [133]. At low temperatures, only a small number of phage particles inject their genetic material into the host cell and, thus, only a few are involved in the phage amplification stage [246]. On the other hand, higher temperatures can lead to the denaturation of proteins from the phage capsid [238,247]. Phages exhibit optimum propagation at temperatures of 37–40 °C [248]. However, several studies confirm that thermal stability is specific to each phage, and it is different depending on the phage isolate. Vörös et al. (2018) showed that high temperatures reduced the infection of the host by phage T7 since the high temperature caused the tail of the phage capsid to break [249]. Other studies suggested that high temperatures enhanced infection since they provided more energy to drive the genome into the host cells [250–252]. Ahmadi et al. (2017) reported that phage A511 was more stable when infecting *L. monocytogenes* at high temperatures (60–80 °C) than the phage P100 due to the slightly higher protein melting point of the tail capsid connector protein of phage A511 [239]. Additionally, some phages are insensitive to temperature. For example, in a study from Taj et al. (2014), the myovirus phage T4 effectively lysed *E. coli* BL21 at temperatures between 15 and 41 °C [253]. Also, the phage P100 in *L. monocytogenes* remained infectious throughout a range of 4–60 °C [131,239].

Regarding pH, in general, most phages are stable within a pH range of 4–11 [254] and their optimum pH conditions are around a neutral pH of 6–8 [133]. Several studies have suggested that the major limitation for the use of phages as natural preservative in fruits is the acidity of the environment (pH < 4) [143,255–257]. Oliveira et al. (2014) tested phage P100 stability on melon, pear, and apple products (juices and slices). Phage P100 was stable at 10 °C for 8 days in melons and pears (pH 4.61 to 5.92), but 2 and 7 log PFU reductions were observed in apple juice and apple slices (pH 3.70 to 3.76), respectively [257]. Also, Komora et al. (2018) showed a lower resistance to the pressure of phage P100 for lowest pH values. An accentuated phage titer reduction in phage P100 was observed in apple juice (pH 3.41) and orange/carrot nectar (pH 3.54) treated with high pressure, which may be related to the acidic pH values of these matrices compared to the other tested matrices, namely UHT whole milk (pH 6.73), "Serra da Estrela" cheese (pH 5.66), and "Alheira", a meat sausage (pH 6.07), which supported phage P100 application in high-pressure processing up to 300 Mpa [143].

Like any other viruses, most phages are also susceptible to UV radiation, which is responsible for intriguing adsorption flaws [258]. Since lethal UV radiation photoproducts are normally thymine dimers, DNA phages are usually more sensitive to UV damage than RNA phages [243] and, generally, phages with double-stranded genomes are more resistant to UV radiation than single-stranded ones [259–261]. UV radiation is more problematic in certain applications, such as crop treatment or aquaculture infection control, where phage inactivation by UV light significantly limits phage efficacy [124,240,262]. However, when directly treating foods at the post-harvest stage, food must also be protected from UV radiation to ensure the effectiveness of the treatment. In a study by J. Zhang et al. (2015), the direct exposure to UV light (302 and 365 nm) significantly reduced the concentration of ten Salmonella phages after 30 s, with no detectable viable phages after 900 s of UV light exposure (302 or 365 nm) [263]. In another study, Hudson et al. (2016) [264] reported the inactivation of *E. coli* O157:H7 NZRM 3614 by untreated and UV-treated phage FAHEc1 in milk and RTE meat. In milk, at 5 °C, a bacterial reduction of about 4.5 log by viable phages was observed after 35 days, whereas UV-treated phages led to a bacterial reduction of only about 1 log. In RTE meat, at 5 °C, higher bacterial reduction was obtained with viable phages (1.5 log) compared with the obtained reduction with UV-treated phages (0.5–1 log), after 24 h. At 24 °C, the reduction produced by the viable phages was once again higher (around 4.5 log) than that produced by UV-treated ones (3.5 log) after 24 h. The results confirmed the impact of UV on phage viability and, consequently, on bacterial reduction levels [264].

Some studies also reported the effect of desiccation on phage survival [132,265–267]. Iriarte et al. (2007) showed the *in vitro* effect of desiccation (at constant darkness, room temperature, and 0% relative humidity) on the survival of phage ΦXacm 2004-16 with reductions of 2.06 log PFU/mL after 60 days [132]. In another study, Rode et al. (2011) reported a 7 log reduction in phage Stx2 titer under desiccation on coupons of stainless steel at 20 °C [266]. More recently, Carrigy et al. (2019) observed phage CP30A titer reductions of 2.0 ± 0.1 log PFU/mL after air drying at room temperature for 48 h in a Petri dish [267].

In addition to environmental factors, other variables in particular the food matrix and the composition of the food surface can also affect the efficacy of phages [137]. The complex microstructures and matrices of food strongly affect the success of phage intervention, both by limiting their diffusion or by specifically interfering with phages, thus reducing their encounter with host bacteria [268]. In a study by Zinno et al. (2014), phage titers increased in all tested foods (whole milk, skimmed milk, energy drink, apple juice) except in liquid egg, in which a decrease probably occurred due to the highly viscous nature of the egg matrix, thus limiting diffusion and homogeneous phage distribution and leading to a decrease in phage numbers [269]. Similarly, in another study, Bao et al. (2015) tested a phage cocktail to reduce *S.* Enteritidis in milk, cabbage, and chicken breast and the highest effect was observed in milk, suggesting that the liquid allowed a greater diffusion of the phages [270]. This is reinforced by the lower phage titer and consequently lower MOI required for inactivation in liquids compared to solids, as observed by Thung et al. (2017), who used an MOI of 10^5 to reduce *S.* Typhimurium load in liquid egg and fruit juice, while a higher MOI of 10^7 was required to achieve a similar reduction in cooked beef and chicken [271].

Many food additives can also impair phage activity [143] with a consequent decrease in treatment effectiveness, namely dairy proteins, organic acids, fatty acids, tea infusions, phenolic compounds, retinoids, high ionic (NaCl), and sucrose concentrations [236,237,272–274]. Guenther et al. (2009) showed that lettuce and cabbage affected the stability of phage A511 by reducing the phage titer from 8 to 7 log PFU/g after 6 d at 6 °C, probably due to the presence of organic acids and tannins [268]. In another study, De Siqueira et al. (2006) found that tea infusions are also capable of reducing the viability of *S.* Typhimurium phages, Felix 01 and P22. Of the nine tea infusions tested, gold leaf tea had the highest antiphage activity with 7 to 8 log PFU/mL titer reductions for both phages [272]. García et al. (2009) also observed a phage titer decrease (about 1 log) throughout the incubation period in semi-skimmed and whole-fat raw milk assays. The authors suggested that fat globules and milk protein might promote phage particle aggregation and render them unable to infect the host under the assay conditions [273]. García-Anaya and colleagues (2019) described the interaction between phage A511 and the protein phases (whey and casein) in both unhomogenized and homogenized milk. The lowest phage affinity (0.06 to 0.48%) occurred in whey from both milk types. Nevertheless, in the casein phase, the affinity increased gradually with the degree of homogenization (from 8 to 20%). The authors proposed that this effect was associated with the exposure of both hydrophobic and hydrophilic sites in casein micelles during the homogenization, suggesting that phage–protein interactions and, consequently, phage inhibition can be modulated by alterations in milk proteins [275].

Different encapsulation strategies, such as extrusion, emulsification, and formation of liposomes, can help protect phages and maintain their viability in adverse environmental conditions, namely acidic environments and high temperatures, by the creation of phage microcapsules. Different polymers have been applied for phage encapsulation, with alginate as one of the most studied polymer materials in this context [276]. Several studies demonstrated the improvement in phage stability after encapsulation. The encapsulation of an *E. coli* phage (UFV-AREG1) in alginate, alginate–carrageenan, and alginate–whey protein spheres by extrusion was shown to increase its survival in simulated gastric fluid, especially in the alginate–whey–protein spheres [277]. Core–shell capsules encapsulating *E. coli* phage T3 by emulsification in a water-in-oil emulsion in the core with either alginate only or Eudragit S100 + alginate shell provided phage protection at pH 1 for 2 h. Also, encapsulated phages were significantly more stable than free phages upon exposure to temperatures up to 95 °C for 120 s [278]. *Salmonella* phages (UAB_Phi20, UAB_Phi78, and UAB_Phi87) encapsulated in liposomes were significantly more stable in simulated gastric fluid (pH 2.8), with losses of 3.7 to 5.4 log, compared to non-encapsulated phages that decreased by 5.7 to 7.8 log [279].

Another strategy of phage protection is its incorporation into active food packaging systems [8].

3.1.4. Phages in Active Food Packaging Systems

The application of phages in their free form directly into food sometimes requires their preservation for longer periods of time in extreme conditions [280], such as those highlighted in the previous section. In most cases, uncontrolled release and fast activity and stability loss during storage may impact the effectiveness of phage treatment in food.

Furthermore, the traditional methods of the application of free phages into food products (usually by immersion, pipetting, or spraying) [281] are often associated with dilution, requiring a high number/quantity of phages to be effective in order to suppress the bacterial growth [281,282].

In recent years, packaging, in addition to its main function of protecting food against external foreign environments and to increase ease of handling, has also started to play an active role in preventing food waste by actively extending its shelf life, thus ensuring quality and security [283]. Packaging systems can be classified into traditional or passive, active, intelligent, and smart [284]. The term "active packaging" is receiving significant attention, and it refers to packaging in which additional functions/ingredients have been purposely added to produce a more efficient packaging system, caring for the quality and safety of food and extending its shelf life, mainly by providing additional protection against the growth of microorganisms or the oxidation of a product [285].

The incorporation of phages as antimicrobial agents in active packaging systems could be explored in order to reduce the effects of external factors and food properties on phage stability and to promote the slower and continuous release of the phages into the food, enhancing food safety, extending food shelf life, and reducing phage waste that occurs with the traditional methods of phage application [8,281]. Moreover, it is expected that this approach will result in targeted and effective antibacterial actions due to phage specificity [11].

In addition, in order to minimize the environmental impact caused by traditional plastic packaging, fully biodegradable biopolymeric-based materials are gaining considerable relevance [286]. Polysaccharides and proteins are the most used biopolymers for food packaging development and production [287]. Water-soluble materials, such as some polysaccharides and proteins, are promising for phage incorporation, since, in general, the stability and viability of phages can be better maintained when they are stored in water-based solutions [288–290]. Plasticizers are commonly added to biopolymers in film-forming solutions in order to increase flexibility and improve the mechanical properties of films and coatings. Glycerol, polyethylene glycol, and sorbitol are among the most widely used plasticizers [291–293]. Glycerol can also act as a phage protectant [294].

Although both active coatings and films can control and prevent foodborne pathogens, these two terms have distinct meanings, as pointed out by García-Anaya et al. (2023a) [22]. Coating formulations are typically film-forming solutions that are applied directly to the food surface, usually by spraying, brushing, or dipping. After drying, they act as barriers that may be removed or consumed as a part of the food to provide protection, enhance the food's appearance, or provide specific properties, such as antibacterial properties when containing phages or other active compounds. Films are thin materials obtained also from a film-forming solution after drying. These pre-formed films are often used as packaging materials to wrap the food or as coatings for other materials [22,295]. Several systems for phage incorporation and delivery, namely biopolymeric edible films and coatings, have been reported [22,295]; these aim to deliver phages to a specific site and allow them to remain intact and viable with prolonged activity and controlled release into food.

Phages can be incorporated into packaging films or coatings through different methods, with each possessing associated advantages and disadvantages [22,295]. The most common methods used to protect phages against adverse environmental and food processing conditions and to ensure their stability when incorporated into different materials rely on some form of encapsulation or simple immobilization into the packaging material. In the encapsulation approach, phages are encapsulated by certain stabilizing agents that confer protection against the external environment and added to a specific packaging material, requiring their release from the capsules in order to contact the target bacterial cells [280,290,296]. Immobilization is based on trapping phages within a matrix. Due to these differences, a small fraction of immobilized phages can be exposed to the external environment, which does not happen with encapsulation [280,297]. Despite the implications of the stability of immobilized phages, immobilization also allows for more direct contact between the phage and the target bacteria [280].

To the best of our knowledge, Table 2 outlines all available studies on phage incorporation into packaging materials, edible coatings, and coated packaging materials against different foodborne bacteria. Table 2 summarizes information regarding the target bacteria, the applied phage(s), the material(s) used for phage incorporation, the tested food products, and the major findings, namely the bacterial reduction, phage behavior, and the effect of phage addition on the properties of the films/coatings. In these studies, the most frequently investigated bacterium is *E. coli* [265,289,294,298–309]. Meat was the most tested type of food [265,289,300–302,307,310–313]. Among all the materials used for phage incorporation, cellulose derivatives [265,289,304,306,314,315], alginate [288,298,302,303,313,315,316] and whey protein [307,308,310,315,317,318] were the most frequently reported.

Although this review considers several studies targeting different bacteria, the focus of this review is on the most important foodborne pathogens, namely *E. coli*, *Salmonella* spp., *L. monocytogenes*, and *Campylobacter* spp., and the following subsections emphasize on the main findings on these specific bacteria. To date, there are no studies available in the literature regarding the incorporation of *Campylobacter* phages into food packaging systems.

Incorporated Phages Targeting *Escherichia coli*

Several *in vitro* [294,298,303,304] and food studies [265,299–302,305–307,309] with incorporated phages targeting *E. coli* have been reported (Table 2).

The majority of studies aimed to combat the serotype *E. coli* O157 [265,294,299–302,307] and most experiments were performed on meat [265,300–302,307], followed by fruits and vegetables [289,299,302,308,309]. Some studies were performed in milk [306] and in medium containing food components [305].

In most studies, a single phage, namely a T-even type phage, was used [300,303–306,308,309], and only a few studies reported the use of phage cocktails [265,289,307].

Cellulose and cellulose derivatives, including modified cellulose membranes, cellulose diacetate, and filter paper functionalized with carboxyl methyl cellulose or chitosan [265,289,294,304,306] were the most explored materials for the incorporation of the phages. Other studies included phage combination with whey protein [307–309], polyethylene oxide [302–304], chitosan [299,301], indium tin oxide [305], polycaprolactone [300], and sodium alginate [298]. Some of these studies reported improved phage stability with the addition of maltose and starch [265] and trehalose [294]. Improved antibacterial activity was also observed in the presence of D-phenylalanine [302]. Higher phage stability and, consequently, higher bacterial reduction were also observed when phages were first encapsulated before their incorporation into different materials [301–303]. The infectivity of the incorporated phages was maintained for 1 week [299,306], 4 weeks [294,308], and 5 weeks [307] with higher stabilities at refrigerated temperatures compared to room and higher temperatures.

Table 2. Phage incorporation in active food packaging.

Target Bacteria (Study)	Phages	Material(s) and Type of Application	Food Products	Findings	
				Bacteria	Phages/Films/Coatings
Escherichia coli					
Escherichia coli O157:H7 (*amp::lux*) (C918) [265]	Cocktail of *E. coli* phages (EcoM-AG2, EcoM-AG3, and EcoM-AG10)	Modified cellulose membranes with polyvinylamine (positively charged cellulose membranes)	Raw beef	In vitro (bioluminescent assays at 25 °C): At higher phage concentrations (10^9, 10^7, and 10^5 PFU/mL)-similar results obtained with modified and unmodified cellulose membranes with almost complete inhibition; At lower phage concentrations (10^3): • Better results obtained with modified cellulose membranes—after 8 h, the bioluminescence diminished until it reached approximately the same values obtained with higher phage concentrations; • Unmodified membranes—bioluminescence patterns are very similar to that of the control.In food: a reduction below the detection limit (aerobic conditions) with at least a 2 log CFU/g reduction at 4 °C for 15 days.	Phage-treated positively charged cellulose membranes showed higher quantities of phage plates than those with the phage-treated unmodified membranes. Drying tests: • Desiccated phages irreversibly lost their activity; • The siphovirus phage was significantly more tolerant to the effect of air drying than myovirus phages; • The addition of maltose or starch significantly improved the tolerance of the T4 phage to air drying; • Freeze-drying was the most effective method to dry phages with little decrease in phage activity.

Table 2. *Cont.*

Target Bacteria (Study)	Phages	Material(s) and Type of Application	Food Products	Findings	
				Bacteria	Phages/Films/Coatings
Escherichia coli ATCC 11303 [303]	Wild type T4 phage	Core (phage suspension)/shell (poly(ethylene oxide)) fibers	N/A	N/A	Better results obtained with coaxial electrospinning (relatively to simple and emulsion electrospinning)—the incorporated T4 phage totally maintained its activity after several weeks at 4 °C.
Escherichia coli [304]	Wild type T4 phage	Core (phage suspension)/shell electrospun fibers made from poly(ethylene oxide), cellulose diacetate, and their blends	N/A	N/A	The phage release rate was dependent on cellulose diacetate/poly(ethylene oxide) ratio and the poly(ethylene oxide) molecular weight. Increasing both parameters resulted in slower phage release.
Escherichia coli BL21 [308]	Coliphage T4	Edible whey protein isolate films	Lettuce leaves (release tests)	In vitro: ~5 log difference between the bacterial control and samples containing phage and ~2 log decrease in the samples from the initial inoculation, after 24 h at 22 °C.	The phages • Were stable in ambient (22 °C and light) and refrigerated (4 °C and dark) conditions without significant loss in infectivity over a period of 1 month; • Were released to a significant degree in an aqueous environment and on a lettuce leaf surface within 3 h; • Showed an homogenous distribution within the film matrix.

Table 2. Cont.

Target Bacteria (Study)	Phages	Material(s) and Type of Application	Food Products	Findings	
				Bacteria	Phages/Films/Coatings
Escherichia coli O104:H4, C1321 [289]	*E. coli* phage cocktail (EcoM-HG2, EcoM-HG7, and EcoM-HG8)	Paper coated with encapsulated phage in alginate beads or Paper impregnated with phage suspension	Alfalfa seed and sprout	At room temperature: • Reduction below the detection limit (~5 log CFU/g) after 1 h in the phage-treated (free, encapsulated, or impregnated phage) seeds relative to controls; • After 5 days, a reduction in sprouts treated with paper and impregnated with phage or with phage microcapsules of 0.6 and 1.3 log CFU/g, respectively, were observed, in comparison with the free phage results (1.5 log CFU/g).	• The phage containing alginate beads (0.5 g containing ~1.22 × 10^{10} PFU) showed a phage release of around 10^7 PFU/mL and around 10^6 PFU/mL, after 1 day at 4 °C or 25 °C, respectively. Similar results were obtained after 7 days at 4 °C or 25 °C; • No significant difference in antibacterial effect between free and immobilized phages on seeds but on sprouts free phage showed greater results, followed by paper coated with phage microcapsules and paper impregnated with phage.

Table 2. *Cont.*

Target Bacteria (Study)	Phages	Material(s) and Type of Application	Food Products	Findings	
				Bacteria	Phages/Films/Coatings
Escherichia coli (EHEC O157:H7 CICC 21530) [301]	*E. coli* O157:H7 phage	Chitosan film containing liposome-encapsulated phage	Beef	Model beef suspension (room temperature with shaking): The antibacterial activity of the chitosan film containing liposome-encapsulated phages was positively associated with its concentration—better results for 400 mg/mL (around 5 log reduction after 7 days in comparison with the bacterial control) were obtained. In food (at 25 °C): The chitosan film containing liposome-encapsulated phage led to a 4.45 log reduction while the chitosan film led to a 0.29 log reduction after 15 days.	• The encapsulation efficiency of phages in the liposome was 57.66 ± 0.12%; • The chitosan film with a volume ratio of 6:4 (liposome: chitosan) was the most suitable; • Phages without liposome protection were unstable and inactive within a short period of time; • After liposome encapsulation, the phage inhibitory effect continued for 15 days; • No impact of chitosan film embedded with liposome-encapsulated phage on the sensory properties of beef.
Escherichia coli O157:H7 [294]	*E. coli* AG10 phages	Pullulan–trehalose films	N/A	N/A	Phage infectivity was maintained for up to 1 month at ambient storage conditions (reduction of 1.90 log PFU/film).

Table 2. *Cont.*

Target Bacteria (Study)	Phages	Material(s) and Type of Application	Food Products	Findings	
				Bacteria	Phages/Films/Coatings
Escherichia coli K12 [306]	T4, T5, and T7 phages	Filter paper with phage addition after functionalization with carboxyl methyl cellulose or chitosan	Milk	At 37 °C: All papers with the T4 phage were able to remove *E. coli* from milk; • Both functionalized papers removed all *E. coli* (9.3 log CFU/mL) in less than 1 h, while non-functionalized papers only reduced *E. coli* by about 4 log with a total reduction only after 3 h.	All papers extended the lifetime of the infective phage by at least a factor of four, with some papers stabilizing phages for up to one week at 37 °C.
Escherichia coli O157:H7 CECT 4075 [299]	vB_EcoMH2W phage	Chitosan-based edible coating	Tomato	A ~3 log bacterial reduction in the samples relative to the controls on tomato surfaces after 6 days at 20 °C.	Phage infectivity was maintained in the film that was applied on the surface of tomatoes for at least 6 days in the presence (higher titer) or absence of *E. coli* at 20 °C.
Escherichia coli BL21 [309]	T7 phage	Edible whey protein isolate-based coating	Apples Cherry tomatoes Cucumbers	After 24 h at 4 °C: A ~2 and 4 log bacterial reduction on cut apples and whole cherry tomatoes, respectively, was observed, while no significant reduction was seen for sliced cucumbers.	Films enhanced phage stability during cold storage (4 °C).

Table 2. *Cont.*

Target Bacteria (Study)	Phages	Material(s) and Type of Application	Food Products	Findings	
				Bacteria	**Phages/Films/Coatings**
Escherichia coli ATCC 11303 [305]	Phage T4 (ATCC 11303-B4)	Phage-conjugated indium tin oxide systems	Food components: starch and casein	In vitro, at 37 °C, after 2 h incubation: • Upon the 2 h eradication of the '1st batch' (about 4 log reduction), the '2nd batch' of *E. coli* concentration was reduced by around 5 log in just 30 min by all of the indium tin oxide/T4 systems at pH 7; • Around 4 log *E. coli* reductionfor all of the indium tin oxide/T4 surfaces at a pH of 7 and 8, after 2 h of incubation; At pH 5, generally no *E. coli* reduction was seen after 2 h of incubation. In vitro (in the presence of food components at pH 7): A 3–4 log bacterial reduction after 2 h was seen, similar to that observed in only medium.	All the indium tin oxide/T4 systems maintained their antimicrobial activity in the presence of model food components (starch and casein), but this activity was still affected by pH.

Table 2. *Cont.*

Target Bacteria (Study)	Phages	Material(s) and Type of Application	Food Products	Findings	
				Bacteria	Phages/Films/Coatings
Escherichia coli (DH5α and O157:H7 STEC strains) [307]	Cocktail of T-even type phages (DT1 to DT6)	Whey protein concentrate films	Meat	In vitro (growth inhibition assay): • At 4 and 37 °C— A reduction to non-detectable levels for DH5α and O157:H7 STEC strains, after 24 h was seen; • At 24 °C—both strains grew; however, the extent of growth was lesser than in the controls.In vitro (inhibition zone assay): At 4 °C, a 1.5 ± 0.1 mm of zone of inhibition for the films with added phages compared to film disk without phages (no inhibition) after 1 week = 1 month. In food: Total elimination (~2 log) of both DH5α and STEC strains of *E. coli* in the meat stored for 24 h at 4 °C and 1 h at 37 °C.	Phage cocktail embedded in the films were stable within 5 weeks of storage (better results at 4 °C than at 24 °C).
Escherichia coli CECT 434 [298]	*Salmonella* phage φ135 and *E. coli* phage vB_EcoS-EC4 (EC4)	Phages and cinnamaldehyde incorporated on sodium alginate emulsion-based films	N/A	In vitro: Better results with a combination of both phages at a higher concentration of cinnamaldehyde (0.4%), with a reduction until reaching the detection limit of the method (~7 log) after 24 h at 20 °C.	The incorporation of phages into the film did not introduce significant changes to the film characteristics, unlike cinnamaldehyde, which increased the roughness, thickness, and swelling ability of films.

Table 2. *Cont.*

Target Bacteria (Study)	Phages	Material(s) and Type of Application	Food Products	Findings	
				Bacteria	Phages/Films/Coatings
Escherichia coli K12 (ATCC 23724) and O157:H7 H1730 [300]	Phage T4 (ATCC 11303)	Polycaprolactone films (phage addition by physical adsorption or chemical functionalization)	Raw beef	*In vitro E. coli* K12 reduction with chemically functionalized film compared to films with no phages: • 37 °C—maximum reduction between 3 and 5 h to about 7 log and 2.44 log after 120 h; • 20 °C—around 6 log maximum bacterial reduction after 24 h and around 2 log after 120 h; • 10 °C—maximum decrease of around 3 log after 120 h; 4 °C—no growth of *E. coli* and no antibacterial activity in the samples treated with films was observed.In food: chemically functionalized film reduced *E. coli* O157:H7 to ~2 log after 120 h at 10 °C.	Chemically functionalized film showed more antibacterial efficacy than the physically adsorbed film.

Table 2. *Cont.*

Target Bacteria (Study)	Phages	Material(s) and Type of Application	Food Products	Findings: Bacteria	Findings: Phages/Films/Coatings
Escherichia coli O157:H7 (EHEC O157:H7 CICC 21530) [302]	*E. coli* O157 phage	Co-encapsulation of phages and D-phenylalanine into sodium alginate/polyethylene oxide nanofibers based-films	Beef Cherry tomatoes Cucumber	In vitro (37 °C): • There was a decrease of about 3 log in planktonic *E. coli* after 8 h; • An around 7 log reduction in the *E. coli* biofilm on stainless steel after 72 h was seen. In food (after 4 days): Beef—reduction of ~2 log of planktonic *E. coli* at both 4 and 25 °C; Cucumber—reduction of ~3 and 2 log of *E. coli* biofilm at 4 and 25 °C, respectively; Cherry tomatoes—reduction of ~2 and 1.5 log on the *E. coli* biofilm at 4 and 25 °C, respectively.	• The films showed good mechanical properties and thermal stability; • Phages were able to remain viable in nanofiber films with high release rate; • The addition of D-phenylalanine significantly enhanced the antimicrobial activity of the nanofiber films.
Salmonella spp.					

Table 2. *Cont.*

Target Bacteria (Study)	Phages	Material(s) and Type of Application	Food Products	Findings	
				Bacteria	Phages/Films/Coatings
Salmonella enterica ser. Typhimurium [314]	Phage cocktail (BFSE16, BFSE18, PaDTA1, PaDTA9, PaDTA10, and PaDTA11)	Cellulose acetate films	N/A	In vitro (diffusion in liquid medium): Increase in the lag phase and slower growth of microorganisms in the sample containing incorporated phages in the films, compared to control, at 150 rpm at 35 ± 2 °C. In vitro (diffusion in solid medium): Halos of films containing phages of 1.23–1.35 cm, larger than that of the pure cellulose acetate film (1 cm), at 35 ± 2 °C for 24 h.	• The mechanical and physical properties of films (thickness, elongation, and puncture resistance) showed no significant differences after phage addition; • The addition of phage altered the film surface, reducing the transparency, tensile resistance, and the modulus of elasticity of the film and increased the porosity; • Phages remained viable for 14 days at 23 ± 2 °C and at a relative humidity 50 ± 10%, and were no longer detected after that time.
Salmonella sp. cocktail (*S.* Typhimurium DT104, 19485A96 SGI1 and ATCC 13311, *S.* Heidelberg ATCC 8326, *S.* Enteritidis ATCC 4931 and *S.* Newport ATCC 6962) [312]	*Salmonella* phage Felix O1	Poly(lactic acid) films with a xanthan coating containing phage	Sliced turkey	In vitro (microtiter plate assay): At 37 °C, *S.* Typhimurium DT104 cultures showed significantly slower growth and lower final bacterial density compared to the control, while at 25 °C, *S.* Typhimurium was reduced in a similar way in the presence and absence of phages over a period of 20 h. In food: Better results under anaerobic packaging with reductions in Salmonella sp. cocktail in about 0.832 and 1.30 log at 4 °C and 10 °C after 30 days.	99.9% of phage released within 30 min into meat.

Table 2. *Cont.*

Target Bacteria (Study)	Phages	Material(s) and Type of Application	Food Products	Findings	
				Bacteria	Phages/Films/Coatings
Salmonella Newport [294]	*Salmonella* CG4-1	Paper coated with pullulan–trehalose mixture containing phages or Pullulan–trehalose films incorporating phages	N/A	In vitro, a 4.59 log CFU/cm^2 bacterial reduction at ambient conditions (~22–25 °C) after 1 month was observed with paper coated with a pullulan–trehalose mixture and containing phages.	Phage infectivity was maintained for up 2 months in the pullulan–trehalose films in ambient storage conditions.
Salmonella Enteritidis EX2 [298]	*Salmonella* phage φ135 and *E. coli* phage vB_EcoS-EC4 (EC4)	Phages and cinnamaldehyde incorporated on sodium alginate emulsion-based films	N/A	In vitro, better results with a combination of both phages at the higher concentration of cinnamaldehyde (0.4%) were obtained, with reduction until the detection limit of the method (~8 log) was reached after 24 h at 20 °C.	The incorporation of phages into the film did not introduce significant changes in its characteristics, unlike cinnamaldehyde, which increased the roughness, thickness, and swelling ability of films.
Salmonella Enteritidis (H40499 SDE) [319]	*Salmonella* Enteritidis phage Felix O1	Phage incorporation into polyvinyl alcohol coatings and fibers deposited by casting and electrospinning on polyhydroxybutyrate/poly-hydroxyvalerate films	N/A	N/A	• Polyvinyl alcohol increased the moisture content, the solubility, and the hydrophilicity of the films; • Phages were successfully incorporated and remained viable (10^6 PFU/mL) after the formation of the coating and nanofibers.

Table 2. *Cont.*

Target Bacteria (Study)	Phages	Material(s) and Type of Application	Food Products	Findings	
				Bacteria	Phages/Films/Coatings
Salmonella mixture [315]	Phage cocktail: *S.* Enteritidis F5–4, *S.* Typhimurium L2–1, and *S.* Typhimurium ICB1–1	Phage-based edible coatings of whey protein concentrate, carboxymethyl cellulose, chitosan, or sodium alginate	Strawberries	The largest antimicrobial effect was observed with a whey protein concentrate coating, with a reduction of 3.1 log CFU/g after 5 days at 4 °C, compared to the other tested polymers.	For a period of 5 days at 4 °C: • Whey protein concentrate coating showedthe smallest escalation in pH and the smallest decrease in the titratable acidity of dip-coated strawberries; During storage, a 0.7 log-unit (PFU/g) reduction in whey protein concentrate coating was observed compared to the phage reduction observed with other polymers (around 1.0).
Salmonella Enteritidis (ATCC 13076) [320]	*Salmonella* Enteritidis phage PBSE191 (BP-1370)	Polyvinyl alcohol film	Chicken eggshell	In vitro: Significant bacterial reduction (2×10^5 CFU/film within 2 h) in the phage-containing films compared to the bacterial control-containing film without phages, at 37 °C. On chicken eggshell surface: An about 2 log CFU reduction within 24 h at 5 °C and 50% relative humidity was observed.	Phages remained stable and were effectively released from the polyvinyl alcohol film without any substantial loss in the phage titer at 5 °C and at 50% relative humidity for 30 days.

Table 2. *Cont.*

Target Bacteria (Study)	Phages	Material(s) and Type of Application	Food Products	Findings	
				Bacteria	Phages/Films/Coatings
Salmonella Enteritidis (ATCC 13076) [311]	*Salmonella* Enteritidis phage PBSE191 (BP-1370)	κ-carrageenan (KC) and konjac glucomannan (KGM) hydrogel film containing adsorbed phages	Raw chicken meat	Around 1 log CFU/mL reduction within 3 h at 25 °C and within 48 h at 5 °C.	• KC/KGM-based hydrogel with a ratio of 7:3 showed the highest compressive strength; • 40% sorbitol containing KC/KGM hydrogel film showed the highest tensile strength; • Phage adsorption significantly increased the tensile strength and water swelling ratio and decreased water solubility and water vapor permeability; • Phage adsorption did not change the chemical structure of the films; • The phage-adsorbed film had an overall smooth and layered morphology.
Salmonella enterica CCCD-S004 [316]	Phage cocktail (SentS01L and SentS01T phages)	Sodium alginate edible coating	Ripened cheese	Coating containing phage cocktail was removed from a cheese matrix sample and positioned in the center of a bacterial lawn of *Salmonella enterica* CCCD-S004. After incubation (24 h at 37 °C), the presence of clear zones of lysis in the bacterial lawn surrounding the coating was observed.	The antibacterial coating showed adequate physicochemical characteristics and zero cytotoxicity in HaCaT and 3T3 cell lines.

Table 2. *Cont.*

Target Bacteria (Study)	Phages	Material(s) and Type of Application	Food Products	Findings	
				Bacteria	Phages/Films/Coatings
Salmonella Enteritidis (MET S1–001) [313]	Phage MET P1-001_43	Layer-by-layer assembly of chitosan/alginate thin films incorporating phages	Chicken meat	Wrapping a *S.* Enteritidis-contaminated chicken piece with aluminum foil whose surface was modified with phage-loaded chitosan/alginate multilayers reduced bacterial numbers in more than 1.0 log CFU/cm^2. Similar results were obtained with free phages.	• The phage-loaded chitosan/alginate multilayers showed antibacterial activity at pH 7, but not in acidic conditions; • Surface roughness decreased distinctly upon treatment of multilayers with phage-containing NaCl solution.
			Listeria monocytogenes		
Listeria monocytogenes C391 [265]	Cocktail of *Listeria* phages (LinM-AG8, LmoM-AG13, and LmoM-AG20)	Modified cellulose membranes with polyvinylamine (positively charged cellulose membranes)	RTE oven-roasted turkey breast	Better results obtained at 4 °C and in vacuum conditions: *Listeria* reduction of 4 log CFU/g after 15 days.	Immobilized phages were able to control the growth of *L. monocytogenes* in meat incubated at different temperatures and under different packaging conditions.

Table 2. Cont.

Target Bacteria (Study)	Phages	Material(s) and Type of Application	Food Products	Findings	
				Bacteria	Phages/Films/Coatings
Listeria monocytogenes cocktail (Li0512, Li0529, ATCC19115 and 08–5578, serotype 1/2b,1/2a, 4b and 1/2a, respectively) [289]	LISTEX™ P100 phage (RTE turkey) or *L. monocytogenes* phage cocktail (LinM-AG8, LmoM-AG13, and LmoM-AG20) (freshly cut cantaloupe)	Modified cellulose membranes with polyvinylamine (positively charged cellulose membranes)	RTE turkey Fresh cut cantaloupe	RTE turkey: Significant *L. monocytogenes* reduction at 4 °C and 10 °C, with around 1 log and 2 log CFU/cm^2 reduction, respectively, after 5 days. Fresh cut cantaloupe: Reduction of ~1 log at the end of different storage conditions (4 and 12 °C for 5 days, and 25 °C for 24 h) in the samples containing the immobilized phage compared to controls; • Better results were obtained using free phages with reductions of 3, 4, and 3 log at 4 °C (after 5 days), 12 °C (after 5 days), and 25 °C (after 24 h), respectively.	RTE turkey: No significant difference in antibacterial effect between free and immobilized phages. Fresh cut cantaloupe: Spray-coating free phages showed greater antibacterial effect than immobilized phages.
Listeria monocytogenes cocktail (FSL F6-367, ATCC 19115, 08e5578, C6-0003, LI 0512) [312]	*Listeria* phage A511	Poly(lactic acid) films with a xanthan coating containing phages	Sliced turkey	In vitro (microtiter plate assay): At 37 °C, a significant decrease in final cell density in the grown cultures of *L. monocytogenes* ATCC 19115 in the presence of the films containing phage, relative to the controls, after 3 h at 25 °C, was observed, with a significant reduction only seen after 21 h. In food: Reductions in *L. monocytogenes* cocktail of 6.31 log at 4 °C and 1.52 log at 10 °C, in anaerobic packaging after 30 days, and reductions of 3.79 log at 4 °C and 2.17 log at 10 °C, after 14 days, in aerobic packaging.	99.9% of both phages were successfully released from the film within 30 min onto the meat samples.

Table 2. *Cont.*

Target Bacteria (Study)	Phages	Material(s) and Type of Application	Food Products	Findings	
				Bacteria	Phages/Films/Coatings
Listeria monocytogenes serotype 1/2a [294]	LISTEX P100	Paper coated with pullulan–trehalose mixture containing phages or Pullulan–trehalose films incorporating phages	N/A	In vitro, more than 2 log CFU/cm^2 reduction at ambient conditions (~22–25 °C) after 6 weeks, with paper coated with pullulan–trehalose mixture containing phages was observed.	Phage infectivity was maintained for up to 2 months in the pullulan–trehalose films at ambient storage conditions.
Listeria monocytogenes serotype 1/2a [321]	LISTEX P100	Phage incorporated into pullulan–trehalose films	N/A	N/A	• Vacuum drying and storing in enclosed containers enhanced the long-term viability of the phage by over 1000-fold; • Main cause of titer reduction in the film was exposure to high humidity.
Listeria monocytogenes (CECT 934, ATCC 19114) [288]	LISTEX™ P100 phage	Films of (1) sodium caseinate, (2) sodium alginate mixed with gelatine, and (3) polyvinyl alcohol	N/A	• All films showed *in vitro* antimicrobial capacity of close to 1 log after 24 h at 30 °C; • The effectiveness of polyvinyl alcohol films was greater at 8 °C, reaching a 2 log reduction after 8 days.	• The incorporation of phages did not alter the morphology, color, opacity, or thermal stability of the films; • Better antibacterial results were obtained with free phage compared to incorporated phages.

Table 2. *Cont.*

Target Bacteria (Study)	Phages	Material(s) and Type of Application	Food Products	Findings	
				Bacteria	Phages/Films/Coatings
Listeria monocytogenes (ATCC 19113) [318]	*Listeria* phage A511	Whey protein concentrate (WPC), pullulan (PULL) films, and their blends	N/A	In vitro: • Inhibition assay by disk diffusion—30WPC:70PULL showed a significant reduction in inhibitory effect after 60 days compared to 50WPC:50PULL and 70WPC:30PULL, which completely lost their inhibitory effects after 60 and 40 days, respectively, at 30 °C; • Growth inhibition assay with 30WPC:70PULL film—4–5 log of *L. monocytogenes* reduction after 24 h at 25 °C.	• Pure films of WPC and PULL were not successful in phage stabilization; • Higher phage recovery in the 30WPC:70PULL film; • Phage distribution in the films was uniform and resulted in enhanced opacity; • In the 30WPC:70PULL film, the elastic modulus and tensile strength were enhanced while the elongation at break diminished after 60 days; • The film 30WPC:70PULL had the best physical, mechanical, and anti-*Listeria* performance during storage (60 days at 25 °C and at a relative humidity of 53%).

Table 2. *Cont.*

Target Bacteria (Study)	Phages	Material(s) and Type of Application	Food Products	Findings: Bacteria	Findings: Phages/Films/Coatings
Listeria monocytogenes (ATCC 19113) [310]	*Listeria* phage A511	Bilayer films of whey protein concentrate/pullulan (WP) containing phage and poly (lactic acid) (PLA)	Chicken breast	In vitro: Disk diffusion assay—The diameter of the inhibition zone in the bilayer and monolayer films were similar and did not differ significantly at 30 °C; • Growth inhibition—The growth of *L. monocytogenes* in the phage-containing film treatment was significantly lower than that in the phage-free film treatment (around 5 log reduction), after 24 h at 25 °C. In food: the phage-containing films inhibited *Listeria* in chicken breast filets in a similar way to free phages with at least a 1.5 log CFU/cm^2 reduction at both 4 and 10 °C after 120 h compared to the control-containing film without phages.	• The addition of 30% thickness ratios of PLA to WP film enhanced the mechanical, barrier, and visual properties of bilayer film; • Among the bilayer films, 30PLA/70WP was the best film, with the highest phage recovery and phage stability; • This film showed a shelf life of 20 days at 25 °C and 50% relative humidity.

Table 2. *Cont.*

Target Bacteria (Study)	Phages	Material(s) and Type of Application	Food Products	Findings	
				Bacteria	Phages/Films/Coatings
Listeria monocytogenes [317]	ATCC A511 phage	Whey protein isolate (WPI)-based coating	Cheese	At the end of storage period (16 days at 4 °C), phages added to water or to WPI reduced bacterial counts in 0.39 and 0.86 log CFU/g, respectively.	• Phages remained stable in the WPI coating for 16 days at 4 °, and had a similar stability to phages in buffer; • A511 phage remained stable on the surface of WPI dip-coated cheese in the presence of *L. monocytogenes*; • In the absence of *L. monocytogenes*, the phage concentration was reduced by 0.8 log PFU/g; • The phage–WPI coating had a significant effect on the color, hardness, and springiness of cheese.
Other bacteria					

Table 2. *Cont.*

Target Bacteria (Study)	Phages	Material(s) and Type of Application	Food Products	Findings	
				Bacteria	Phages/Films/Coatings
Vibrio parahaemolyticus ATCC 17802 [322]	Phage isolated from raw bonito fishes (*Sarda sarda*)	Edible methylcellulose films coated with capsules of sodium alginate containing phages	Raw fish filets	In vitro (after 24 h at room temperature): • 1.27 and 3.99 log of bacterial reduction in the treated samples with film containing phages compared to the initial bacterial inoculum and the bacterial control, respectively. In food (after 14 days at 4 °C): • A 2.65 and 6.46 log reduction in bacteria in the treated samples with films containing phages compared to the initial bacterial inoculum and the bacterial control, respectively.	Phage stability: • In both tested conditions (darkness at 4 °C and illuminated room at 22 °C), the most significant decrease in the encapsulated phage stability was observed on the 7th and 14th days; • Lower phage stability at 22 °C in a lit environment with a decrease of 4 log after 14 days. • Phage release from the film into water: 1.5×10^5 PFU/mL during the first 30 min and in total 1.6×10^7 PFU/mL after 5 h of incubation.
Pseudomonas fluorescens PF7A [323]	Phage ϕIBB-PF7A	Sodium alginate-based films	Chicken breast filet	In vitro (at 4 °C): A significant decrease in *P. fluorescens* growth in films that incorporated phages (1.9 and 4 log reductions after 24 h) when in direct contact with *P. fluorescens* or when completely immersed on a bacterial suspension, respectively. In food (at 4 °C): Films incorporating phages reduced *P. fluorescens* counts by 2 log after 2 days and maintained a significant reduction for the next 5 days (1 log).	• Phages were homogeneously distributed inside the films; • A decrease in phage viability by ~2.4 log was detected after 8 weeks at 4 °C, while phages were inactivated after 15 s and 30 s of exposure at 77.5 and 67.5 °C, respectively.

Table 2. *Cont.*

Target Bacteria (Study)	Phages	Material(s) and Type of Application	Food Products	Findings	
				Bacteria	Phages/Films/Coatings
Clavibacter michiganensis subsp. *nebraskensis* (Cmn-91R) [324]	Phage CN8	Coatings based on phages, polymers (polyvinylpyrrolidone, polyvinyl alcohol, or poly(methyl vinyl ether)), and stabilizers (whey protein isolate, skim milk, sucrose, maltodextrin, or D-mannitol)	Maize seeds	Coating of polyvinyl alcohol combined with whey protein isolate containing phage significantly reduced target bacterial cells up to 3.8×10^3 CFU/seed after four months at 10 °C, without affecting seed germination.	• Polyvinyl alcohol offered the greatest stability for CN8 phages on seeds when coatings did not contain a stabilizer; • Polyvinyl alcohol combined with whey protein isolate stabilizer maintained CN8 phage activity for seven months and four months at 10 and 26 °C, respectively.
Staphylococcus aureus IPLA1 [325]	Phage phiIPLA-RODI	Edible gelatine films/coatings	Cottage cheese	In vitro: Reductions of 5 and 7 log for the films with the lowest and the highest phage concentrations, respectively, after 17 h at 37 °C and at 250 rpm. On cheese (at 4 °C): • A lower reduction than that obtained *in vitro* (1–2 log) was observed; • Best results (except for the highest phage concentration) were observed when the cheese was immersed in the film-forming solution containing phages and the coating was directly formed on the surface of the cheese.	The edible films/coatings were not physically impacted by the addition of phages.

N/A—not available; KC—κ-carrageenan; KGM—konjac glucomannan; RTE—ready-to-eat; WPC—whey protein concentrate; PULL—pullulan; WP—whey protein concentrate/pullulan; PLA—poly (lactic acid); WPI—whey protein isolate.

Although refrigerated temperatures were better for maintaining phage stability, higher and/or faster bacterial reductions were observed at room and higher temperatures of 37 °C [300,307]. In the majority of the studies, higher bacterial reductions were obtained with free phages; however, in some cases, similar results between free and incorporated phages were observed [289]. In general, higher and faster bacterial reductions were obtained *in vitro* compared to in food assays.

Incorporated Phages Targeting *Salmonella* spp.

Regarding the incorporation of *Salmonella* phages into different systems (Table 2), both *in vitro* [294,298,314,319] and food studies [311–313,315,316,320] are also reported in the literature.

Studies targeting *S.* Enteritidis are the most frequent [298,311,313,319,320], but some examine other individual *Salmonella* strains [294,314,316] or mixtures of different *Salmonella* strains in bacterial cocktails [312,315].

Similarly to those on *E. coli*, most studies used a single phage [294,298,311,313,320], namely the commercially available phage Felix O1 [312,319], but also phage cocktails are investigated [314–316].

Food studies were performed on meat [311–313], fruit [315], cheese [316], and eggshells [320].

Salmonella phages were incorporated into coatings/films of chitosan [315], alginate [298,315,316], chitosan–alginate [313], cellulose derivatives [314,315], whey protein [315], polyvinyl alcohol [319,320], pullulan–trehalose [294], and poly(lactic acid) films coated with xanthan [312]. Also, phage incorporation into a hydrogel film was studied [311].

In a study with different polymers (whey protein concentrate, carboxymethyl cellulose, chitosan, and sodium alginate), films of whey protein concentrate showed the greatest results in terms of phage stability and antibacterial effects [315]. The addition of cinnamaldehyde to sodium alginate films also resulted in higher bacterial reduction; however, alterations in film properties were reported after the addition of this compound [298].

In some cases, phage addition did not result in changes to film properties [298]. However, in others [314], phage addition led to modifications in the film surface, reducing the transparency, tensile resistance, and modulus of elasticity of the film and increasing the porosity.

The stability of *Salmonella* phages in the different produced films/coatings varied from 14 days in cellulose acetate at room temperature [314] to 30 days in polyvinyl alcohol at 5 °C [320] and 60 days in pullulan–trehalose-coated paper at room temperature [294]. Similar antibacterial effects were observed with the layer-by-layer assembly of chitosan/alginate thin films that incorporated phages and free phages [313].

Incorporated Phages Targeting *Listeria monocytogenes*

Table 2 also includes several *in vitro* and food studies with incorporated phages targeting a specific *L. monocytogenes* bacterial strain [265,288,294,310,317,318,321] or a mixture of strains in a bacterial cocktail [289,312].

A majority of studies utilized phage A511 [310,312,317,318] and the commercially available phage LISTEX™ P100 [288,289,294,321]. Other studies also used phage cocktails [265,289].

From the available studies that utilized foods, meat was the most tested type [265,289,310,312]. However, there are also studies on fruit [289] and cheese [317].

Listeria phages were incorporated into whey protein [317] or into films composed of whey protein and pullulan [310,318], in positively charged cellulose membranes [265,289], xanthan coatings on poly(lactic acid) films [312], a pullulan–trehalose mixture [294,321], a pullulan–trehalose mixture coated onto packaging paper [294], sodium caseinate, and sodium alginate mixed with gelatine and polyvinyl alcohol [288].

Despite free phages in some cases showing a higher antibacterial effect than immobilized phages [288,289], in others, no significant differences in antibacterial effect between free and immobilized phages were observed [289]. Also, although some studies reported better results at refrigerated temperatures [288], in general, bacterial reduction was temperature-dependent with higher bacterial reductions obtained at room temperature [289] and at temperatures higher than 37 °C [312].

The stability of the incorporated *Listeria* phages varied from 16 days at 4 °C [317], to 20 days [310] and 60 days [294,318] at 25 °C.

Regarding film properties, some studies reported that phage addition altered film properties, such as opacity [318], while in others, the addition of phages did not cause changes in their properties [288].

In general, better results in terms of bacterial inactivation were obtained *in vitro* compared to the results obtained in food. Also, in food, the bacterial reductions were matrix-dependent. The lowest *L. monocytogenes* reduction was obtained in cheese with a long incubation period of 16 days at 4 °C [317].

Regarding phage stability, Leung et al. (2018) reported that drying method and humidity significantly affected the long-term viability of LISTEX P100 phages in pullulan/trehalose films, with trehalose also playing an important role in the enhancement of phage stability [321]. Also, other authors reported better results in terms of phage stability, bacterial reduction, and film properties in whey protein concentrate and pullulan blends instead of the corresponding pure films [310,318].

4. Summary and Future Perspectives

Phages in their free form or incorporated in active food packaging systems have proven to be an eco-friendly strategy to effectively decontaminate food and improve food safety. With several commercial phage products already approved and numerous studies demonstrating their antibacterial efficacy against problematic foodborne pathogens, phages hold great promise for future use in common food safety practices. However, there are significant regulatory, industrial, and implementation challenges that must be addressed to scale them up effectively.

Regulatory frameworks remain one of the key obstacles of phage use in the food industry. The approval process for phages as food additives, biocontrol agents, or components of packaging materials requires clear guidelines on their safety, efficacy, impact on food quality and environmental impact. There is a need for continued collaboration between scientists, food producers, and regulatory bodies to create a consistent regulatory framework that addresses specific concerns such as dosage, phage stability, and potential resistance development.

Industrial scalability also presents challenges, whether for direct phage application or incorporation into packaging. In both cases, ensuring the sustainable production of phages at a large scale is critical. Phage production for large-scale use in food safety applications needs to be both economically viable and able to meet the high-volume demands of the food industry. For active food packaging, phage incorporation must not only be cost-effective but also retain phage viability and efficacy under real-world conditions. For direct application to food, additional complexities include maintaining phage stability during transportation, storage, and application, especially when dealing with perishable items. Researchers are exploring methods such as encapsulation or the use of stabilizing agents (e.g., trehalose) to improve phage retention and activity, whether in packaging or in food applications.

Barriers to implementation are equally significant. Consumer acceptance remains a critical issue, especially given concerns about the safety of consuming phages, despite their natural occurrence in food. It is essential to educate both consumers and food producers about the safety, natural ubiquity, and efficacy of phages to overcome these barriers. This requires careful coordination between food producers, packaging companies, and regulators to create solutions that are both effective and practical at an industrial scale.

Further research to refine application techniques, ensure consistent performance across different food matrices, and evaluate the long-term impact of phage use is needed. For instance, understanding how phages interact with non-targeted commensal bacteria and ensuring that they do not disrupt either food microbiota or consumer microbiota or cause unwanted resistance in other microbial populations are important areas for future investigation. Phage combination with other decontamination methods, including application timings, also needs more research in order to maximize their effect and reproducibility.

Supplementary Materials: The following supporting information can be downloaded at: https://www.mdpi.com/article/10.3390/microorganisms13030515/s1. Table S1: Results of recent studies on the application of free phages directly to food.

Author Contributions: M.B.: conceptualization; writing—original draft, review and editing. C.P., C.S.R.F., and A.A.: conceptualization; supervision; writing—review and editing. C.S.R.F. and A.A.: funding acquisition. All authors have read and agreed to the published version of the manuscript.

Funding: This work was supported by FCT/MECTES (PIDDAC), which provided financial support to CESAM (UIDP/50017/2020 + UIDB/50017/2020 + LA/P/0094/2020) and the CICECO-Aveiro Institute of Materials (UIDB/50011/2020 (DOI 10.54499/UIDB/50011/2020), UIDP/50011/2020 (DOI 10.54499/UIDP/50011/2020) and LA/P/0006/2020 (DOI 10.54499/LA/P/0006/2020)), through national funds. FCT also supported this work through the attribution of a PhD grant to Márcia Braz (DOI 10.54499/2020.06571.BD) and research contracts to Carla Pereira and Carmen S. R. Freire (DOI 10.54499/CEECIND/03974/2017/CP1459/CT0022 and CEECIND/00464/2017, respectively).

Institutional Review Board Statement: Not applicable.

Data Availability Statement: No new data were created or analyzed in this study.

Acknowledgments: The authors thank the Department of Biology, Department of Chemistry, and the University of Aveiro, where this research was carried out.

Conflicts of Interest: The authors declare no conflicts of interest.

References

1. World Health Organization (WHO). Food Safety Fact Sheet. Available online: https://www.who.int/news-room/fact-sheets/detail/food-safety (accessed on 16 January 2025).
2. European Food Safety Authority (EFSA); European Centre for Disease Prevention and Control (ECDC). The European Union One Health 2021 Zoonoses Report. *EFSA J.* **2022**, *20*, e07666. [CrossRef]
3. Chacha, J.S.; Zhang, L.; Ofoedu, C.E.; Suleiman, R.A.; Dotto, J.M.; Roobab, U.; Agunbiade, A.O.; Duguma, H.T.; Mkojera, B.T.; Hossaini, S.M.; et al. Revisiting non-thermal food processing and preservation methods—Action mechanisms, pros and cons: A technological update (2016–2021). *Foods* **2021**, *10*, 1430. [CrossRef]
4. Deng, L.Z.; Sutar, P.P.; Mujumdar, A.S.; Tao, Y.; Pan, Z.; Liu, Y.H.; Xiao, H.W. Thermal decontamination technologies for microorganisms and mycotoxins in low-moisture foods. *Annu. Rev. Food Sci. Technol.* **2021**, *12*, 287–305. [CrossRef]
5. Mendoza, I.C.; Luna, E.O.; Pozo, M.D.; Vásquez, M.V.; Montoya, D.C.; Moran, G.C.; Romero, L.G.; Yépez, X.; Salazar, R.; Romero-Peña, M.; et al. Conventional and non-conventional disinfection methods to prevent microbial contamination in minimally processed fruits and vegetables. *LWT—Food Sci. Technol.* **2022**, *165*, 113714. [CrossRef]
6. Rendueles, C.; Duarte, A.C.; Escobedo, S.; Fernández, L.; Rodríguez, A.; García, P.; Martínez, B. Combined use of bacteriocins and bacteriophages as food biopreservatives. A review. *Int. J. Food Microbiol.* **2022**, *368*, 109611. [CrossRef]

7. Sillankorva, S.M.; Oliveira, H.; Azeredo, J. Bacteriophages and their role in food safety. *Int. J. Microbiol.* **2012**, *2012*, 863945. [CrossRef]
8. Deka, D.; Annapure, U.S.; Shirkole, S.S.; Thorat, B.N. Bacteriophages: An organic approach to food decontamination. *J. Food Process. Preserv.* **2021**, *46*, e16101. [CrossRef]
9. Li, J.; Zhao, F.; Zhan, W.; Li, Z.; Zou, L.; Zhao, Q. Challenges for the application of bacteriophages as effective antibacterial agents in the food industry. *J. Sci. Food Agric.* **2022**, *102*, 461–471. [CrossRef]
10. Alomari, M.M.M.; Dec, M.; Urban-Chmiel, R. Bacteriophages as an alternative method for control of zoonotic and foodborne pathogens. *Viruses* **2021**, *13*, 2348. [CrossRef]
11. Endersen, L.; Coffey, A. The use of bacteriophages for food safety. *Curr. Opin. Food Sci.* **2020**, *36*, 1–8. [CrossRef]
12. Ge, H.; Fu, S.; Guo, H.; Hu, M.; Xu, Z.; Zhou, X. Application and challenge of bacteriophage in the food protection. *Int. J. Food Microbiol.* **2022**, *380*, 109872. [CrossRef]
13. Hyla, K.; Dusza, I.; Skaradzińska, A. Recent advances in the application of bacteriophages against common foodborne pathogens. *Antibiotics* **2022**, *11*, 1536. [CrossRef]
14. Lavilla, M.; Domingo-Calap, P.; Sevilla-Navarro, S.; Lasagabaster, A. Natural Killers: Opportunities and challenges for the use of bacteriophages in microbial food safety from the one health perspective. *Foods* **2023**, *12*, 552. [CrossRef]
15. Puligundla, P.; Lim, S. Biocontrol approaches against *Escherichia coli* O157:H7 in foods. *Foods* **2022**, *11*, 756. [CrossRef]
16. Almutairi, M.; Imam, M.; Alammari, N.; Hafiz, R.; Patel, F.; Alajel, S. Using phages to reduce Salmonella prevalence in chicken meat: A systematic review. *PHAGE Ther. Appl. Res.* **2022**, *3*, 15–27. [CrossRef]
17. Kawacka, I.; Olejnik-Schmidt, A.; Schmidt, M.; Sip, A. Effectiveness of phage-based inhibition of *Listeria monocytogenes* in food products and food processing environments. *Microorganisms* **2020**, *8*, 1764. [CrossRef]
18. Ushanov, L.; Lasareishvili, B.; Janashia, I.; Zautner, A.E. Application of *Campylobacter jejuni* phages: Challenges and perspectives. *Animals* **2020**, *10*, 279. [CrossRef]
19. Abbas, R.Z.; Alsayeqh, A.F.; Aqib, A.I. Role of bacteriophages for optimized health and production of poultry. *Animals* **2022**, *12*, 3378. [CrossRef]
20. Islam, M.R.; Martinez-Soto, C.E.; Lin, J.T.; Khursigara, C.M.; Barbut, S.; Anany, H. A systematic review from basics to omics on bacteriophage applications in poultry production and processing. *Crit. Rev. Food Sci. Nutr.* **2023**, *63*, 3097–3129. [CrossRef]
21. Shahdadi, M.; Safarirad, M.; Berizi, E.; Hosseinzadeh, S. Investigating the effect of phage on reducing *Salmonella* spp. in poultry meat: A systematic review and meta-analysis. *Food Control* **2024**, *160*, 110380. [CrossRef]
22. García-Anaya, M.C.; Sepulveda, D.R.; Zamudio-Flores, P.B.; Acosta-Muñiz, C.H. Bacteriophages as additives in edible films and coatings. *Trends Food Sci. Technol.* **2023**, *132*, 150–161. [CrossRef]
23. Aaliya, B.; Sunooj, K.V.; Navaf, M.; Akhila, P.P.; Sudheesh, C.; Mir, S.A.; Sabu, S.; Sasidharan, A.; Hlaing, M.T.; George, J. Recent trends in bacterial decontamination of food products by hurdle technology: A synergistic approach using thermal and non-thermal processing techniques. *Food Res. Int.* **2021**, *147*, 110514. [CrossRef] [PubMed]
24. Fang, J.; Liu, C.; Law, C.L.; Mujumdar, A.S.; Xiao, H.W.; Zhang, C. Superheated steam processing: An emerging technology to improve food quality and safety. *Crit. Rev. Food Sci. Nutr.* **2022**, *63*, 8720–8736. [CrossRef] [PubMed]
25. Liao, X.; Cullen, P.J.; Muhammad, A.I.; Jiang, Z.; Ye, X.; Liu, D.; Ding, T. Cold plasma–based hurdle interventions: New strategies for improving food safety. *Food Eng. Rev.* **2020**, *12*, 321–332. [CrossRef]
26. Qiu, L.; Zhang, M.; Tang, J.; Adhikari, B.; Cao, P. Innovative technologies for producing and preserving intermediate moisture foods: A review. *Food Res. Int.* **2019**, *116*, 90–102. [CrossRef]
27. Khan, I.; Tango, C.N.; Miskeen, S.; Lee, B.H.; Oh, D.H. Hurdle technology: A novel approach for enhanced food quality and safety—A review. *Food Control* **2017**, *73*, 1426–1444. [CrossRef]
28. Alfy, A.; Kiran, B.V.; Jeevitha, G.C.; Hebbar, H.U. Recent developments in superheated steam processing of foods—A review. *Crit. Rev. Food Sci. Nutr.* **2016**, *56*, 2191–2208. [CrossRef]
29. Rifna, E.J.; Singh, S.K.; Chakraborty, S.; Dwivedi, M. Effect of thermal and non-thermal techniques for microbial safety in food powder: Recent advances. *Food Res. Int.* **2019**, *126*, 108654. [CrossRef]
30. Aboud, S.A.; Altemimi, A.B.; Al-Hilphy, A.R.S.; Yi-Chen, L.; Cacciola, F. A comprehensive review on infrared heating applications in food processing. *Molecules* **2019**, *24*, 4125. [CrossRef]
31. Manyatsi, T.S.; Al-Hilphy, A.R.; Majzoobi, M.; Farahnaky, A.; Gavahian, M. Effects of infrared heating as an emerging thermal technology on physicochemical properties of foods. *Crit. Rev. Food Sci. Nutr.* **2022**, *63*, 6840–6859. [CrossRef]
32. Khampakool, A.; Soisungwan, S.; Park, S.H. Potential application of infrared-assisted freeze drying (IRAFD) for banana snacks: Drying kinetics, energy consumption, and texture. *LWT—Food Sci. Technol.* **2019**, *99*, 355–363. [CrossRef]
33. Mohammadi, Z.; Kashaninejad, M.; Ziaiifar, A.M.; Ghorbani, M. Peeling of kiwifruit using infrared heating technology: A feasibility and optimization study. *LWT—Food Sci. Technol.* **2019**, *99*, 128–137. [CrossRef]
34. Smith, D.L.; Atungulu, G.G.; Wilson, S.; Shad, Z.M. Deterrence of *Aspergillus flavus* regrowth and aflatoxin accumulation on shelled corn using infrared heat treatments. *Appl. Eng. Agric.* **2020**, *36*, 151–158. [CrossRef]

35. Wilson, S.A.; Mohammadi Shad, Z.; Oduola, A.A.; Zhou, Z.; Jiang, H.; Carbonero, F.; Atungulu, G.G. Decontamination of mycotoxigenic fungi on shelled corn using selective infrared heating technique. *Cereal Chem.* **2021**, *98*, 31–43. [CrossRef]
36. Alkaya, G.B.; Erdogdu, F.; Ekiz, H.I. Comparison of conventional far-infrared (IR) heating to continuous IR heating-cooling for surface pasteurization of shell eggs contaminated by *Salmonella enterica* serotype Enteritidis. *J. Food Process. Preserv.* **2022**, *46*, e16168. [CrossRef]
37. Rosenthal, I.; Rosen, B.; Bernstein, S. Surface pasteurization of cottage cheese. *Milchwissenschaft-Milk Sci. Int.* **1996**, *51*, 198–201.
38. Krishnamurthy, K.; Jun, S.; Irudayaraj, J.; Demirci, A. Efficacy of infrared heat treatment for inactivation of *Staphylococcus aureus* in milk. *J. Food Process Eng.* **2008**, *31*, 798–816. [CrossRef]
39. Lee, E.-H. A review on applications of infrared heating for food processing in comparison to other industries. In *Innovative Food Processing Technologies: A Comprehensive Review*; Elsevier: Amsterdam, The Netherlands, 2021; pp. 431–455. [CrossRef]
40. Stratakos, A.C.; Koidis, A. Suitability, efficiency, and microbiological safety of novel physical technologies for the processing of ready-to-eat meals, meats, and pumpable products. *Int. J. Food Sci. Technol.* **2015**, *50*, 1283–1302. [CrossRef]
41. Verma, T.; Subbiah, J. Conical twin-screw extrusion is an effective inactivation process for *Salmonella* in low-moisture foods at temperatures above 65 °C. *LWT—Food Sci. Technol.* **2019**, *114*, 108369. [CrossRef]
42. Goodburn, C.; Wallace, C.A. The microbiological efficacy of decontamination methodologies for fresh produce: A review. *Food Control* **2013**, *32*, 418–427. [CrossRef]
43. Smaoui, S.; Agriopoulou, S.; D'Amore, T.; Tavares, L.; Mousavi Khaneghah, A. The control of *Fusarium* growth and decontamination of produced mycotoxins by lactic acid bacteria. *Crit. Rev. Food Sci. Nutr.* **2022**, *63*, 11125–11152. [CrossRef] [PubMed]
44. Trias, R.; Bañeras, L.; Badosa, E.; Montesinos, E. Bioprotection of golden delicious apples and iceberg lettuce against foodborne bacterial pathogens by lactic acid bacteria. *Int. J. Food Microbiol.* **2008**, *123*, 50–60. [CrossRef] [PubMed]
45. Corrêa, A.N.R.; Ferreira, C.D. Essential oil for the control of fungi, bacteria, yeasts and viruses in food: An overview. *Crit. Rev. Food Sci. Nutr.* **2022**, *63*, 8960–8974. [CrossRef] [PubMed]
46. Kaavya, R.; Pandiselvam, R.; Abdullah, S.; Sruthi, N.U.; Jayanath, Y.; Ashokkumar, C.; Chandra Khanashyam, A.; Kothakota, A.; Ramesh, S.V. Emerging non-thermal technologies for decontamination of *Salmonella* in food. *Trends Food Sci. Technol.* **2021**, *112*, 400–418. [CrossRef]
47. Gracy, T.K.R.; Sharanyakanth, P.S.; Radhakrishnan, M. Non-thermal technologies: Solution for hazardous pesticides reduction in fruits and vegetables. *Crit. Rev. Food Sci. Nutr.* **2022**, *62*, 1782–1799. [CrossRef]
48. Mandal, R.; Singh, A.; Pratap Singh, A. Recent developments in cold plasma decontamination technology in the food industry. *Trends Food Sci. Technol.* **2018**, *80*, 93–103. [CrossRef]
49. Turantaş, F.; Kiliç, G.B.; Kiliç, B. Ultrasound in the meat industry: General applications and decontamination efficiency. *Int. J. Food Microbiol.* **2015**, *198*, 59–69. [CrossRef]
50. Ochoa-Velasco, C.E.; Díaz-Lima, M.C.; Ávila-Sosa, R.; Ruiz-López, I.I.; Corona-Jiménez, E.; Hernández-Carranza, P.; López-Malo, A.; Guerrero-Beltrán, J.A. Effect of UV-C light on *Lactobacillus rhamnosus*, *Salmonella Typhimurium*, and *Saccharomyces cerevisiae* kinetics in inoculated coconut water: Survival and residual effect. *J. Food Eng.* **2018**, *223*, 255–261. [CrossRef]
51. Mandal, R.; Mohammadi, X.; Wiktor, A.; Singh, A.; Singh, A.P. Applications of pulsed light decontamination technology in food processing: An overview. *Appl. Sci.* **2020**, *10*, 3606. [CrossRef]
52. Kuan, Y.H.; Bhat, R.; Patras, A.; Karim, A.A. Radiation processing of food proteins—A review on the recent developments. *Trends Food Sci. Technol.* **2013**, *30*, 105–120. [CrossRef]
53. Harper, D.R.; Parracho, H.M.R.T.; Walker, J.; Sharp, R.; Hughes, G.; Werthén, M.; Lehman, S.; Morales, S. Bacteriophages and biofilms. *Antibiotics* **2014**, *3*, 270–284. [CrossRef]
54. O'Sullivan, L.; Bolton, D.; McAuliffe, O.; Coffey, A. Bacteriophages in food applications: From foe to friend. *Annu. Rev. Food Sci. Technol.* **2019**, *10*, 151–172. [CrossRef] [PubMed]
55. Yan, J.; Mao, J.; Xie, J. Bacteriophage polysaccharide depolymerases and biomedical applications. *BioDrugs* **2014**, *28*, 265–274. [CrossRef] [PubMed]
56. Park, I.K.; Ha, J.W.; Kang, D.H. Investigation of optimum ohmic heating conditions for inactivation of *Escherichia coli* O157:H7, *Salmonella enterica* serovar Typhimurium, and *Listeria monocytogenes* in apple juice. *BMC Microbiol.* **2017**, *17*, 117. [CrossRef]
57. Yildiz, H.; Bozkurt, H.; Icier, F. Ohmic and conventional heating of pomegranate juice: Effects on rheology, color, and total phenolics. *Food Sci. Technol. Int.* **2009**, *15*, 503–512. [CrossRef]
58. Alkanan, Z.T.; Altemimi, A.B.; Al-Hilphy, A.R.S.; Watson, D.G.; Pratap-Singh, A. Ohmic heating in the food industry: Developments in concepts and applications during 2013–2020. *Appl. Sci.* **2021**, *11*, 2507. [CrossRef]
59. Michalak, J.; Czarnowska-Kujawska, M.; Klepacka, J.; Gujska, E.Z. Effect of microwave heating on the acrylamide formation in foods. *Molecules* **2020**, *25*, 4140. [CrossRef]

0. Tavman, S.; Otles, S.; Glaue, S.; Gogus, N. Food preservation technologies. In *Saving Food: Production, Supply Chain, Food Waste and Food Consumption*; Galanakis, C.M., Ed.; Academic Press: Cambridge, MA, USA; Elsevier Ltd.: Amsterdam, The Netherlands, 2019; pp. 117–140. [CrossRef]
1. Lazou, A.E. Food extrusion: An advanced process for innovation and novel product development. *Crit. Rev. Food Sci. Nutr.* **2022**, *64*, 4532–4560. [CrossRef]
2. Praeger, U.; Herppich, W.B.; Hassenberg, K. Aqueous chlorine dioxide treatment of horticultural produce: Effects on microbial safety and produce quality—A review. *Crit. Rev. Food Sci. Nutr.* **2018**, *58*, 318–333. [CrossRef]
3. Deng, L.Z.; Mujumdar, A.S.; Pan, Z.; Vidyarthi, S.K.; Xu, J.; Zielinska, M.; Xiao, H.W. Emerging chemical and physical disinfection technologies of fruits and vegetables: A comprehensive review. *Crit. Rev. Food Sci. Nutr.* **2020**, *60*, 2481–2508. [CrossRef]
4. Reddy, S.V.R.; Rao, D.V.S.; Sharma, R.R.; Preethi, P.; Pandiselvam, R. Role of ozone in post-harvest disinfection and processing of horticultural crops: A review. *Ozone Sci. Eng.* **2021**, *44*, 127–146. [CrossRef]
5. Xiang, Q.; Fan, L.; Li, Y.; Dong, S.; Li, K.; Bai, Y. A review on recent advances in plasma-activated water for food safety: Current applications and future trends. *Crit. Rev. Food Sci. Nutr.* **2022**, *62*, 2250–2268. [CrossRef] [PubMed]
6. Barbosa, A.A.T.; Mantovani, H.C.; Jain, S. Bacteriocins from lactic acid bacteria and their potential in the preservation of fruit products. *Crit. Rev. Biotechnol.* **2017**, *37*, 852–864. [CrossRef] [PubMed]
7. Verma, A.K.; Banerjee, R.; Dwivedi, H.P.; Juneja, V.K. Bacteriocins: Potential in food preservation. In *Encyclopedia of Food Microbiology*, 2nd ed.; Batt, C.A., Tortorello, M.L., Eds.; Academic Press: Cambridge, MA, USA; Elsevier Ltd.: Amsterdam, The Netherlands, 2014; pp. 180–186. [CrossRef]
8. Cotter, P.D.; Hill, C.; Ross, R.P. Bacteriocins: Developing innate immunity for food. *Nat. Rev. Microbiol.* **2005**, *3*, 777–788. [CrossRef]
9. Chikindas, M.L.; Weeks, R.; Drider, D.; Chistyakov, V.A.; Dicks, L.M. Functions and emerging applications of bacteriocins. *Curr. Opin. Biotechnol.* **2018**, *49*, 23–28. [CrossRef]
0. López-Cuellar, M.R.; Rodríguez-Hernández, A.I.; Chavarría-Hernández, N. LAB bacteriocin applications in the last decade. *Biotechnol. Biotechnol. Equip.* **2016**, *30*, 1039–1050. [CrossRef]
1. Da Costa, R.J.; Voloski, F.L.S.; Mondadori, R.G.; Duval, E.H.; Fiorentini, Â.M. Preservation of meat products with bacteriocins produced by lactic acid bacteria isolated from meat. *J. Food Qual.* **2019**, *2019*, 4726510. [CrossRef]
2. Silva, C.C.G.; Silva, S.P.M.; Ribeiro, S.C. Application of bacteriocins and protective cultures in dairy food preservation. *Front. Microbiol.* **2018**, *9*, 594. [CrossRef]
3. Darbandi, A.; Asadi, A.; Mahdizade Ari, M.; Ohadi, E.; Talebi, M.; Halaj Zadeh, M.; Darb Emamie, A.; Ghanavati, R.; Kakanj, M. Bacteriocins: Properties and potential use as antimicrobials. *J. Clin. Lab. Anal.* **2022**, *36*, e24093. [CrossRef]
4. Li, Y.X.; Erhunmwunsee, F.; Liu, M.; Yang, K.; Zheng, W.; Tian, J. Antimicrobial mechanisms of spice essential oils and application in food industry. *Food Chem.* **2022**, *382*, 132312. [CrossRef]
5. Salanță, L.C.; Cropotova, J. An update on effectiveness and practicability of plant essential oils in the food industry. *Plants* **2022**, *11*, 2488. [CrossRef] [PubMed]
6. Dini, H.; Fallah, A.A.; Bonyadian, M.; Abbasvali, M.; Soleimani, M. Effect of edible composite film based on chitosan and cumin essential oil-loaded nanoemulsion combined with low-dose gamma irradiation on microbiological safety and quality of beef loins during refrigerated storage. *Int. J. Biol. Macromol.* **2020**, *164*, 1501–1509. [CrossRef] [PubMed]
7. Iseppi, R.; Sabia, C.; de Niederhäusern, S.; Pellati, F.; Benvenuti, S.; Tardugno, R.; Bondi, M.; Messi, P. Antibacterial activity of *Rosmarinus officinalis* L. and *Thymus vulgaris* L. essential oils and their combination against food-borne pathogens and spoilage bacteria in ready-to-eat vegetables. *Nat. Prod. Res.* **2019**, *33*, 3568–3572. [CrossRef] [PubMed]
8. Ozdikmenli, S.; Demirel Zorba, N.N. Evaluation of usage of essential oils instead of spices in meat ball formulation for controlling *Salmonella* spp. *Food Sci. Technol. Int.* **2016**, *22*, 93–101. [CrossRef]
9. Hernández, H.; Fraňková, A.; Klouček, P.; Banout, J. The effect of the application of thyme essential oil on microbial load during meat drying. *J. Vis. Exp.* **2018**, *2018*, e57054. [CrossRef]
0. Fancello, F.; Petretto, G.L.; Marceddu, S.; Venditti, T.; Pintore, G.; Zara, G.; Mannazzu, I.; Budroni, M.; Zara, S. Antimicrobial activity of gaseous *Citrus limon* var. pompia leaf essential oil against *Listeria monocytogenes* on ricotta salata cheese. *Food Microbiol.* **2020**, *87*, 103386. [CrossRef]
1. Roobab, U.; Abida, A.; Chacha, J.S.; Athar, A.; Madni, G.M.; Ranjha, M.M.A.N.; Rusu, A.V.; Zeng, X.A.; Aadil, R.M.; Trif, M. Applications of innovative non-thermal pulsed electric field technology in developing safer and healthier fruit juices. *Molecules* **2022**, *27*, 4031. [CrossRef]
2. Mohamed, M.E.A.; Eissa, A.H.A. Pulsed electric fields for food processing technology. In *Structure and Function of Food Engineering*; Eissa, A.A., Ed.; InTech: Rijeka, Croatia, 2012; pp. 275–306. [CrossRef]
3. Alirezalu, K.; Munekata, P.E.S.; Parniakov, O.; Barba, F.J.; Witt, J.; Toepfl, S.; Wiktor, A.; Lorenzo, J.M. Pulsed electric field and mild heating for milk processing: A review on recent advances. *J. Sci. Food Agric.* **2020**, *100*, 16–24. [CrossRef]

84. Arshad, R.N.; Abdul-Malek, Z.; Munir, A.; Buntat, Z.; Ahmad, M.H.; Jusoh, Y.M.M.; Bekhit, A.E.D.; Roobab, U.; Manzoor, M.F.; Aadil, R.M. Electrical systems for pulsed electric field applications in the food industry: An engineering perspective. *Trends Food Sci. Technol.* **2020**, *104*, 1–13. [CrossRef]
85. Song, Q.; Li, R.; Song, X.; Clausen, M.P.; Orlien, V.; Giacalone, D. The effect of high-pressure processing on sensory quality and consumer acceptability of fruit juices and smoothies: A review. *Food Res. Int.* **2022**, *157*, 111250. [CrossRef]
86. Bhargava, N.; Mor, R.S.; Kumar, K.; Sharanagat, V.S. Advances in application of ultrasound in food processing: A review. *Ultrason. Sonochem.* **2021**, *70*, 105293. [CrossRef] [PubMed]
87. Gabriel, A.A. Inactivation behaviors of foodborne microorganisms in multi-frequency power ultrasound-treated orange juice. *Food Control* **2014**, *46*, 189–196. [CrossRef]
88. Moosavi, M.H.; Khaneghah, A.M.; Javanmardi, F.; Hadidi, M.; Hadian, Z.; Jafarzadeh, S.; Huseyn, E.; Sant'Ana, A.S. A review of recent advances in the decontamination of mycotoxin and inactivation of fungi by ultrasound. *Ultrason. Sonochem.* **2021**, *79*, 105755. [CrossRef] [PubMed]
89. Monteiro, S.H.M.C.; Silva, E.K.; Alvarenga, V.O.; Moraes, J.; Freitas, M.Q.; Silva, M.C.; Raices, R.S.L.; Sant'Ana, A.S.; Meireles, M.A.A.; Cruz, A.G. Effects of ultrasound energy density on the non-thermal pasteurization of chocolate milk beverage. *Ultrason. Sonochem.* **2018**, *42*, 1–10. [CrossRef]
90. Techathuvanan, C.; D'Souza, D.H. High intensity ultrasound for *Salmonella* Enteritidis inactivation in culture and liquid whole eggs. *J. Food Sci.* **2018**, *83*, 1733–1739. [CrossRef]
91. Pandiselvam, R.; Barut Gök, S.; Yüksel, A.N.; Tekgül, Y.; Koç, G.Ç.; Kothakota, A. Evaluation of the impact of UV radiation on rheological and textural properties of food. *J. Texture Stud.* **2022**, *53*, 800–808. [CrossRef]
92. Dhar, R.; Basak, S.; Chakraborty, S. Pasteurization of fruit juices by pulsed light treatment: A review on the microbial safety, enzymatic stability, and kinetic approach to process design. *Compr. Rev. Food Sci. Food Saf.* **2022**, *21*, 499–540. [CrossRef]
93. Cheigh, C.I.; Hwang, H.J.; Chung, M.S. Intense pulsed light (IPL) and UV-C treatments for inactivating *Listeria monocytogenes* on solid medium and seafoods. *Food Res. Int.* **2013**, *54*, 745–752. [CrossRef]
94. Ganan, M.; Hierro, E.; Hospital, X.F.; Barroso, E.; Fernández, M. Use of pulsed light to increase the safety of ready-to-eat cured meat products. *Food Control* **2013**, *32*, 512–517. [CrossRef]
95. Aguiló-Aguayo, I.; Charles, F.; Renard, C.M.G.C.; Page, D.; Carlin, F. Pulsed light effects on surface decontamination, physical qualities and nutritional composition of tomato fruit. *Postharvest Biol. Technol.* **2013**, *86*, 29–36. [CrossRef]
96. Salehi, F. Application of pulsed light technology for fruits and vegetables disinfection: A review. *J. Appl. Microbiol.* **2022**, *132*, 2521–2530. [CrossRef] [PubMed]
97. Agüero, M.V.; Jagus, R.J.; Martín-Belloso, O.; Soliva-Fortuny, R. Surface decontamination of spinach by intense pulsed light treatments: Impact on quality attributes. *Postharvest Biol. Technol.* **2016**, *121*, 118–125. [CrossRef]
98. Kasahara, I.; Carrasco, V.; Aguilar, L. Inactivation of *Escherichia coli* in goat milk using pulsed ultraviolet light. *J. Food Eng.* **2015**, *152*, 43–49. [CrossRef]
99. Sharifi-Yazdi, M.K.; Darghahi, H. Inactivation of pathogenic bacteria using pulsed UV-light and its application in water disinfection and quality control. *Acta Med. Iran.* **2006**, *44*, 305–308.
100. Fernandes, Â.; Antonio, A.L.; Oliveira, M.B.P.P.; Martins, A.; Ferreira, I.C.F.R. Effect of gamma and electron beam irradiation on the physico-chemical and nutritional properties of mushrooms: A review. *Food Chem.* **2012**, *135*, 641–650. [CrossRef]
101. Rosario, D.K.A.; Rodrigues, B.L.; Bernardes, P.C.; Conte-Junior, C.A. Principles and applications of non-thermal technologies and alternative chemical compounds in meat and fish. *Crit. Rev. Food Sci. Nutr.* **2021**, *61*, 1163–1183. [CrossRef]
102. Ranathunga, N.S.; Wijayasekara, K.N.; Abeyrathne, E.D.N.S. Application of bio-preservation to enhance food safety: A review. *Korean J. Food Preserv.* **2023**, *30*, 179–189. [CrossRef]
103. Sankar, K.; Baranitharan, M.; Shreyasi, E.; Keerthish, P.N.; Swethaa, R. Biopreservative technologies of food: An alternative to chemical preservation and recent developments. *Food Sci. Biotechnol.* **2023**, *32*, 1337–1350. [CrossRef]
104. Zapaśnik, A.; Sokołowska, B.; Bryła, M. Role of lactic acid bacteria in food preservation and safety. *Foods* **2022**, *11*, 1283. [CrossRef]
105. Muccilli, S.; Restuccia, C. Bioprotective role of yeasts. *Microorganisms* **2015**, *3*, 588–611. [CrossRef]
106. Ge, H.; Lin, C.; Xu, Y.; Hu, M.; Xu, Z.; Geng, S.; Jiao, X.; Chen, X. A phage for the controlling of *Salmonella* in poultry and reducing biofilms. *Vet. Microbiol.* **2022**, *269*, 109432. [CrossRef] [PubMed]
107. Poojari, K.; Akhila, D.S.; Raj, M.J.R.; Santhosh, K.S.; Kenjar, A.; Ashwath, P. Biocontrol of *Escherichia coli* and *Salmonella* in poultry meat using phage cocktail. *Iran. J. Vet. Res.* **2022**, *23*, 270–274. [CrossRef] [PubMed]
108. Witte, S.; Huijboom, L.; Klamert, S.; van de Straat, L.; Hagens, S.; Fieseler, L.; de Vegt, B.T.; van Mierlo, J.T. Application of bacteriophages EP75 and EP335 efficiently reduces viable cell counts of *Escherichia coli* O157 on beef and vegetables. *Food Microbiol.* **2022**, *104*, 103978. [CrossRef] [PubMed]
109. Vikram, A.; Tokman, J.I.; Woolston, J.; Sulakvelidze, A. Phage biocontrol improves food safety by significantly reducing the level and prevalence of *Escherichia coli* O157:H7 in various foods. *J. Food Prot.* **2020**, *83*, 668–676. [CrossRef]

110. Wang, L.; Qu, K.; Li, X.; Cao, Z.; Wang, X.; Li, Z.; Song, Y.; Xu, Y. Use of bacteriophages to control *Escherichia coli* O157:H7 in domestic ruminants, meat products, and fruits and vegetables. *Foodborne Pathog. Dis.* **2017**, *14*, 483–493. [CrossRef]
111. Lopes, R.P.; Mota, M.J.; Gomes, A.M.; Delgadillo, I.; Saraiva, J.A. Application of high pressure with homogenization, temperature, carbon dioxide, and cold plasma for the inactivation of bacterial spores: A review. *Compr. Rev. Food Sci. Food Saf.* **2018**, *17*, 532–555. [CrossRef]
112. Kuchment, A. The fading of phage therapy. In *The Forgotten Cure: The Past and Future of Phage Therapy*, 1st ed.; Springer: New York, NY, USA, 2012; pp. 35–42. [CrossRef]
113. Ranveer, S.A.; Dasriya, V.; Ahmad, F.; Dhillon, H.S.; Samtiya, M.; Shama, E. Positive and negative aspects of bacteriophages and their immense role in the food chain. *npj Sci. Food* **2024**, *8*, 1. [CrossRef]
114. Nanda, A.M.; Thormann, K.; Frunzke, J. Impact of spontaneous prophage induction on the fitness of bacterial populations and host-microbe interactions. *J. Bacteriol.* **2015**, *197*, 410–419. [CrossRef]
115. Chen, Q.; Dharmaraj, T.; Cai, P.C.; Burgener, E.B.; Haddock, N.L.; Spakowitz, A.J.; Bollyky, P.L. Bacteriophage and bacterial susceptibility, resistance, and tolerance to antibiotics. *Pharmaceutics* **2022**, *14*, 1425. [CrossRef]
116. Harada, L.K.; Silva, E.C.; Campos, W.F.; Del Fiol, F.S.; Vila, M.; Dąbrowska, K.; Krylov, V.N.; Balcão, V.M. Biotechnological applications of bacteriophages: State of the art. *Microbiol. Res.* **2018**, *212–213*, 38–58. [CrossRef]
117. Pereira, C.; Costa, P.; Duarte, J.; Balcão, V.M.; Almeida, A. Phage therapy as a potential approach in the biocontrol of pathogenic bacteria associated with shellfish consumption. *Int. J. Food Microbiol.* **2021**, *338*, 108995. [CrossRef] [PubMed]
118. Rios, A.C.; Moutinho, C.G.; Pinto, F.C.; Del Fiol, F.S.; Jozala, A.; Chaud, M.V.; Vila, M.M.D.C.; Teixeira, J.A.; Balcão, V.M. Alternatives to overcoming bacterial resistances: State-of-the-art. *Microbiol. Res.* **2016**, *191*, 51–80. [CrossRef] [PubMed]
119. Garvey, M. Bacteriophages and the one health approach to combat multidrug resistance: Is this the way? *Antibiotics* **2020**, *9*, 414. [CrossRef] [PubMed]
120. Sevilla-Navarro, S.; Catalá-Gregori, P.; García, C.; Cortés, V.; Marin, C. *Salmonella* Infantis and *Salmonella* Enteritidis specific bacteriophages isolated from poultry faeces as a complementary tool for cleaning and disinfection against *Salmonella*. *Compar. Immunol. Microbiol. Infect. Dis.* **2020**, *68*, 101405. [CrossRef]
121. Tang, S.S.; Biswas, S.K.; Tan, W.S.; Saha, A.K.; Leo, B.F. Efficacy and potential of phage therapy against multidrug resistant *Shigella* spp. *PeerJ* **2019**, *7*, e6225. [CrossRef]
122. Moye, Z.D.; Woolston, J.; Sulakvelidze, A. Bacteriophage applications for food production and processing. *Viruses* **2018**, *10*, 205. [CrossRef]
123. Vikram, A.; Woolston, J.; Sulakvelidze, A. Phage biocontrol applications in food production and processing. *Curr. Issues Mol. Biol.* **2021**, *40*, 267–302. [CrossRef]
124. Jones, J.B.; Vallad, G.E.; Iriarte, F.B.; Obradović, A.; Wernsing, M.H.; Jackson, L.E.; Balogh, B.; Hong, J.C.; Momol, M.T. Considerations for using bacteriophages for plant disease control. *Bacteriophage* **2012**, *2*, e23857. [CrossRef]
125. Almeida, A.; Cunha, Â.; Gomes, N.C.M.; Alves, E.; Costa, L.; Faustino, M.A.F. Phage therapy and photodynamic therapy: Low environmental impact approaches to inactivate microorganisms in fish farming plants. *Mar. Drugs* **2009**, *7*, 268–313. [CrossRef]
126. Jassim, S.A.A.; Limoges, R.G. *Bacteriophages: Practical Applications for Nature's Biocontrol*; Springer: Cham, Switzerland, 2017.
127. Ruban, S.; Kannan, P.; Quintoil, M. Bacteriophage: A novel biocontrol agent against food pathogens: Myth or reality. In *Applied Food Science and Engineering with Industrial Applications*, 1st ed.; Aguilar, C.N., Carvajal-Millan, E., Eds.; Apple Academic Press: New York, NY, USA, 2019; pp. 1–15.
128. Guenther, S.; Herzig, O.; Fieseler, L.; Klumpp, J.; Loessner, M.J. Biocontrol of *Salmonella* Typhimurium in RTE foods with the virulent bacteriophage FO1-E2. *Int. J. Food Microbiol.* **2012**, *154*, 66–72. [CrossRef]
129. Nobrega, F.L.; Costa, A.R.; Kluskens, L.D.; Azeredo, J. Revisiting phage therapy: New applications for old resources. *Trends Microbiol.* **2015**, *23*, 185–191. [CrossRef] [PubMed]
130. Colavecchio, A.; Cadieux, B.; Lo, A.; Goodridge, L.D. Bacteriophages contribute to the spread of antibiotic resistance genes among foodborne pathogens of the *Enterobacteriaceae* family—A review. *Front. Microbiol.* **2017**, *8*, 1108. [CrossRef] [PubMed]
131. Fister, S.; Robben, C.; Witte, A.K.; Schoder, D.; Wagner, M.; Rossmanith, P. Influence of environmental factors on phage-bacteria interaction and on the efficacy and infectivity of phage P100. *Front. Microbiol.* **2016**, *7*, 1152. [CrossRef] [PubMed]
132. Iriarte, F.B.; Balogh, B.; Momol, M.T.; Smith, L.M.; Wilson, M.; Jones, J.B. Factors affecting survival of bacteriophage on tomato leaf surfaces. *Appl. Environ. Microbiol.* **2007**, *73*, 1704–1711. [CrossRef]
133. Jonczyk, E.; Kłak, M.; Międzybrodzki, R.; Górski, A. The influence of external factors on bacteriophages-review. *Folia Microbiol.* **2011**, *56*, 191–200. [CrossRef]
134. Ly-Chatain, M.H. The factors affecting effectiveness of treatment in phage therapy. *Front. Microbiol.* **2014**, *5*, 51. [CrossRef]
135. Wessels, K.; Rip, D.; Gouws, P. The effect of spray parameters on the survival of bacteriophages. *Processes* **2022**, *10*, 673. [CrossRef]
136. De Melo, A.G.; Levesque, S.; Moineau, S. Phages as friends and enemies in food processing. *Curr. Opin. Biotechnol.* **2018**, *49*, 185–190. [CrossRef]

137. Yen, M.; Cairns, L.S.; Camilli, A. A cocktail of three virulent bacteriophages prevents *Vibrio cholerae* infection in animal models. *Nat. Commun.* **2017**, *8*, 14187. [CrossRef]
138. Garvey, M. Bacteriophages and food production: Biocontrol and bio-preservation options for food safety. *Antibiotics* **2022**, *11*, 1324. [CrossRef]
139. Phothaworn, P.; Supokaivanich, R.; Lim, J.; Klumpp, J.; Imam, M.; Kutter, E.; Galyov, E.E.; Dunne, M.; Korbsrisate, S. Development of a broad-spectrum *Salmonella* phage cocktail containing Viunalike and Jerseylike viruses isolated from Thailand. *Food Microbiol.* **2020**, *92*, 103586. [CrossRef] [PubMed]
140. Tomat, D.; Casabonne, C.; Aquili, V.; Balagué, C.; Quiberoni, A. Evaluation of a novel cocktail of six lytic bacteriophages against Shiga toxin-producing *Escherichia coli* in broth, milk and meat. *Food Microbiol.* **2018**, *76*, 434–442. [CrossRef] [PubMed]
141. Komora, N.; Maciel, C.; Pinto, C.A.; Ferreira, V.; Brandão, T.R.S.; Saraiva, J.M.A.; Castro, S.M.; Teixeira, P. Non-thermal approach to *Listeria monocytogenes* inactivation in milk: The combined effect of high pressure, pediocin PA-1 and bacteriophage P100. *Food Microbiol.* **2020**, *86*, 103315. [CrossRef] [PubMed]
142. Komora, N.; Maciel, C.; Amaral, R.A.; Fernandes, R.; Castro, S.M.; Saraiva, J.A.; Teixeira, P. Innovative hurdle system towards *Listeria monocytogenes* inactivation in a fermented meat sausage model—High pressure processing assisted by bacteriophage P100 and bacteriocinogenic *Pediococcus acidilactici*. *Food Res. Int.* **2021**, *148*, 110628. [CrossRef]
143. Komora, N.; Bruschi, C.; Ferreira, V.; Maciel, C.; Brandão, T.R.S.; Fernandes, R.; Saraiva, J.A.; Castro, S.M.; Teixeira, P. The protective effect of food matrices on *Listeria* lytic bacteriophage P100 application towards high pressure processing. *Food Microbiol.* **2018**, *76*, 416–425. [CrossRef]
144. Ishaq, A.; Syed, Q.A.; Ebner, P.D.; Ubaid ur Rahman, H. Multiple hurdle technology to improve microbial safety, quality and oxidative stability of refrigerated raw beef. *LWT—Food Sci. Technol.* **2021**, *138*, 110529. [CrossRef]
145. Monferrer, E.; Domingo-Calap, P. Virus-host coevolution as a tool for controlling bacterial resistance to phage therapy. *J. Biotechnol. Biomed.* **2019**, *2*, 96–104. [CrossRef]
146. Peng, H.; Chen, I.A.; Qimron, U. Engineering phages to fight multidrug-resistant bacteria. *Chem. Rev.* **2025**, *125*, 933–971. [CrossRef]
147. Cristobal-Cueto, P.; García-Quintanilla, A.; Esteban, J.; García-Quintanilla, M. Phages in food industry biocontrol and bioremediation. *Antibiotics* **2021**, *10*, 786. [CrossRef]
148. Suja, E.; Gummadi, S.N. Advances in the applications of bacteriophages and phage products against food-contaminating bacteria. *Crit. Rev. Microbiol.* **2023**, *50*, 702–727. [CrossRef]
149. Centers for Disease Control and Prevention (CDC). Salmonella. Available online: https://www.cdc.gov/salmonella/general/index.html#two (accessed on 16 January 2025).
150. Popa, G.L.; Popa, M.I. *Salmonella* spp. Infection—A continuous threat worldwide. *Germs* **2021**, *11*, 88–96. [CrossRef] [PubMed]
151. Trampel, D.W.; Holder, T.G.; Gast, R.K. Integrated farm management to prevent *Salmonella* Enteritidis contamination of eggs. *J. Appl. Poultry Res.* **2014**, *23*, 353–365. [CrossRef]
152. Paudyal, N.; Anihouvi, V.; Hounhouigan, J.; Matsheka, M.I.; Sekwati-Monang, B.; Amoa-Awua, W.; Atter, A.; Ackah, N.B.; Mbugua, S.; Asagbra, A.; et al. Prevalence of foodborne pathogens in food from selected African countries—A meta-analysis. *Int. J. Food Microbiol.* **2017**, *249*, 35–43. [CrossRef] [PubMed]
153. Al-Hindi, R.R.; Alharbi, M.G.; Alotibi, I.; Azhari, S.A.; Algothmi, K.M.; Esmael, A. Application of a novel lytic Jerseyvirus phage LPSent1 for the biological control of the multidrug-resistant *Salmonella* Enteritidis in foods. *Front. Microbiol.* **2023**, *14*, 1135806. [CrossRef] [PubMed]
154. Brenner, T.; Schultze, D.M.; Mahoney, D.; Wang, S. Reduction of nontyphoidal *Salmonella enterica* in broth and on raw chicken breast by a broad-spectrum bacteriophage cocktail. *J. Food Prot.* **2024**, *87*, 100207. [CrossRef]
155. Han, S.; Choi, M.W.; Byun, K.H.; Kim, B.H.; Song, M.S.; Kang, I.; Ha, S.D. Characterization of *Salmonella* ser. Enteritidis-specific bacteriophages and biocontrol strategy to reduce *S.* Enteritidis on egg products using bacteriophages and essential oil compounds. *Food Control* **2024**, *160*, 110304. [CrossRef]
156. He, J.; Wong, C.W.Y.; Schultze, D.M.; Wang, S. Inactivation of *Salmonella* Enteritidis in liquid egg yolk and egg white using bacteriophage cocktails. *Curr. Res. Food Sci.* **2024**, *8*, 100703. [CrossRef]
157. Kokab, A.; Sheikh, A.A.; Rabbani, M.; Shehzad, W.; Riaz, M.I.; Raza, S.; Durrani, R.H. Physiological profiling of lytic bacteriophages and their efficiency against *Salmonella* Enteritidis on chicken breast cuts. *Pak. J. Zool.* **2024**, *56*, 375–384. [CrossRef]
158. Choi, Y.; Kwak, M.J.; Kang, M.G.; Kang, A.N.; Lee, W.; Mun, D.; Choi, H.; Park, J.; Eor, J.Y.; Song, M.; et al. Molecular characterization and environmental impact of newly isolated lytic phage SLAM_phiST1N3 in the *Cornellvirus* genus for biocontrol of a multidrug-resistant *Salmonella* Typhimurium in the swine industry chain. *Sci. Total Environ.* **2024**, *922*, 171208. [CrossRef]
159. Evran, E.; Dasan, B.G.; Tayyarcan, E.K.; Boyaci, I.H. Effect of sequential treatment of plasma activated water and bacteriophage on decontamination of *Salmonella* Typhimurium in lettuce. *Food Bioprocess Technol.* **2024**, *17*, 3790–3799. [CrossRef]
160. Jung, S.J.; Ashrafudoulla, M.; Kang, I.; Ha, S.D. Isolation and characterization of multidrug-resistant *Salmonella*-specific bacteriophages and their antibacterial efficiency in chicken breast. *Poultry Sci.* **2023**, *102*, 103073. [CrossRef] [PubMed]

61. Unverdi, A.; Erol, H.B.; Kaskatepe, B.; Babacan, O. Characterization of Salmonella phages isolated from poultry coops and its effect with nisin on food bio-control. *Food Sci. Nutr.* **2024**, *12*, 2760–2771. [CrossRef] [PubMed]
62. Gonmei, L.; Inbaraj, S.; Geyi, D.; Prakashan, L.; Dhiman, H.; Athira, V.; Thomas, P. Evaluation of bacteriophage cocktail as biopreservatives against Salmonella enterica serovar Typhimurium in chicken meat. *Food Biosci.* **2023**, *56*, 103290. [CrossRef]
63. Pan, H.; Shu, M.; Li, T.; Shen, K.; Zhao, Y.; Liao, N.; Zhong, C.; Wu, G. Isolation and characterization of two lytic phages against multidrug-resistant Salmonella and their application as a cocktail for biocontrol in foods. *LWT—Food Sci. Technol.* **2023**, *184*, 115184. [CrossRef]
64. Torkashvand, N.; Kamyab, H.; Reza, A.; Mohammad, S.; Khoshayand, R. Isolation, characterization, and genome analysis of a broad host range Salmonella phage vB_SenS_TUMS_E4: A candidate bacteriophage for biocontrol. *Vet. Res. Commun.* **2023**, *47*, 1493–1503. [CrossRef]
65. Zhao, J.; Lin, Y.; Wang, C.; Zayda, M.; Thida, A.; Noor, T.; Minh, H.; Yu, P.; Ma, M.; Gong, D.; et al. Biocontrol of Salmonella Typhimurium in milk, lettuce, raw pork meat and ready-to-eat steamed-chicken breast by using a novel bacteriophage with broad host range. *Int. J. Food Microbiol.* **2023**, *402*, 110295. [CrossRef]
66. Erol, H.B.; Kaskatepe, B. Effect of phage and rhamnolipid on Salmonella Infantis biofilm removal and biological control of phage on food deterioration. *Int. J. Food Sci. Technol.* **2024**, *59*, 120–128. [CrossRef]
67. Ashrafudoulla, M.; Kim, H.J.; Her, E.; Shaila, S.; Park, S.H.; Ha, S.D. Characterization of Salmonella Thompson-specific bacteriophages and their preventive application against Salmonella Thompson biofilm on eggshell as a promising antimicrobial agent in the food industry. *Food Control* **2023**, *154*, 110008. [CrossRef]
68. Guo, Y.; Li, J.; Islam, M.S.; Yan, T.; Zhou, Y.; Liang, L.; Connerton, I.F.; Deng, K.; Li, J. Application of a novel phage vB_SalS-LPSTLL for the biological control of Salmonella in foods. *Food Res. Int.* **2021**, *147*, 110492. [CrossRef]
69. Lu, Z.; Marchant, J.; Thompson, S.; Melgarejo, H.; Ignatova, D.; Kopić, S.; Damaj, R.; Trejo, H.; Paramo, R.; Reed, A.; et al. Bacteriophages isolated from turkeys infecting diverse Salmonella serovars. *Front. Microbiol.* **2022**, *13*, 933751. [CrossRef]
70. Thung, T.Y.; Lee, E.; Mahyudin, N.A.; Anuradha, K.; Mazlan, N.; Kuan, C.H.; Pui, C.F.; Ghazali, F.M.; Mahmud Ab Rashid, N.K.; Rollon, W.D.; et al. Evaluation of a lytic bacteriophage for bio-control of Salmonella Typhimurium in different food matrices. *LWT—Food Sci. Technol.* **2019**, *105*, 211–214. [CrossRef]
71. Li, T.; Chen, H.; Zhao, J.; Tao, Z.; Lan, W.; Zhao, Y.; Sun, X. Characterization of phage vB_SalM_SPJ41 and the reduction of risk of antibiotic-resistant Salmonella enterica contamination in two ready-to-eat foods. *Antibiotics* **2023**, *12*, 364. [CrossRef] [PubMed]
72. Xu, D.; Jiang, Y.; Wang, L.; Yao, L.; Li, F.; Zhai, Y.; Zhang, Y. Biocontrol of Salmonella Typhimurium in raw salmon fillets and scallop adductors by using bacteriophage SLMP1. *J. Food Prot.* **2018**, *81*, 1304–1312. [CrossRef] [PubMed]
73. Le, T.S.; Southgate, P.C.; O'Connor, W.; Poole, S.; Kurtböke, D.I. Bacteriophages as biological control agents of enteric bacteria contaminating edible oysters. *Curr. Microbiol.* **2018**, *75*, 611–619. [CrossRef] [PubMed]
74. Esmael, A.; Azab, E.; Gobouri, A.A.; Nasr-Eldin, M.A.; Moustafa, M.M.A.; Mohamed, S.A.; Badr, O.A.M.; Abdelatty, A.M. Isolation and characterization of two lytic bacteriophages infecting a multi-drug resistant Salmonella typhimurium and their efficacy to combat salmonellosis in ready-to-use foods. *Microorganisms* **2021**, *9*, 423. [CrossRef]
75. Kim, S.H.; Lee, H.; Park, M.K. Isolation, characterization, and application of a novel, lytic phage vB_SalA_KFSST3 with depolymerase for the control of Salmonella and its biofilm on cantaloupe under cold temperature. *Food Res. Int.* **2023**, *172*, 113062. [CrossRef]
76. Wong, C.W.Y.; Delaquis, P.; Goodridge, L.; Lévesque, R.C.; Fong, K.; Wang, S. Inactivation of Salmonella enterica on post-harvest cantaloupe and lettuce by a lytic bacteriophage cocktail. *Curr. Res. Food Sci.* **2020**, *2*, 25–32. [CrossRef]
77. Azari, R.; Yousefi, M.H.; Taghipour, Z.; Wagemans, J.; Lavigne, R.; Hosseinzadeh, S.; Mazloomi, S.M.; Vallino, M.; Khalatbari-Limaki, S.; Berizi, E. Application of the lytic bacteriophage Rostam to control *Salmonella enteritidis* in eggs. *Int. J. Food Microbiol.* **2023**, *389*, 110097. [CrossRef]
78. Zhang, X.; Niu, Y.D.; Nan, Y.; Stanford, K.; Holley, R.; McAllister, T.; Narváez-Bravo, C. SalmoFresh™ effectiveness in controlling Salmonella on romaine lettuce, mung bean sprouts and seeds. *Int. J. Food Microbiol.* **2019**, *305*, 108250. [CrossRef]
79. Li, S.; Konoval, H.M.; Marecek, S.; Lathrop, A.A.; Pokharel, S. Salmonella spp. response to lytic bacteriophage and lactic acid on marinated and tenderized raw pork loins. *Foods* **2022**, *11*, 879. [CrossRef]
80. Shebs-Maurine, E.L.; Giotto, F.M.; Laidler, S.T.; de Mello, A.S. Effects of bacteriophages and peroxyacetic acid applications on beef contaminated with Salmonella during different grinding stages. *Meat Sci.* **2021**, *173*, 108407. [CrossRef] [PubMed]
81. Walker, N.; Li, S.; Strauss, H.; Pokharel, S. Salmonella Typhimurium DT 104 response to lytic bacteriophage and lactobionic acid on raw chicken breast. *Food Microbiol.* **2021**, *100*, 103862. [CrossRef] [PubMed]
82. Rani, A.; Ravindran, V.B.; Surapaneni, A.; Mantri, N.; Ball, A.S. Review: Trends in point-of-care diagnosis for *Escherichia coli* O157:H7 in food and water. *Int. J. Food Microbiol.* **2021**, *349*, 109233. [CrossRef] [PubMed]
83. Liu, B.; Wang, J.; Wang, L.; Ding, P.; Yang, P.; Yang, B. Transcriptional activator OvrA encoded in O island 19 modulates virulence gene expression in enterohemorrhagic *Escherichia coli* O157:H7. *J. Infect. Dis.* **2020**, *221*, 820–829. [CrossRef]

184. Li, Y.K.; Chen, H.; Shu, M.; Zhong, C.; Bi, Y.; Yang, H.H.; Wu, G.P. Isolation, characterization and application of an alkaline resistant virulent bacteriophage JN01 against *Escherichia coli* O157:H7 in milk and beef. *LWT—Food Sci. Technol.* **2021**, *144*, 111266 [CrossRef]
185. Salaheen, S.; Kim, S.W.; Karns, J.S.; Hovingh, E.; Haley, B.J.; Van Kessel, J.A.S. Metagenomic analysis of the fecal microbiomes from *Escherichia coli* O157:H7-shedding and non-shedding cows on a single dairy farm. *Food Control* **2019**, *102*, 76–80. [CrossRef]
186. Kintz, E.; Byrne, L.; Jenkins, C.; McCarthy, N.; Vivancos, R.; Hunter, P. Outbreaks of Shiga toxin–producing *Escherichia coli* linked to sprouted seeds, salad, and leafy greens: A systematic review. *J. Food Prot.* **2019**, *82*, 1950–1958. [CrossRef]
187. Mok, J.H.; Niu, Y.; Yousef, A.; Zhao, Y.; Sastry, S.K. Spatial persistence of *Escherichia coli* O157:H7 flowing on micropatterned structures inspired by stomata and microgrooves of leafy greens. *Innov. Food Sci. Emerg. Technol.* **2022**, *75*, 102889. [CrossRef]
188. Yeni, F.; Yavaş, S.; Alpas, H.; Soyer, Y. Most common foodborne pathogens and mycotoxins on fresh produce: A review of recent outbreaks. *Crit. Rev. Food Sci. Nutr.* **2016**, *56*, 1532–1544. [CrossRef]
189. Choo, K.W.; Mao, L.; Mustapha, A. CAM-21, a novel lytic phage with high specificity towards *Escherichia coli* O157:H7 in food products. *Int. J. Food Microbiol.* **2023**, *386*, 110026. [CrossRef]
190. Ding, Y.; Nan, Y.; Qiu, Y.; Niu, D.; Holley, R.; McAllister, T.; Narváez-Bravo, C.; Stanford, K. Use of a phage cocktail to reduce the numbers of seven *Escherichia coli* strains belonging to different STEC serogroups applied to fresh produce and seeds. *J. Food Saf.* **2023**, *43*, e13044. [CrossRef]
191. Li, S.; Konoval, H.M.; Marecek, S.; Lathrop, A.A.; Feng, S.; Pokharel, S. Control of *Escherichia coli* O157:H7 using lytic bacteriophage and lactic acid on marinated and tenderized raw pork loins. *Meat Sci.* **2023**, *196*, 109030. [CrossRef] [PubMed]
192. Oh, M.; Cevallos-Urena, A.; Kim, B.S. Bacteriophages PECP14, PECP20, and their endolysins as effective biocontrol agents for *Escherichia coli* O157:H7 and other foodborne pathogens. *Int. J. Food Microbiol.* **2024**, *409*, 110460. [CrossRef] [PubMed]
193. Yesil, M.; Kasler, D.R.; Huang, E.; Yousef, A.E. Lytic *Escherichia* phage OSYSP acts additively and synergistically with gaseous ozone against *Escherichia coli* O157:H7 on spinach leaves. *Sci. Rep.* **2023**, *13*, 10706. [CrossRef]
194. Artawinata, P.C.; Lorraine, S.; Waturangi, D.E. Isolation and characterization of bacteriophages from soil against food spoilage and foodborne pathogenic bacteria. *Sci. Rep.* **2023**, *13*, 9282. [CrossRef]
195. Wang, Y.; Li, J.; Wang, Y.; Wu, J.; Wang, X.; Xue, F.; Ren, J.; Dai, J.; Tang, F. Sterilizing effect of phage cocktail against Shiga toxin-producing *Escherichia coli* O157:H7 in foods. *Food Biosci.* **2023**, *51*, 102282. [CrossRef]
196. Zhang, H.Z.; Shu, M.; Yang, W.Y.; Pan, H.; Tang, M.X.; Zhao, Y.Y.; Zhong, C.; Wu, G.P. Isolation and characterization of a novel Salmonella bacteriophage JNwz02 capable of lysing *Escherichia coli* O157:H7 and its antibacterial application in foods. *LWT—Food Sci. Technol.* **2023**, *173*, 114251. [CrossRef]
197. L.A, L.A.; Waturangi, D.E. Application of BI-EHEC and BI-EPEC bacteriophages to control enterohemorrhagic and enteropathogenic *Escherichia coli* on various food surfaces. *BMC Res. Notes* **2023**, *16*, 102. [CrossRef]
198. El-Nour, S.A.A.; Hammad, A.A.; Fathy, R.; Eid, A.S. Application of coliphage as biocontrol agent in combination with gamma irradiation to eliminate multi-drug-resistant *E. coli* in minimally processed vegetables. *Environ. Sci. Pollut. Res. Int.* **2023**, *30*, 123907–123924. [CrossRef]
199. Kuek, M.; Mclean, S.K.; Palombo, E.A. Control of *Escherichia coli* in fresh-cut mixed vegetables using a combination of bacteriophage and carvacrol. *Antibiotics* **2023**, *12*, 1579. [CrossRef]
200. Fawzy, S.; El-shenawy, M.A.; Senna, A.S.A.; El-shibiny, A.; Industries, F.; Therapy, P. The application of bacteriophage ZCEC5 to control multi-drug resistant *E. coli* in dairy products. *Egypt. J. Chem.* **2023**, *66*, 369–382. [CrossRef]
201. Duc, H.M.; Son, H.M.; Yi, H.P.S.; Sato, J.; Ngan, P.H.; Masuda, Y.; Honjoh, K.-i.; Miyamoto, T. Isolation, characterization and application of a polyvalent phage capable of controlling *Salmonella* and *Escherichia coli* O157:H7 in different food matrices. *Food Res. Int.* **2020**, *131*, 108977. [CrossRef] [PubMed]
202. Dewanggana, M.N.; Evangeline, C.; Ketty, M.D.; Waturangi, D.E.; Yogiara; Magdalena, S. Isolation, characterization, molecular analysis and application of bacteriophage DW-EC to control Enterotoxigenic *Escherichia coli* on various foods. *Sci. Rep.* **2022**, *12*, 495. [CrossRef] [PubMed]
203. Zhu, L.; Hou, Y.; Huang, X.; Wang, S.; Xie, R.; Yang, J.; Lv, Q.; Hua, L.; Liang, W.; Peng, Z.; et al. Isolation of three coliphages and the evaluation of their phage cocktail for biocontrol of Shiga toxin-producing *Escherichia coli* O157 in milk. *Curr. Microbiol.* **2022**, *79*, 216. [CrossRef]
204. Shebs, E.L.; Lukov, M.J.; Giotto, F.M.; Torres, E.S.; de Mello, A.S. Efficacy of bacteriophage and organic acids in decreasing STEC O157:H7 populations in beef kept under vacuum and aerobic conditions: A simulated high event period scenario. *Meat Sci.* **2020**, *162*, 108023. [CrossRef]
205. Stratakos, A.C.; Grant, I.R. Evaluation of the efficacy of multiple physical, biological and natural antimicrobial interventions for control of pathogenic *Escherichia coli* on beef. *Food Microbiol.* **2018**, *76*, 209–218. [CrossRef]
206. Zhou, Y.; Wan, Q.; Bao, H.; Guo, Y.; Zhu, S.; Zhang, H.; Pang, M.; Wang, R. Application of a novel lytic phage vB_EcoM_SQ17 for the biocontrol of enterohemorrhagic *Escherichia coli* O157:H7 and enterotoxigenic *E. coli* in food matrices. *Front. Microbiol.* **2022**, *13*, 929005. [CrossRef]

207. Centers for Disease Control and Prevention (CDC). Listeria Infection (Listeriosis). Available online: https://www.cdc.gov/listeria/index.html (accessed on 16 January 2025).
208. Barancelli, G.V.; Camargo, T.M.; Reis, C.M.F.; Porto, E.; Hofer, E.; Oliveira, C.A.F. Incidence of *Listeria monocytogenes* in cheese manufacturing plants from the northeast region of São Paulo, Brazil. *J. Food Prot.* **2011**, *74*, 816–819. [CrossRef]
209. Buchanan, R.L.; Gorris, L.G.M.; Hayman, M.M.; Jackson, T.C.; Whiting, R.C. A review of *Listeria monocytogenes*: An update on outbreaks, virulence, dose-response, ecology, and risk assessments. *Food Control* **2017**, *75*, 1–13. [CrossRef]
210. Garner, D.; Kathariou, S. Fresh produce-associated listeriosis outbreaks, sources of concern, teachable moments, and insights. *J. Food Prot.* **2016**, *79*, 337–344. [CrossRef]
211. Kurpas, M.; Wieczorek, K.; Osek, J. Ready-to-eat meat products as a source of *Listeria monocytogenes*. *J. Vet. Res.* **2018**, *62*, 49–55. [CrossRef] [PubMed]
212. Saraiva, C.; García-Díez, J.; da Conceição Fontes, M.; Esteves, A. Modeling the behavior of *Listeria monocytogenes* in meat. In *Listeria Monocytogenes*, 1st ed.; Nyila, M.A., Ed.; InTech: Rijeka, Croatia, 2018; pp. 154–196. [CrossRef]
213. Ahmadi, H.; Barbut, S.; Lim, L.T.; Balamurugan, S. Examination of the use of bacteriophage as an additive and determining its best application method to control *Listeria monocytogenes* in a cooked-meat model system. *Front. Microbiol.* **2020**, *11*, 779. [CrossRef] [PubMed]
214. Ayaz, N.D.; Onaran, B.; Cufaoglu, G.; Goncuoglu, M.; Ormanci, F.S.; Erol, I. Prevalence and characterization of *Listeria monocytogenes* isolated from beef and sheep carcasses in Turkey with characterization of locally isolated Listeriophages as a control measure. *J. Food Prot.* **2018**, *81*, 2045–2053. [CrossRef] [PubMed]
215. Byun, K.H.; Han, S.H.; Choi, M.W.; Kim, B.H.; Ha, S.D. Efficacy of disinfectant and bacteriophage mixture against planktonic and biofilm state of *Listeria monocytogenes* to control in the food industry. *Int. J. Food Microbiol.* **2024**, *413*, 110587. [CrossRef]
216. Colás-Medà, P.; Viñas, I.; Alegre, I. Evaluation of commercial anti-Listerial Products for improvement of food safety in ready-to-eat meat and dairy products. *Antibiotics* **2023**, *12*, 414. [CrossRef]
217. Hluchanova, L.; Korena, K.; Juricova, H. Vacuum-packed steak tartare: Prevalence of *Listeria monocytogenes* and evaluation of efficacy of ListexTM P100. *Foods* **2022**, *11*, 533. [CrossRef]
218. Ishaq, A.; Ebner, P.D.; Syed, Q.A.; Ubaid ur Rahman, H. Employing list-shield bacteriophage as a bio-control intervention for *Listeria monocytogenes* from raw beef surface and maintain meat quality during refrigeration storage. *LWT—Food Sci. Technol.* **2020**, *132*, 109784. [CrossRef]
219. Byun, K.H.; Han, S.H.; Choi, M.W.; Park, S.H.; Ha, S.D. Isolation, characterization, and application of bacteriophages to reduce and inhibit *Listeria monocytogenes* in celery and enoki mushroom. *Food Control* **2022**, *135*, 108826. [CrossRef]
220. Gómez-Galindo, M.; Truchado, P.; Allende, A.; Gil, M.I. Optimization of the use of a commercial phage-based product as a control strategy of *Listeria monocytogenes* in the fresh-cut industry. *Foods* **2023**, *12*, 3171. [CrossRef]
221. Lewis, R.; Bolocan, A.S.; Draper, L.A.; Ross, P.; Hill, C. The effect of a commercially available bacteriophage and bacteriocin on *Listeria monocytogenes* in coleslaw. *Viruses* **2019**, *11*, 977. [CrossRef]
222. Stone, E.; Lhomet, A.; Neve, H.; Grant, I.R.; Campbell, K.; McAuliffe, O. Isolation and characterization of *Listeria monocytogenes* phage vB_LmoH_P61, a phage with biocontrol potential on different food matrices. *Front. Sustain. Food Syst.* **2020**, *4*, 521645. [CrossRef]
223. Truchado, P.; Elsser-Gravesen, A.; Gil, M.I.; Allende, A. Post-process treatments are effective strategies to reduce *Listeria monocytogenes* on the surface of leafy greens: A pilot study. *Int. J. Food Microbiol.* **2020**, *313*, 108390. [CrossRef] [PubMed]
224. Axelsson, L.; Bjerke, G.A.; McLeod, A.; Berget, I.; Holck, A.L. Growth behavior of *Listeria monocytogenes* in a traditional Norwegian fermented fish product (Rakfisk), and its inhibition through bacteriophage addition. *Foods* **2020**, *9*, 119. [CrossRef] [PubMed]
225. Heir, E.; Jensen, M.R.; Aasli, A.W.; Berget, I.; Holck, A.L. Reduction and growth inhibition of *Listeria monocytogenes* by use of anti-listerial nisin, P100 phages and buffered dry vinegar fermentates in standard and sodium-reduced cold-smoked salmon. *Foods* **2023**, *12*, 4391. [CrossRef]
226. Zhou, C.; Zhu, M.; Wang, Y.; Yang, Z.; Ye, M.; Wu, L.; Bao, H.; Pang, M.; Zhou, Y.; Wang, R.; et al. Broad host range phage vB-LmoM-SH3-3 reduces the risk of *Listeria* contamination in two types of ready-to-eat food. *Food Control* **2020**, *108*, 106830. [CrossRef]
227. Henderson, L.O.; Cabrera-Villamizar, L.A.; Skeens, J.; Kent, D.; Murphy, S.; Wiedmann, M.; Guariglia-Oropeza, V. Environmental conditions and serotype affect *Listeria monocytogenes* susceptibility to phage treatment in a laboratory cheese model. *J. Dairy Sci.* **2019**, *102*, 9674–9688. [CrossRef]
228. Schamp, C.N.; Dhowlaghar, N.; Hudson, L.K.; Bryan, D.W.; Zhong, Q.; Fozo, E.M.; Gaballa, A.; Wiedmann, M.; Denes, T.G. Selection of mutant *Listeria* phages under food-relevant conditions can enhance application potential. *Appl. Environ. Microbiol.* **2023**, *89*, e01007-23. [CrossRef]
229. World Health Organization (WHO). Campylobacter. Available online: https://www.who.int/news-room/fact-sheets/detail/campylobacter (accessed on 16 January 2025).

230. Deng, W.; Dittoe, D.K.; Pavilidis, H.O.; Chaney, W.E.; Yang, Y.; Ricke, S.C. Current perspectives and potential of probiotics to limit foodborne *Campylobacter* in poultry. *Front. Microbiol.* **2020**, *11*, 583429. [CrossRef]
231. Mota-Gutierrez, J.; Lis, L.; Lasagabaster, A.; Nafarrate, I.; Ferrocino, I.; Cocolin, L.; Rantsiou, K. *Campylobacter* spp. prevalence and mitigation strategies in the broiler production chain. *Food Microbiol.* **2022**, *104*, 103998. [CrossRef]
232. Umar, S.; Maiyah, A.T.; Mushtaq, A. *Campylobacter* infections in poultry: Update on challenges and potential immune interventions. *World's Poult. Sci. J.* **2016**, *72*, 381–390. [CrossRef]
233. Thung, T.Y.; Lee, E.; Mahyudin, N.A.; Wan Mohamed Radzi, C.W.J.; Mazlan, N.; Tan, C.W.; Radu, S. Partial characterization and in vitro evaluation of a lytic bacteriophage for biocontrol of *Campylobacter jejuni* in mutton and chicken meat. *J. Food Saf.* **2020**, *40*, e12770. [CrossRef]
234. Zhang, X.; Tang, M.; Zhou, Q.; Lu, J.; Zhang, H.; Tang, X.; Ma, L.; Zhang, J.; Chen, D.; Gao, Y. A broad host phage, CP6, for combating multidrug-resistant *Campylobacter* prevalent in poultry meat. *Poult. Sci.* **2024**, *103*, 103548. [CrossRef] [PubMed]
235. Jäckel, C.; Hammerl, J.A.; Hertwig, S. *Campylobacter* phage isolation and characterization: What we have learned so far. *Methods Protoc.* **2019**, *2*, 18. [CrossRef] [PubMed]
236. García-Anaya, M.C.; Sepúlveda, D.R.; Rios-Velasco, C.; Zamudio-Flores, P.B.; Sáenz-Mendoza, A.I.; Acosta-Muñiz, C.H. The role of food compounds and emerging technologies on phage stability. *Innov. Food Sci. Emerg. Technol.* **2020**, *64*, 102436. [CrossRef]
237. García-Anaya, M.C.; Sepulveda, D.R.; Sáenz-Mendoza, A.I.; Rios-Velasco, C.; Zamudio-Flores, P.B.; Acosta-Muñiz, C.H. Phages as biocontrol agents in dairy products. *Trends Food Sci. Technol.* **2020**, *95*, 10–20. [CrossRef]
238. Jonczyk-Matysiak, E.; Łodej, N.; Kula, D.; Owczarek, B.; Orwat, F.; Międzybrodzki, R.; Neuberg, J.; Bagińska, N.; Weber-Dąbrowska, B.; Górski, A. Factors determining phage stability/activity: Challenges in practical phage application. *Expert Rev. Anti-Infect. Ther.* **2019**, *17*, 583–606. [CrossRef]
239. Ahmadi, H.; Radford, D.; Kropinski, A.M.; Lim, L.T.; Balamurugan, S. Thermal-stability and reconstitution ability of *Listeria* phages P100 and A511. *Front. Microbiol.* **2017**, *8*, 2375. [CrossRef]
240. Duarte, J.; Pereira, C.; Moreirinha, C.; Salvio, R.; Lopes, A.; Wang, D.; Almeida, A. New insights on phage efficacy to control *Aeromonas salmonicida* in aquaculture systems: An in vitro preliminary study. *Aquaculture* **2018**, *495*, 970–982. [CrossRef]
241. Ackermann, H.-W.; Tremblay, D.; Moineau, S. Long-term bacteriophage preservation. *WFCC Newsl.* **2004**, *38*, 35–40.
242. Vandenheuvel, D.; Lavigne, R.; Brüssow, H. Bacteriophage therapy: Advances in formulation strategies and human clinical trials. *Annu. Rev. Virol.* **2015**, *2*, 599–618. [CrossRef]
243. Mojica, K.D.A.; Brussaard, C.P.D. Factors affecting virus dynamics and microbial host-virus interactions in marine environments. *FEMS Microbiol. Ecol.* **2014**, *89*, 495–515. [CrossRef] [PubMed]
244. Pinheiro, L.A.M.; Pereira, C.; Frazão, C.; Balcão, V.; Almeida, A. Efficiency of phage φ6 for biocontrol of *Pseudomonas syringae* pv. *syringae: An in vitro preliminary study. Microorganisms* **2019**, *7*, 286. [CrossRef]
245. Silva, Y.J.; Costa, L.; Pereira, C.; Cunha, Â.; Calado, R.; Gomes, N.C.M.; Almeida, A. Influence of environmental variables in the efficiency of phage therapy in aquaculture. *Microb. Biotechnol.* **2014**, *7*, 401–413. [CrossRef] [PubMed]
246. Tey, B.T.; Ooi, S.T.; Yong, K.C.; Ng, M.Y.T.; Ling, T.C.; Tan, W.S. Production of fusion m13 phage bearing the di-sulphide constrained peptide sequence (C-WSFFSNI-C) that interacts with hepatitis B core antigen. *Afr. J. Biotechnol.* **2009**, *8*, 268–273.
247. Kowalska, J.D.; Kazimierczak, J.; Sowińska, P.M.; Wójcik, E.A.; Siwicki, A.K.; Dastych, J. Growing trend of fighting infections in aquaculture environment—Opportunities and challenges of phage therapy. *Antibiotics* **2020**, *9*, 301. [CrossRef]
248. Qadir, M.I.; Chauhdary, Z. Antibacterial activity of novel strains of bacteriophages: An experimental approach. *Crit. Rev. Eukaryot. Gene Expr.* **2018**, *28*, 1–12. [CrossRef]
249. Vörös, Z.; Sevcsik, G.; Csík, G.; Herényi, L.; Kellermayer, M.S. Temperature-dependent nanomechanics and topography of bacteriophage T7. *Biophys. J.* **2018**, *114*, 354A. [CrossRef]
250. Broeker, N.K.; Kiele, F.; Casjens, S.R.; Gilcrease, E.B.; Thalhammer, A.; Koetz, J.; Barbirz, S. In vitro studies of lipopolysaccharide-mediated DNA release of podovirus HK620. *Viruses* **2018**, *10*, 289. [CrossRef]
251. Liu, T.; Sae-Ueng, U.; Li, D.; Lander, G.C.; Zuo, X.; Jönsson, B.; Rau, D.; Shefer, I.; Evilevitch, A.; Reiss, H. Solid-to-fluid-like DNA transition in viruses facilitates infection. *Proc. Natl. Acad. Sci. USA* **2014**, *111*, 14675–14680. [CrossRef]
252. Zhang, C.-Y.; Zhang, N.-H. Size dependent correlation between structure and apparent stiffness of viral DNA during temperature variation. *J. Mech. Phys. Solids* **2021**, *154*, 104501. [CrossRef]
253. Taj, M.K.; Ling, J.X.; Bing, L.L.; Qi, Z.; Taj, I.; Hassani, T.M.; Samreen, Z.; Yunlin, W. Effect of dilution, temperature and pH on the lysis activity of T4 phage against *E. coli* BL21. *J. Anim. Plant Sci.* **2014**, *24*, 1252–1255.
254. Ravi, Y.; Pooja, M.K.; Krushna Yadav, D.K. Review—Bacteriophages in food preservation. *Int. J. Pure Appl. Biosci.* **2017**, *5*, 197–205. [CrossRef]
255. Leverentz, B.; Conway, W.S.; Alavidze, Z.; Janisiewicz, W.J.; Fuchs, Y.; Camp, M.J.; Chighladze, E.; Sulakvelidze, A. Examination of bacteriophage as a biocontrol method for *Salmonella* on fresh-cut fruit: A model study. *J. Food Prot.* **2001**, *64*, 1116–1121. [CrossRef]

256. Leverentz, B.; Conway, W.S.; Camp, M.J.; Janisiewicz, W.J.; Abuladze, T.; Yang, M.; Saftner, R.; Sulakvelidze, A. Biocontrol of *Listeria monocytogenes* on fresh cut produce by treatment with lytic bacteriophages and a bacteriocin. *Appl. Environ. Microbiol.* **2003**, *69*, 4519–4526. [CrossRef] [PubMed]
257. Oliveira, M.; Viñas, I.; Colàs, P.; Anguera, M.; Usall, J.; Abadias, M. Effectiveness of a bacteriophage in reducing *Listeria monocytogenes* on fresh-cut fruits and fruit juices. *Food Microbiol.* **2014**, *38*, 137–142. [CrossRef] [PubMed]
258. Buttimer, C.; McAuliffe, O.; Ross, R.P.; Hill, C.; O'Mahony, J.; Coffey, A. Bacteriophages and bacterial plant diseases. *Front. Microbiol.* **2017**, *8*, 34. [CrossRef]
259. Turgeon, N.; Toulouse, M.J.; Martel, B.; Moineau, S.; Duchaine, C. Comparison of five bacteriophages as models for viral aerosol studies. *Appl. Environ. Microbiol.* **2014**, *80*, 4242–4250. [CrossRef]
260. Vasickova, P.; Kovarcik, K. Natural persistence of food- and waterborne viruses. In *Viruses in Food and Water: Risks, Surveillance and Control*; Cook, N., Ed.; Woodhead Publishing Limited: Cambridge, UK, 2013; pp. 179–204. [CrossRef]
261. Verreault, D.; Marcoux-Voiselle, M.; Turgeon, N.; Moineau, S.; Duchaine, C. Resistance of aerosolized bacterial viruses to relative humidity and temperature. *Appl. Environ. Microbiol.* **2015**, *81*, 7305–7311. [CrossRef]
262. Sharma, S.; Chatterjee, S.; Datta, S.; Prasad, R.; Dubey, D.; Prasad, R.K.; Vairale, M.G. Bacteriophages and its applications: An overview. *Folia Microbiol.* **2017**, *62*, 17–55. [CrossRef]
263. Zhang, J.; Hong, Y.; Fealey, M.; Singh, A.; Walton, K.; Martin, C.; Harman, N.J.; Mahlie, J.; Ebner, P.D. Physiological and molecular characterization of *Salmonella* bacteriophages previously used in phage therapy. *J. Food Prot.* **2015**, *78*, 2143–2149. [CrossRef]
264. Hudson, J.A.; Billington, C.; Premaratne, A.; On, S.L.W. Inactivation of *Escherichia coli* O157:H7 using ultraviolet light-treated bacteriophages. *Food Sci. Technol. Int.* **2016**, *22*, 3–9. [CrossRef] [PubMed]
265. Anany, H.; Chen, W.; Pelton, R.; Griffiths, M.W. Biocontrol of *Listeria monocytogenes* and *Escherichia coli* O157:H7 in meat by using phages immobilized on modified cellulose membranes. *Appl. Environ. Microbiol.* **2011**, *77*, 6379–6387. [CrossRef] [PubMed]
266. Rode, T.M.; Axelsson, L.; Granum, P.E.; Heir, E.; Holck, A.; L'Abée-Lund, T.M. High stability of Stx2 phage in food and under food-processing conditions. *Appl. Environ. Microbiol.* **2011**, *77*, 5336–5341. [CrossRef] [PubMed]
267. Carrigy, N.B.; Liang, L.; Wang, H.; Kariuki, S.; Nagel, T.E.; Connerton, I.F.; Vehring, R. Spray-dried anti-Campylobacter bacteriophage CP30A powder suitable for global distribution without cold chain infrastructure. *Int. J. Pharm.* **2019**, *569*, 118601. [CrossRef]
268. Guenther, S.; Huwyler, D.; Richard, S.; Loessner, M.J. Virulent bacteriophage for efficient biocontrol of *Listeria monocytogenes* in ready-to-eat foods. *Appl. Environ. Microbiol.* **2009**, *75*, 93–100. [CrossRef]
269. Zinno, P.; Devirgiliis, C.; Ercolini, D.; Ongeng, D.; Mauriello, G. Bacteriophage P22 to challenge *Salmonella* in foods. *Int. J. Food Microbiol.* **2014**, *191*, 69–74. [CrossRef]
270. Bao, H.; Zhang, P.; Zhang, H.; Zhou, Y.; Zhang, L.; Wang, R. Bio-control of *Salmonella* Enteritidis in foods using bacteriophages. *Viruses* **2015**, *7*, 4836–4853. [CrossRef]
271. Thung, T.Y.; Krishanthi Jayarukshi Kumari Premarathne, J.M.; San Chang, W.; Loo, Y.Y.; Chin, Y.Z.; Kuan, C.H.; Tan, C.W.; Basri, D.F.; Jasimah Wan Mohamed Radzi, C.W.; Radu, S. Use of a lytic bacteriophage to control *Salmonella* Enteritidis in retail food. *LWT—Food Sci. Technol.* **2017**, *78*, 222–225. [CrossRef]
272. De Siqueira, R.S.; Dodd, C.E.R.; Rees, C.E.D. Evaluation of the natural virucidal activity of teas for use in the phage amplification assay. *Int. J. Food Microbiol.* **2006**, *111*, 259–262. [CrossRef]
273. García, P.; Madera, C.; Martínez, B.; Rodríguez, A.; Suárez, E. Prevalence of bacteriophages infecting *Staphylococcus aureus* in dairy samples and their potential as biocontrol agents. *J. Dairy Sci.* **2009**, *92*, 3019–3026. [CrossRef]
274. García, P.; Rodríguez, L.; Rodríguez, A.; Martínez, B. Food biopreservation: Promising strategies using bacteriocins, bacteriophages and endolysins. *Trends Food Sci. Technol.* **2010**, *21*, 373–382. [CrossRef]
275. García-Anaya, M.C.; Sepulveda, D.R.; Rios-Velasco, C.; Zamudio-Flores, P.B.; Acosta-Muñiz, C.H. Effect of homogenization on binding-affinity of bacteriophage A511 in bovine milk fractions. *J. Food Eng.* **2019**, *244*, 73–79. [CrossRef]
276. Liu, S.; Quek, S.Y.; Huang, K. Advanced strategies to overcome the challenges of bacteriophage-based antimicrobial treatments in food and agricultural systems. *Crit. Rev. Food Sci. Nutr.* **2023**, *64*, 12574–12598. [CrossRef]
277. Batalha, L.S.; Gontijo, M.T.P.; Teixeira, A.V.N.C.; Boggione, D.M.G.; Lopez, M.E.S.; Eller, M.R.; Mendonça, R.C.S. Encapsulation in alginate-polymers improves stability and allows controlled release of the UFV-AREG1 bacteriophage. *Food Res. Int.* **2021**, *139*, 109947. [CrossRef] [PubMed]
278. Richards, K.; Malik, D.J. Bacteriophage encapsulation in pH-responsive core-shell capsules as an animal feed additive. *Viruses* **2021**, *13*, 1131. [CrossRef] [PubMed]
279. Colom, J.; Cano-Sarabia, M.; Otero, J.; Cortés, P.; Maspoch, D.; Llagostera, M. Liposome-encapsulated bacteriophages for enhanced oral phage therapy against *Salmonella* spp. *Appl. Environ. Microbiol.* **2015**, *81*, 4841–4849. [CrossRef]
280. Rosner, D.; Clark, J. Formulations for bacteriophage therapy and the potential uses of immobilization. *Pharmaceuticals* **2021**, *14*, 359. [CrossRef]

281. Appendini, P.; Hotchkiss, J.H. Review of antimicrobial food packaging. *Innov. Food Sci. Emerg. Technol.* **2002**, *3*, 113–126. [CrossRef]
282. Quintavalla, S.; Vicini, L. Antimicrobial food packaging in meat industry. *Meat Sci.* **2002**, *62*, 373–380. [CrossRef]
283. Tan, C.; Han, F.; Zhang, S.; Li, P.; Shang, N. Novel bio-based materials and applications in antimicrobial food packaging: Recent advances and future trends. *Int. J. Mol. Sci.* **2021**, *22*, 9663. [CrossRef]
284. Amin, U.; Khan, M.K.I.; Maan, A.A.; Nazir, A.; Riaz, S.; Khan, M.U.; Sultan, M.; Munekata, P.E.S.; Lorenzo, J.M. Biodegradable active, intelligent, and smart packaging materials for food applications. *Food Packag. Shelf Life* **2022**, *33*, 100903. [CrossRef]
285. Duda-Chodak, A.; Tarko, T.; Petka-Poniatowska, K. Antimicrobial compounds in food packaging. *Int. J. Mol. Sci.* **2023**, *24*, 2457. [CrossRef] [PubMed]
286. Wang, J.; Euring, M.; Ostendorf, K.; Zhang, K. Biobased materials for food packaging. *J. Bioresour. Bioprod.* **2022**, *7*, 1–13. [CrossRef]
287. Fernández-Pan, I.; Ignacio, J.; Caballero, M. Biopolymers for edible films and coatings in food applications. In *Biopolymers—New Materials for Sustainable Films and Coatings*; Plackett, D., Ed.; John Wiley & Sons, Ltd.: Hoboken, NJ, USA, 2011; pp. 233–254. [CrossRef]
288. De Dicastillo, C.L.; Settier-Ramírez, L.; Gavara, R.; Hernández-Muñoz, P.; Carballo, G.L. Development of biodegradable films loaded with phages with antilisterial properties. *Polymers* **2021**, *13*, 327. [CrossRef] [PubMed]
289. Lone, A.; Anany, H.; Hakeem, M.; Aguis, L.; Avdjian, A.C.; Bouget, M.; Atashi, A.; Brovko, L.; Rochefort, D.; Griffiths, M.W. Development of prototypes of bioactive packaging materials based on immobilized bacteriophages for control of growth of bacterial pathogens in foods. *Int. J. Food Microbiol.* **2016**, *217*, 49–58. [CrossRef]
290. Malik, D.J.; Sokolov, I.J.; Vinner, G.K.; Mancuso, F.; Cinquerrui, S.; Vladisavljevic, G.T.; Clokie, M.R.J.; Garton, N.J.; Stapley, A.G.F.; Kirpichnikova, A. Formulation, stabilisation and encapsulation of bacteriophage for phage therapy. *Adv. Colloid Interface Sci.* **2017**, *249*, 100–133. [CrossRef]
291. Bolívar-Monsalve, J.; Ramírez-Toro, C.; Bolívar, G.; Ceballos-González, C. Mechanisms of action of novel ingredients used in edible films to preserve microbial quality and oxidative stability in sausages—A review. *Trends Food Sci. Technol.* **2019**, *89*, 100–109. [CrossRef]
292. Jarray, A.; Gerbaud, V.; Hemati, M. Polymer-plasticizer compatibility during coating formulation: A multi-scale investigation. *Prog. Org. Coat.* **2016**, *101*, 195–206. [CrossRef]
293. Malik, G.K.; Khuntia, A.; Mitra, J. Comparative effect of different plasticizers on barrier, mechanical, optical, and sorption properties of hydroxypropyl methylcellulose (HPMC)–based edible film. *J. Biosyst. Eng.* **2022**, *47*, 93–105. [CrossRef]
294. Leung, V.; Szewczyk, A.; Chau, J.; Hosseinidoust, Z.; Groves, L.; Hawsawi, H.; Anany, H.; Griffiths, M.W.; Monsur Ali, M.; Filipe, C.D.M. Long-term preservation of bacteriophage antimicrobials using sugar glasses. *ACS Biomater. Sci. Eng.* **2018**, *4*, 3802–3808. [CrossRef]
295. Costa, M.J.; Pastrana, L.M.; Teixeira, J.A.; Sillankorva, S.M.; Cerqueira, M.A. Bacteriophage delivery systems for food applications: Opportunities and perspectives. *Viruses* **2023**, *15*, 1271. [CrossRef] [PubMed]
296. Choińska-Pulit, A.; Mituła, P.; Śliwka, P.; Łaba, W.; Skaradzińska, A. Bacteriophage encapsulation: Trends and potential applications. *Trends Food Sci. Technol.* **2015**, *45*, 212–221. [CrossRef]
297. O'Connell, L.; Marcoux, P.R.; Roupioz, Y. Strategies for surface immobilization of whole bacteriophages: A review. *ACS Biomater. Sci. Eng.* **2021**, *7*, 1987–2014. [CrossRef] [PubMed]
298. Alves, D.; Cerqueira, M.A.; Pastrana, L.M.; Sillankorva, S. Entrapment of a phage cocktail and cinnamaldehyde on sodium alginate emulsion-based films to fight food contamination by *Escherichia coli* and *Salmonella* Enteritidis. *Food Res. Int.* **2020**, *128*, 108791. [CrossRef]
299. Amarillas, L.; Lightbourn-Rojas, L.; Angulo-Gaxiola, A.K.; Basilio Heredia, J.; González-Robles, A.; León-Félix, J. The antibacterial effect of chitosan-based edible coating incorporated with a lytic bacteriophage against *Escherichia coli* O157:H7 on the surface of tomatoes. *J. Food Saf.* **2018**, *38*, e12571. [CrossRef]
300. Choi, I.; Yoo, D.S.; Chang, Y.; Kim, S.Y.; Han, J. Polycaprolactone film functionalized with bacteriophage T4 promotes antibacterial activity of food packaging toward *Escherichia coli*. *Food Chem.* **2021**, *346*, 128883. [CrossRef]
301. Cui, H.; Yuan, L.; Lin, L. Novel chitosan film embedded with liposome-encapsulated phage for biocontrol of *Escherichia coli* O157:H7 in beef. *Carbohydr. Polym.* **2017**, *177*, 156–164. [CrossRef]
302. Cui, H.; Yang, X.; Li, C.; Ye, Y.; Chen, X.; Lin, L. Enhancing anti-*E. coli* O157:H7 activity of composite phage nanofiber film by D-phenylalanine for food packaging. *Int. J. Food Microbiol.* **2022**, *376*, 109762. [CrossRef]
303. Korehei, R.; Kadla, J. Incorporation of T4 bacteriophage in electrospun fibres. *J. Appl. Microbiol.* **2013**, *114*, 1425–1434. [CrossRef]
304. Korehei, R.; Kadla, J.F. Encapsulation of T4 bacteriophage in electrospun poly(ethylene oxide)/cellulose diacetate fibers. *Carbohydr. Polym.* **2014**, *100*, 150–157. [CrossRef]
305. Liana, A.E.; Marquis, C.P.; Gunawan, C.; Gooding, J.J.; Amal, R. Antimicrobial activity of T4 bacteriophage conjugated indium tin oxide surfaces. *J. Colloid Interface Sci.* **2018**, *514*, 227–233. [CrossRef] [PubMed]

306. Meyer, A.; Greene, M.; Kimmelshue, C.; Cademartiri, R. Stabilization of T4 bacteriophage at acidic and basic pH by adsorption on paper. *Colloids Surf. B Biointerfaces* **2017**, *160*, 169–176. [CrossRef] [PubMed]
307. Tomat, D.; Soazo, M.; Verdini, R.; Casabonne, C.; Aquili, V.; Balagué, C.; Quiberoni, A. Evaluation of an WPC edible film added with a cocktail of six lytic phages against foodborne pathogens such as enteropathogenic and Shigatoxigenic *Escherichia coli*. *LWT—Food Sci. Technol.* **2019**, *113*, 108316. [CrossRef]
308. Vonasek, E.; Le, P.; Nitin, N. Encapsulation of bacteriophages in whey protein films for extended storage and release. *Food Hydrocoll.* **2014**, *37*, 7–13. [CrossRef]
309. Vonasek, E.L.; Choi, A.H.; Sanchez, J.; Nitin, N. Incorporating phage therapy into WPI dip coatings for applications on fresh whole and cut fruit and vegetable surfaces. *J. Food Sci.* **2018**, *83*, 1871–1879. [CrossRef]
310. Kamali, S.; Yavarmanesh, M.; Habibi Najafi, M.B.; Koocheki, A. Poly (lactic acid) and whey protein/pullulan composite bilayer film containing phage A511 as an anti-*Listeria* packaging for chicken breast at refrigerated temperatures. *LWT—Food Sci. Technol.* **2022**, *170*, 114085. [CrossRef]
311. Kim, S.; Kim, B.S.; Bai, J.; Chang, Y. Antibacterial κ-carrageenan/konjac glucomannan-based edible hydrogel film containing *Salmonella* phage PBSE191 and its application in chicken meat. *LWT—Food Sci. Technol.* **2023**, *180*, 114707. [CrossRef]
312. Radford, D.; Guild, B.; Strange, P.; Ahmed, R.; Lim, L.T.; Balamurugan, S. Characterization of antimicrobial properties of *Salmonella* phage Felix O1 and *Listeria* phage A511 embedded in xanthan coatings on poly(lactic acid) films. *Food Microbiol.* **2017**, *66*, 117–128. [CrossRef]
313. Tidim, G.; Guzel, M.; Soyer, Y.; Erel-Goktepe, I. Layer-by-layer assembly of chitosan/alginate thin films containing *Salmonella enterica* bacteriophages for antibacterial applications. *Carbohydr. Polym.* **2024**, *328*, 121710. [CrossRef]
314. Gouvêa, D.M.; Mendonça, R.C.S.; Soto, M.L.; Cruz, R.S. Acetate cellulose film with bacteriophages for potential antimicrobial use in food packaging. *LWT—Food Sci. Technol.* **2015**, *63*, 85–91. [CrossRef]
315. Sezer, B.; Kubra, E.; Hakki, I. The use of bacteriophage-based edible coatings for the biocontrol of *Salmonella* in strawberries. *Food Control* **2022**, *135*, 108812. [CrossRef]
316. Vila, M.M.D.C.; Cinto, E.C.; Pereira, A.O.; Baldo, D.Á.; Oliveira, J.M., Jr.; Balcão, V.M. An edible antibacterial coating integrating lytic bacteriophage ripened cheese. *Polymers* **2024**, *16*, 680. [CrossRef] [PubMed]
317. García-Anaya, M.C.; Sepulveda, D.R.; Rios-Velasco, C.; Acosta-Muñiz, C.H. Incorporation of A511 bacteriophage in a whey protein isolate-based edible coating for the control of *Listeria monocytogenes* in cheese. *Food Packag. Shelf Life* **2023**, *37*, 101095. [CrossRef]
318. Kamali, S.; Yavarmanesh, M.; Habibi Najafi, M.B.; Koocheki, A. Development of whey protein concentrate/pullulan composite films containing bacteriophage A511: Functional properties and anti-*Listeria* effects during storage. *Food Packag. Shelf Life* **2022**, *33*, 100902. [CrossRef]
319. Costa, M.J.; Pastrana, L.M.; Teixeira, J.A.; Sillankorva, S.M.; Cerqueira, M.A. Characterization of PHBV films loaded with FO1 bacteriophage using polyvinyl alcohol-based nanofibers and coatings: A comparative study. *Innov. Food Sci. Emerg. Technol.* **2021**, *69*, 102646–102656. [CrossRef]
320. Kim, S.; Chang, Y. Anti-*Salmonella* polyvinyl alcohol coating containing a virulent phage PBSE191 and its application on chicken eggshell. *Food Res. Int.* **2022**, *162*, 111971. [CrossRef]
321. Leung, V.; Groves, L.; Szewczyk, A.; Hosseinidoust, Z.; Filipe, C.D.M. Long-term antimicrobial activity of phage-sugar glasses is closely tied to the processing conditions. *ACS Omega* **2018**, *3*, 18295–18303. [CrossRef]
322. Kalkan, S. *Vibrio parahaemolyticus* ATCC 17802 inactivation by using methylcellulose films containing encapsulated bacteriophages. *Turk. J. Vet. Anim. Sci.* **2018**, *42*, 480–485. [CrossRef]
323. Alves, D.; Marques, A.; Milho, C.; Costa, M.J.; Pastrana, L.M.; Cerqueira, M.A.; Sillankorva, S.M. Bacteriophage ϕIBB-PF7A loaded on sodium alginate-based films to prevent microbial meat spoilage. *Int. J. Food Microbiol.* **2019**, *291*, 121–127. [CrossRef]
324. Kimmelshue, C.; Goggi, A.S.; Cademartiri, R. The use of biological seed coatings based on bacteriophages and polymers against *Clavibacter michiganensis* subsp. *nebraskensis in maize seeds. Sci. Rep.* **2019**, *9*, 17950. [CrossRef]
325. Weng, S.; López, A.; Sáez-Orviz, S.; Marcet, I.; García, P.; Rendueles, M.; Díaz, M. Effectiveness of bacteriophages incorporated in gelatine films against *Staphylococcus aureus*. *Food Control* **2021**, *121*, 107666. [CrossRef]